CONCISE THIRD EDITION

Successful Writing at Work

Philip C. Kolin
University of Southern Mississippi

WADSWORTH
CENGAGE Learning™

Australia · Brazil · Japan · Korea · Mexico · Singapore · Spain · United Kingdom · United States

WADSWORTH
CENGAGE Learning

Successful Writing at Work, Concise Third Edition
Philip C. Kolin

Senior Publisher: Lyn Uhl

Publisher: Michael Rosenberg

Development Editor: Ed Dodd

Assistant Editor: Jillian D'Urso

Editorial Assistant: Erin Pass

Media Editor: Janine Tangney

Marketing Manager: Jason Sakos

Marketing Coordinator: Ryan Ahern

Marketing Communications Manager: Stacey Purviance

Senior Content Project Manager: Michael Lepera

Art Director: Marissa Falco

Print Buyer: Sue Spencer

Senior Rights Acquisition Specialist, Image: Jennifer Meyer Dare

Senior Rights Acquisition Specialist, Text: Katie Huha

Production Service: Integra Software Services, Inc.

Text Designer: Glenna Collett

Cover Designer: Anne Katzeff

Cover Image: © Getty Images/Jupiterimages © iStockphoto © JLP/Sylvia Torres/ CORBIS

Compositor: Integra Software Services, Inc.

Image for title page and chapter openers courtesy of Glenna Collett.

For product information and technology assistance, contact us at **Cengage Learning Customer & Sales Support, 1-800-354-9706**

For permission to use material from this text or product, submit all requests online at **www.cengage.com/permissions**. Further permissions questions can be emailed to **permissionrequest@cengage.com**

Library of Congress Control Number: 2010934801

ISBN-13: 978-0-495-90194-5

ISBN-10: 0-495-90194-6

Wadsworth
20 Channel Center Street
Boston, MA 02210
USA

Cengage Learning is a leading provider of customized learning solutions with office locations around the globe, including Singapore, the United Kingdom, Australia, Mexico, Brazil and Japan. Locate your local office at **international.cengage.com/region**

Cengage Learning products are represented in Canada by Nelson Education, Ltd.

For your course and learning solutions, visit **www.cengage.com**.

Purchase any of our products at your local college store or at our preferred online store **www.cengagebrain.com**.

To Kristin, Eric, and Theresa
Evan Philip and Megan Elise
Julie and Loretta
Diane
and
MARY

Printed in the United States of America
2 3 4 5 14 13 12 11

Contents

Chapter 8: Writing Effective Short Reports and Proposals   280

Preface

Overview

Successful Writing at Work, Concise Third Edition, is a practical introductory text for students in business, professional, and occupational writing courses. As readers of the full-length edition of this text have found, *Successful Writing at Work* clearly helps students develop and master key communication skills vital for success in the global workplace. The *Concise Third Edition* serves the same purpose, but it is designed for those readers who prefer a more compact text, one that covers nearly as many business writing topics but is more streamlined and focuses on the most essential skills and strategies for writing successfully on the job. Whereas the full-length edition includes 17 chapters, the *Concise Edition* contains 10 chapters, yet these fully cover a range of workplace communications technologies and a variety of e-communications from essential considerations such as audience analysis and ethics, to writing increasingly more complex business documents (memos through long reports and websites), to making presentations, to preparing a resume and interviewing for a job.

This compact edition has also been designed for a variety of educational settings where business writing is taught and practiced. It is versatile enough for a full semester or trimester course, or it can be used successfully in a shorter course, such as on a quarter system. It can also meet the diverse goals of varied educational settings, including online, distance education, continuing education, and week-long intensive courses, as well as in-house training programs, workshops, and conferences.

Successful Writing at Work, Concise Third Edition, provides students with easy-to-understand guidelines for writing and designing clear, well-organized, and readable documents. Along with user-friendly guidelines, this edition provides students with realistic models of the precise kinds of documents they will be asked to write on the job. In addition, this text can serve as a ready reference that readers can easily carry with them to the workplace. Students will quickly find that this book includes many practical applications, which are useful to those who have little or no job experience as well as those with years of experience in the world of work.

Distinctive Features of *Successful Writing at Work, Concise Third Edition*

The distinctive features that have made *Successful Writing at Work, Concise Edition,* a user-friendly text in the contemporary workplace continue to be emphasized in this new, third edition. These features, emphasizing up-to-date approaches to teaching business writing, can be found throughout the book:

1. **Approaching writing as a problem-solving activity.** The *Concise Third Edition* continues to approach writing not merely as a set of rules and formats but as a problem-solving activity in which employees meet the needs of their employers, co-workers, customers, clients, community groups, and vendors worldwide by getting to the bottom line. This approach to writing, introduced in Chapter 1 and carried throughout the text, helps students to think through the writing process by asking the key questions of *who* (who is the audience?), *why* (why do they need this document?), *what* (what is the message?), and *how* (how can the writer present the most appropriate style, tone, and format?). As in earlier editions, this new edition teaches students how to develop the critical skills necessary for planning, drafting, revising, editing, and formatting a variety of documents from e-mails, blogs, letters, instructions, and proposals, to long reports. In addition, case studies and numerous figures demonstrate how writers answer these key questions above to solve problems in the business world.

2. **Writing for the global marketplace.** In today's international workplace, effective employees must be consistently aware of how to write for a variety of readers, both in the United States and across the globe. Consequently, almost every chapter in this new *Concise Third Edition* includes increased coverage of writing for international readers and non-native speakers of English. The needs and expectations of these international audiences receives special attention starting with Chapter 1 in a much expanded section "Writing for the Global Marketplace" and continues with coverage of writing letters for international speakers of English in Chapter 4, designing appropriate visuals and documents for this audience in Chapter 6, preparing clear instructions in Chapter 7, and making presentations for global audiences in Chapter 10. Especially important is the long report in Chapter 9 on the role international workers play in a corporation that must meet their needs.

3. **Viewing student readers as business professionals.** To encourage students in their job-related writing, this new *Concise Third Edition* treats them as professionals seeking success at different phases of their business careers. Students are asked to place themselves in the workplace setting (or, in the case of Chapter 5, in the role of job seekers) as they approach each topic, to better understand the differences between workplace and academic writing. In Chapter 1, they are given the kinds of orientation to company culture and protocols that they might find in the early days of their employment. Students are then asked to see themselves as members of a collaborative team drafting and developing an important workplace document (Chapter 2); workers writing routine hardcopy and e-documents (Chapters 3 and 4); employees designing and writing more complex documents, instructions, proposals, and reports (Chapters 7 to 9); co-workers designing documents and websites (Chapter 6); and as company representatives making presentations before co-workers and potential clients worldwide (Chapter 10).

4. **Using the latest workplace technologies.** This new edition offers the most current coverage of communication technologies for writing successfully in the rapidly changing world of work, including the Internet, e-mail, instant messaging, wikis, and document tracking systems used to collaboratively draft, revise, and edit reports, business blogs,

video conferencing tools, and presentation software. Coverage of these technologies is integrated into each chapter. Easy-to-understand explanations and annotated models throughout this edition assist students to discover the *hows* as well as the *whys* of writing for the digital world of work.

5. **Being an ethical employee.** Companies and agencies expect their employees to behave and write ethically. As in earlier editions, the *Concise Third Edition* reinforces the importance of ethical workplace writing. Beginning with enhanced coverage of ethical writing and solving ethical dilemmas at work, Chapter 1 further stresses ethics in the workplace with a new section titled "Successful Employees Are Successful Writers." Special attention to ethics continues in sections of Chapter 2 on avoiding sexism and biased language in the workplace while Chapter 3 draws students' attention to the ethical choices they have to make when writing e-communications, including e-mail and blogs, drafting diplomatic letters in Chapter 4; preparing honest and realistic resumes and webfolios in Chapter 5; constructing unbiased and unaltered visuals and websites in Chapter 6; preparing safe and effective instructions in Chapter 7; writing truthful proposals and reports in Chapters 8 and 9; and making clear and accurate presentations in Chapter 10.

New and Updated Material in the Third Concise Edition

As in the earlier *Concise* editions, this new Third Edition continues to offer students a streamlined alternative to the full edition of *Successful Writing at Work* while still providing many important new additions. Throughout this new Third Edition, you will find strengthened coverage of key topics; updated guidelines; and a wealth of new annotated examples of workplace documents, case studies, and exercises to make the teaching and learning of workplace writing more relevant and current. Highlights of this new edition include:

- **New and updated material on collaborative writing and meetings at work in Chapter 2 and collaborative exercises throughout.** Because a great deal of workplace writing is done collaboratively, this new edition emphasizes this topic more than in earlier editions. In addition to streamlined and updated guidelines for setting up, conducting, and avoiding conflicts in group settings, Chapter 2 now includes a section on how to be a better team player, a revised discussion of collaborating electronically with new figures showing how documents are collaboratively drafted, revised, and edited using a document tracking system as well as writing with wikis and further guidelines on planning virtual and face-to-face meetings. Exercises on preparing collaborative documents—from e-communications to letters to reports to proposals and to websites—are now included in each chapter.
- **New and expanded attention to workplace technologies.** Along with document tracking systems in Chapter 2, discussions of business communication technologies are enhanced in many chapters, especially

in Chapter 3 on e-communication at work, which includes new sections on (and examples of) instant messaging and external business blogs as well as updated coverage of writing and organizing e-mail. Chapter 5, "How to Get a Job," features updated coverage of online resumes plus valuable new sections on developing and designing career portfolios/webfolios. Chapter 6 — on designing documents, visuals, and websites — contains a new streamlined discussion of writing and formatting text and visuals for an online environment. Chapter 10 helps students understand how to be better, more persuasive speakers using PowerPoint technology and includes a revised sample slide presentation.

- **Updated and increased coverage of employment documents.** Already praised for its helpful coverage of the job search, a much-revised Chapter 5 on "How to Get a Job" offers the most current advice on searching for and applying for a job in this extremely competitive market. It includes a new section on helping students prepare for their careers while they are still in college; identify their most marketable skills; updated advice on searching for a job; streamlined discussions and numerous annotated and redesigned examples of various types of resumes and letters of application; cutting-edge coverage of online resumes; and an annotated sample webfolio. Reflecting changes in how companies interview and hire candidates, the chapter closes with new, highly practical advice on interview strategies and finding pay scales.

- **Greatly expanded material on using the Internet in the world of work.** Chapter 6 has been retitled "Designing Successful Documents, Visuals, and Websites" to include new material on how to write text for the web and how to incorporate such key web elements as color, art, animation, and space to convey the most appropriate message for key audience(s). Numerous examples and guidelines prepare students for web authorship. This Third Edition also offers expanded coverage of how to best use e-mail and IM in the workplace as well as detailed guidelines for writing company blogs, including a sample, annotated business blog post.

- **New and enhanced discussion of workplace correspondence.** Chapters 3 and 4 on workplace correspondence contain new and updated material to help students become more proficient and diplomatic writers. Chapter 3 on routine correspondence, for example, gives students updated advice and new annotated models of memos, e-mail, IMs, and blogs to show them how to be clear, concise, and ethical employees. Chapter 4, on writing letters, supplies updated guidelines for writing a variety of business letters as well as an enhanced discussion of international business correspondence showing both bad and good examples of such correspondence.

- **Most current guidelines on using MLA and APA documentation.** The section on documentation in Chapter 9 has been completely revised, reformatted, and updated to show students how to document their work using MLA and APA guidelines. Clear and easy-to-follow examples of the most frequently cited works, both print and e-documents, including websites, podcasts, blogs, and media programs, help students to document these sources accurately and completely. A helpful new feature of this chapter is the table that clearly and concisely spells out the different key information students need to cite according to MLA or APA. In addition to a wide range of sample entries, the chapter concludes with a revised and fully annotated and collaboratively written long report using an using an APA-documented References list.

■ **New and stronger emphasis on greening the workplace.** This Third Edition gives greater attention to the importance of protecting and preserving the environment, both in the workplace and at off-site locations. Chapter 1 includes a major example of how a power company and its employees safeguard the natural resources their customers need and then describes the ethical responsibilities companies and their employees have to respect the environment. Subsequent chapters offer a draft of a report on recycling, a progress report from a contractor rehabbing an office space to save energy, and an e-exchange about an environmental impact report. Most important, many new exercises have been added to get students "thinking green" as they produce eco-sensitive workplace documents.

■ **Complete annotations of model documents.** Every document in this new edition, including all the visuals and web homepage in Chapter 6, has been thoroughly annotated to better help students understand the choices writers make in selecting their words and visuals, organizing their documents, and formatting them for their audiences. Many chapters now show students both an ineffective document and a revised, much more effective version, illustrating the value of revision as well as the benefits of careful, often collaborative, workplace writing.

■ **Updated figures and exercises.** Most figures have been updated in this new edition, not only for currency but also to show students the importance of including the most persuasive and relevant information and graphics in their own work. In addition, new exercises have been included in every chapter, offering students greater opportunities to develop their writing skills for the world work, either alone or as part of a collaborative team.

Supplements

The *Successful Writing at Work, Concise Third Edition*, **Companion Website** (www.cengage.com/english/kolin/writingatworkconcise3e) includes the following resources for students:

■ **Improve Your Grade.** *Online exercises* for each chapter are designed to help students simultaneously practice chapter skills and become effective writers using the latest technologies—from word processing features such as report templates and document tracking, to presentation software, to Internet technologies like mind-mapping software and resume, survey, and blog builders. In addition, annotated *Web links* accompanying every chapter enable students to explore chapter topics even further.

■ **ACE the Test.** Two gradable 10-question *ACE self-tests* per chapter are provided to help students test their full understanding of chapter topics.

The Instructor Companion Website (www.cengage.com/english/kolin/writingat workconcise3e) provides plentiful material for instructors looking for ideas and aids to teach the course:

- **Correlation Guide.** For those instructors transitioning from either *Successful Writing at Work, Ninth Edition*, or the *Concise Second Edition*, this guide provides side-by-side content comparisons for easy updating of course syllabi.
- **Sample Syllabi.** Two syllabi are provided, one for a 15-week course incorporating research and long reports, and one for a shorter 10-week course. Both syllabi provide course goals and week-by-week strategies, but they can also be downloaded and adapted to the particular needs of teachers and students in various courses.
- **Some Suggestions on How to Teach Job-Related Writing.** This helpful guide provides ideas for simulating real-world experience in the classroom; enhancing classes by bringing in outside speakers and examples; and highlighting the crucial topics of ethics, global audience, technology, and collaboration.
- **PowerPoint Slides.** Slide shows for each chapter thoroughly cover all chapter topics and allow for enhanced classroom presentation.
- **Suggested Approaches to Exercises.** Because most of the exercises in *Successful Writing at Work, Concise Third Edition,* are designed to elicit a variety of responses from students, suggested approaches to evaluating and grading exercises are provided, rather than "right" or "wrong" answers.

The InSite online writing and research tool includes electronic peer review, an originality checker, an assignment library, help with common grammar and writing errors, and access to InfoTrac® College Edition. Portfolio management gives you the ability to grade papers, run originality reports, and offer feedback in an easy-to-use online course management system. Using InSite's peer review feature, students can easily review and respond to their classmates' work. Other features include fully integrated discussion boards, streamlined assignment creation, and more. Visit **www.academic.cengage.com/insite** to view a demonstration.

Acknowledgments

In a very real sense, the *Concise Third Edition* has profited from the collaboration of various reviewers with me as I worked on this new edition. I am, therefore, honored to thank the following reviewers who have joined with me to create this new edition.

Eileen M. Finelli, *Northampton Community College*
Christy L. Kinnion, *Wake Technical Community College*
Mary Mullaly, *Washtenaw Community College*
Becky Newman, *Dixie Applied Technology College*
Linda Nicole Patino, *Surry Community College*
Catherine Ramsden, *DePaul University*
Carol Whittaker, *Pennsylvania State University*

I am also deeply grateful to the following individuals at the University of Southern Mississippi for their help—Danielle Sypher-Haley, Penny White, Cecily Hill (Department of English), David Tisdale (University Communications), Mary Beth Applin and Sherry Laughlin (Information Services, Cook Library), Mary Lux (Department of Medical Technology), and Cliff Burgess (Department of Computer Science). I am especially grateful to Denise von Herrmann, Dean of the College of Arts and Letters, for her continued appreciation of my work.

I am also grateful to Terri Smith Ruckel, Jianqing Zheng at Mississippi Valley State University, and Erin Smith at the University of Tennessee—Knoxville.

Several individuals from the business world also gave me wise counsel, for which I am thankful. They include Sally Eddy at Georgia Pacific; Kirk Woodward at Visiting Nurses Services of New York; Jimmy Stockstill at Petro Automotive; Nancy Steen from Adelman and Steen; Theresa Rogers and Rachel Sullivan at Regents Bank, Inc.; and Sgt. Scott Jamison of the U.S. Army.

I am also especially grateful to Father Michael Tracey for his counsel and contributions to Chapter 6 on document design, particularly on websites.

My thanks go to the team at Cengage Learning for their assistance, encouragement, and friendship—Michael Rosenberg, Michael Lepera, Jillian D'Urso, Megan Garvey, Erin Pass, Janine Tangney, Jason Sakos, Stacey Purviance, and to freelance development editor Ed Dodd for his excellent assistance. I want to thank Integra Software Services, Inc. and their production manager Katie Ostler for their hard work, and Cindy Gierhart for her copyediting assistance on this edition. I am also grateful to Katie Huha and Jen Meyer Dare, who handled the permissions for this concise edition of *Successful Writing at Work*.

I thank my extended family—Margie and Al Parish, Sister Carmelita Stinn, Mary and Ralph Torrelli, and Lois and Norman Dobson—for their prayers and love.

Finally, I am deeply grateful to my son, Eric, and my daughter-in-law, Theresa, for their enthusiastic and invaluable assistance as I prepared Chapter 8; to my grandson, Evan Philip, and granddaughter, Megan Elise, for their love and encouragement. My daughter, Kristin, also merits loving praise for her help throughout this new edition by doing various searches and revisions and by offering practical advice on successful writing at work. And to Diane Dobson, my wife, I say thank you for bringing so much peace, music, and love into my life.

P.C.K.
January 2011

1 Getting Started
Writing and Your Career

Writing—An Essential Job Skill

Visit www.
cengage.com/
english/kolin/
writingatwork
concise3e for
this chapter's
online exercises,
ACE quizzes,
and web links.

Writing is a part of every job, from your initial letter of application conveying first impressions to memos, e-mail, blogs, letters, websites, proposals, instructions, and reports. Writing keeps businesses moving. It allows employees to communicate with one another, with management, and with the customers, clients, and agencies a company must serve to stay in business.

Clearly, then, writing is an essential skill for employees and employers alike. According to Don Bagin, a communications consultant, most people need an hour or more to write a typical business letter. If an employer is paying someone $30,000 a year, one letter costs $14 of that employee's time; for someone who earns $50,000 a year, the cost for the average letter jumps to $24. Mistakes in letters are costly for workers as well for as employers. As David Noble cautions in his book *Gallery of Best Cover Letters*, "The cost of a cover letter (in applying for a job, for instance) might be as much of a third of a million dollars—even more if you figure the amount of income and benefits you don't receive, say, in a 10-year period for a job you don't get because of an error that got you screened out."

Unfortunately, as the Associated Press (AP) reported in a recent survey, "Most American businesses say workers need to improve their writing . . . skills." Yet that same report cited a survey of more than 400 companies that identified writing as "the most valuable skill employees can have." In fact, the employers polled in that AP survey indicated that 80 percent of their work force needed to improve their writing. Beyond a doubt, your success as an employee will depend on your success as a writer. The higher you advance in an organization, the more and better writing you will be expected to do. Promotions, and other job recognitions, are often based on an employee's writing skills. This book will show you, step by step, how to write clearly and efficiently the job-related communications you need for success in the world of work.

Chapter 1 gives you some basic information about writing in the global marketplace and raises major questions you need to ask yourself to make the writing process easier and the results more effective. It also describes the basic functions of on-the-job writing and introduces you to one of the most important requirements in the business world—writing ethically.

Writing for the Global Marketplace

Visit www.
cengage.com/
english/kolin/
writingatwork
concise3e for an
online exercise,
"Exploring the
Online Global
Marketplace."

The Internet, e-mail, teleconferencing, blogging, and e-commerce have shrunk the world into a global village. Accordingly, it is no longer feasible to think of business in exclusively regional or even national terms. Many companies are multinational corporations with offices throughout the world. In fact, many U.S. businesses are branches of international firms. A large, multinational corporation may have its products designed in Japan; manufactured in Bangladesh; and sold in Detroit, Atlanta, and Los Angeles. Its stockholders may be in Mexico City as well as Saudi Arabia—in fact, anywhere. In our global economy, every country is affected by every other one.

Competing for International Business

Companies must compete for international sales to stay in business. Every business, whether large or small, has to appeal to diverse international markets to be competitive. Each year a larger share of the U.S. gross national product (GNP) depends on global markets in China, Saudi Arabia, India, Eastern Europe, and elsewhere. Some U.S. firms estimate that 40 to 50 percent of their business is conducted outside of the United States. Wal-Mart, for example, has expanded into hundreds of stores in mainland China, and General Electric has opened plants in more than 60 countries. Jupiter Research estimates that 75 percent of the global Internet population lives outside the United States. A large corporation such as Citigroup, for instance, is eager to promote its image of helping customers worldwide, as Figure 1.1 illustrates. If your company, however small, has a website, then it is an international business.

Communicating with Global Audiences

To be a successful employee in this highly competitive global market, you have to communicate clearly and diplomatically with a host of readers from different cultural backgrounds. As a result, don't presume that you will be writing only to native speakers of American English. As a vital part of your job, you may be communicating with readers in Singapore, Jamaica, and South Africa, for example, who speak varieties of English quite different from American English. You will also very likely be writing to readers for whom English is not their first (or native) language. These individuals, who may reside either in the United States or in a foreign country, will constitute a large and important audience for your work.

Seeing the World Through Their Eyes

Writing to these international readers with proper business etiquette means first learning about their cultural values and assumptions—what they value and also what they regard as communication taboos. They may not conduct business exactly

How Citigroup Meets Banking Needs Around the World

WITH A BANKING EMPIRE that spans more than 100 countries, Citigroup is experienced at meeting the diverse financial services needs of businesses, in-dividuals, customers, and governments. The bank is headquartered in New York City but has offices in Africa, Asia, Central and South America, Europe, the Middle East, as well as throughout North America. Live or work in Japan? You can open a checking account at Citigroup's Citibank branch in downtown Tokyo. How about Mexico? Visit a Grupo Financiero Banamex-Accival branch, owned by Citigroup. Citigroup owns European American Bank and has even bought a stake in a Shanghai-based bank with an eye toward attracting more of China's $1 trillion in bank deposits. Between acquisitions and long-established branches, Citigroup covers the globe from the Atlantic to the Pacific and the Indian Oceans.

Source: From Pride, Hughes, and Kapoor, *Business*, 8th ed., p.587. Copyright © 2005 by Houghton Mifflin Company. Used by permission. Photo by Greg Baker/AP Images.

the way it is done in the United States, and to think they should is wrong. Your international audience is likely to have different expectations of how they want a letter addressed or written to them, whether they allow you to use their first name, how they prefer a business meeting to be conducted, or how they think questions should be framed and asked and agreements reached. Their concepts of time, family, money, the environment, managers, and communication itself may be nothing like those in the United States. Visuals, including icons, which are easily understood in the United States, may be baffling elsewhere in the world. If you misunderstand your audience's culture by inadvertently writing, creating, saying or illustrating something inappropriate can cost your company a contract and you your job.

Cultural Diversity at Home

Cultural diversity exists inside as well as outside the company you work for. Don't conclude that your boss or co-workers are all native speakers of English, either, or that they come from the same cultural background that you do. In the next decade, as much as 40 to 50 percent of the U.S. skilled work force may be composed of recent immigrants who bring their own business traditions and languages with them. These are highly educated, multicultural, and multinational individuals who have acquired English as a second language.

For the common good of your company, then respect these international colleagues. In fact, multinational employees can be tremendously important for your company in making contacts in their native country and in helping your firm understand and appreciate ethical/cultural differences among customers. The long report in Chapter 9 (pages 338–354) describes some ways in which a company can both acknowledge and respect the different cultural traditions of its international employees. Businesses want to emphasize their commitments to globalization.

Using International English

Whether your international readers are customers or colleagues, you need to adapt your writing to respect their language needs and cultural protocols. To communicate with non-native speakers, use "international English," a way of writing that is easily understood, culturally appropriate, and diplomatic. International English is user friendly in terms of the words, sentences, formats, and visuals you choose.

To write international English means you re-examine your own writing. The words, idioms, phrases, and sentences you select instinctively for U.S. readers may not be appropriate for an audience for whom English is a second, or even a third, language. If you find a set of instructions accompanying your computer software package confusing, imagine how much more intimidating such a document would be for non-native speakers of English. You can eliminate such confusion by making your message clear, straightforward, and appropriately polite for readers who are not native speakers.

Here are some basic guidelines to help you write international English:

- Use clear, easy-to-understand sentences, not rambling, complex ones. That does not mean you write insultingly short and simple ones but that you take into account that readers will find your message easier to translate if your sentences do not exceed 15–20 words. Moreover, do not try to pack too much information in a single sentence; consider using two or more sentences instead. See pages 42–47.

- Avoid jargon, idioms (e.g., "to line one's pockets"), and abbreviations (e.g. "FEMA") that international readers may not know.
- Choose clear, commonly used words that unambiguously translate into the non-native speaker's language. Avoid flowery or pretentious ("amend" for "change") language.
- Select visuals and icons that are free from cultural bias, or that are taboo in the non-native speaker's country. (For more on this, see pages 235–237.)
- When in doubt, consult someone from the native speaker's country—a co-worker, an instructor.

Because it is so important, international English is discussed in greater detail on pages 134–139. Later chapters of this book will also give you additional practical guidelines on writing correspondence, instructions, proposals, reports, websites, and other work-related documents suitable for a global audience.

Four Keys to Effective Writing

Effective writing on the job is carefully planned, thoroughly researched, and clearly presented. Its purpose is always to accomplish a specific goal and to be as persuasive as possible. Whether you send a routine e-mail to a co-worker in Cincinnati or in Shanghai or a commissioned report to the president of the company, your writing will be more effective if you ask yourself these four questions:

1. *Who* will read what I write? (Identify your *audience*.)
2. *Why* should they read what I write? (Establish your *purpose*.)
3. *What* do I have to say to them? (Formulate your *message*.)
4. *How* can I best communicate? (Select an appropriate *style* and *tone*.)

The questions *who*? *why*? *what*? and *how*? do not function independently; they are all related. You write (1) for a specific audience (2) with a clearly defined purpose in mind (3) about a topic your readers need to understand (4) in language appropriate for the occasion. Once you answer the first question, you are off to a good start toward answering the other three. Now let's examine each of the four questions in detail.

Identifying Your Audience

Knowing *who* makes up your audience is one of your most important responsibilities as a writer. Keep in mind that you are not writing for yourself but for a specific reader or group of readers. Expect to analyze your audience throughout the composing process.

Look at the advertisements in Figures 1.2, 1.3, and 1.4. The main purpose of all three documents is the same: to discourage people from smoking. The essential message in each ad—smoking is dangerous to your health—is also the same. But note

Figure 1.2   **No-smoking advertisement aimed at fathers who smoke.**

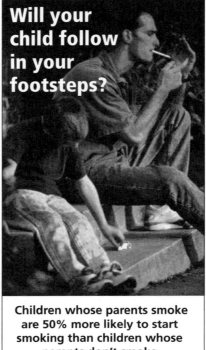

Photo by Peter Poulides/Getty Images.

how the different details—words, photographs, situations—have been selected to appeal to three different audiences.

The advertisement in Figure 1.2 is aimed at fathers who smoke. As you can see, it shows an image of a father smoking next to his son, who is reaching for his pack of cigarettes. Note how the caption "Will your child follow in your footsteps?" plays on the fact that the father and son are literally sitting on steps, but at the same time implies that the son will imitate his father's behavior as a smoker. The statistic at the bottom of the advertisement reinforces both the photo caption and the image, hitting home the point that parental behavior strongly influences children's behavior. The child in the photograph already is following his father by showing a clear interest in smoking.

The advertisement in Figure 1.3, on the other hand, is aimed at an audience of pregnant women and appropriately shows a member of this audience with a lit cigarette. The words on the advertisement appeal to a mother's sense of responsibility as the reason to stop smoking, a reason to which this audience would be most likely to respond—smoking can harm the unborn child.

No-smoking advertisement directed at pregnant women.   Figure 1.3

Photo by Bill Crump/Brand X Pictures/Fotosearch/Royalty-Free Image.

Figure 1.4 is directed toward still another audience, young athletes. The word *smoke* in this advertisement is aimed directly at their game and their goal. In fact, the writer aptly made the goal the same for the game as for the players' lives. Note, too, how this image with its four photos is suitable for an international audience.

The copywriters who created these advertisements have chosen appropriate details—words, pictures, captions, and so on—to persuade each audience not to smoke. With their careful choices, they successfully answered the question "How can we best communicate with each audience?" Note that details relevant for one audience (athletes, for example) could not be used as effectively for another audience (such as fathers).

The three advertisements in Figures 1.2, 1.3 and 1.4 illustrate some fundamental points you need to keep in mind when identifying your audience.

- Members of each audience differ in backgrounds, experiences, and needs.
- How you picture your audience will determine what you say to them.
- Viewing something from the audience's perspective will help you to select the most relevant details for that audience.

Figure 1.4   **No-smoking advertisement appealing to young athletes.**

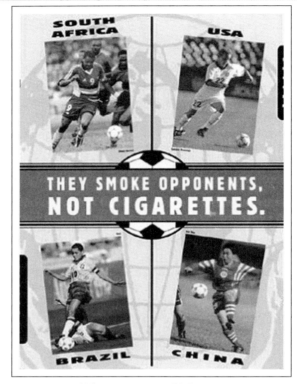

CDC, Tobacco Free Sports Initiative

Some Questions to Ask About Your Audience

You can form a fairly accurate picture of your audience by asking yourself key questions *before* you write. For each audience you need to reach, consider the following questions.

1. **Who is my audience?** What individual(s) will most likely be reading my work?

   If you are writing for colleagues/managers at work:

   - What is my reader's job title? Co-worker? Immediate supervisor? Vice president?
   - What kind of job experience, education, and interests does my reader have?

   If you are writing for clients or consumers (a very large, often-times diverse audience):

   - How can I find out about their interest in my product or service?
   - How much will this audience know about my company? About me?

2. **How many people will make up my audience?**

   - Will just one individual read what I write (the nurse on the next shift, the production manager) or will many people read it (all the consumers of my company's product or service)?

- Will my boss want to see my work (say, a letter to a consumer in response to a complaint) to approve it?
- Will I be sending my message to a large group of people sharing a similar interest in my topic, such as a listserv?

3. **How well does my audience understand English?**

- Are all my readers native speakers of English?
- Will I be communicating with people around the globe?
- Will some of my readers speak English as a second or even third language and thereby require extra sensitivity on my part to their needs as non-native speakers of English?
- Will some of my readers speak no English but instead use an English grammar book and foreign language dictionary to understand what I've written?

4. **How much does my audience already know about my writing topic?**

- Will my audience know as much as I do about the particular problem or issue, or will they need to be briefed, be given background information, or be updated?
- Are my readers familiar with, and do they expect me to use, technical terms and descriptions, or will I have to provide definitions and easy-to-understand and nontechnical wording and visuals?

5. **What is my audience's reason for reading my work?**

- Is my communication part of their routine duties, or are they looking for information to solve a problem or make a decision?
- Am I writing to describe benefits that another writer or company cannot offer?
- Will my readers expect complete details, or will a short summary be enough?
- Are they looking at my work to make an important decision affecting a co-worker, a client, a community, or the environment?
- Are they reading something I write because they must (a legal notification, an incident report, for instance)?

6. **What are my audience's expectations about my written work?**

- Do they want an e-mail, or will they expect a formal letter?
- Will they expect me to follow a company format and style?
- Are they looking for a one-page memo or for a comprehensive report?
- Should I use a formal tone or a more relaxed and conversational style?

7. **What is my audience's attitude toward me and my work?**

- Will I be writing to a group of disgruntled and angry customers or vendors about a sensitive issue (a product recall, discontinuation of a service, a refusal of credit, or a shipment delay)?
- Will I have to be sympathetic while at the same time give firm, convincing reasons for my company's (or my) decision?
- Will my readers be skeptical, indifferent, or friendly about what I write?
- Will my readers feel guilty that they have not answered an earlier message of mine, not paid a bill now overdue, or not kept a promise or commitment?

8. **What do I want my audience to do after reading my work?**

   - Do I want my readers to purchase something from me, approve my plan, or send me additional documentation?
   - Do I expect my readers to acknowledge my message, save it for future reference, or review and e-mail it to another individual or office?
   - Do my readers have to take immediate action, or do they have several days or weeks to respond?
   - Do I simply want my readers to get my message and not respond at all?

As your answers to these questions will show, you may have to communicate with many different audiences on your job. Each group of readers will have different expectations and requirements; you need to understand those audience differences if you want to supply relevant information.

Case Study: Writing to Different Audiences in a Large Corporation

To examine more closely the way that your audience influences the writing process, consider the following situation.

Let's say you work for a company that designs and produces heavy-duty earth moving equipment and that you have to write for many individuals in that organization. Here are the priorities of five different audiences at your firm, along with appropriate information you need to give each one to answer their questions:

Audience	Information to Communicate
Owner or principal executive	Stress financial benefits, indicating that the equipment is a "money-maker" and is compatible with other existing (and competing) models.
Production engineer	Emphasize "state-of the-art" transmissions, productivity, availability of parts.
Operator	Focus on information about how easy and safe it is to run.
Maintenance worker	Provide key details about routine maintenance as well as troubleshooting advice on problems.
Production supervisor	Concentrate on the speed and efficiency the machine offers.

As these examples show, to succeed in the world of work, give each reader the details he or she needs to accomplish a given job. Each specific audience has very different needs and questions you will be expected to answer.

Establishing Your Purpose

By knowing *why* you are writing, you will communicate better and find writing itself to be an easier process. Make sure you follow the most important rule in occupational writing: **Get to the point right away.** At the start of your message, state your goal clearly.

```
I want to teach new employees the security procedures for
logging onto and off the company computer.
```

Because your purpose controls the amount and order of information you include, state it clearly at the beginning of every e-mail, memo, letter, and report.

```
This e-mail will acquaint new employees with the security
measures they need to take when logging onto and off the
company computer.
```

In the opening purpose statement that follows, note how the author clearly informs the reader as to what the report will and will not cover.

```
As you requested at last week's organizational meeting, I have
surveyed how well our websites promote our services. This
report describes and priortizes respondents' assessments.
```

Formulating Your Message

Your message is the sum of *what* facts, responses, and recommendations you put into writing. A message includes the scope and details of your communication.

- *Scope* refers to how much information you give readers about the key details.
- The *details* are those key points you think readers need to know to perform their jobs.

Some messages will consist of only one or two sentences: "Do not touch; wet paint." "Order #756 was sent this afternoon by Federal Express. It should arrive at your office on March 22." At the other extreme, messages may extend over 20 or 30 pages or more. Messages can carry good news or bad news. They may deal with routine matters; or they may handle changes in policy, special situations, or problems.

Keep in mind that you need to adapt the message to fit your audience. For technical audiences, such as engineers, you may have to supply a complete report with statistical or other mathematical data. For other readers—busy decision makers, for example—a short discussion or summary of the financial or managerial significance will be enough. See page 339 for an example of an abstract.

Selecting Your Style and Tone

Style

Style is *how* something is written rather than what is written. Style helps to determine how well you communicate with an audience, how well your readers understand and receive your message. It involves the choices you make about

- the construction of your paragraphs
- the length and patterns of your sentences
- your choice of words

You will have to adapt your style to take into account different messages, different purposes, and different audiences. Your words, for example, will certainly vary with your audience. If all potential readers are specialists in your field, you may safely use the technical language and symbols of your profession. Nonspecialists, however, will be confused and annoyed if you write to them in the same way. The average consumer, for example, will not know what a *potentiometer* is; by writing "volume control on a radio," you will be using words that the general public can understand. As we saw, when writing to an international audience, you have to take into account their proficiency in English and so choose your words and sentences with their needs in mind.

Tone

Tone in writing, like tone of voice, expresses your attitude toward a topic and toward your audience. In general, your tone can range from formal and impersonal (a scientific report) to informal and conversational (an e-mail or IM to a colleague). It can be unprofessionally sarcastic or diplomatically agreeable.

Tone, like style, is signaled in part by the words you choose. For example, saying that someone is "interested in details" conveys a more positive tone than saying the person is a "nitpicker." The word *economical* is more positive than *stingy* or *cheap; assertive* sounds better than *rude* or *aggressive.*

The tone of your writing is especially important in occupational writing because it reflects the image you project to your readers and thus determines how they will respond to you, your work, and your company. Your tone can be informal or formal. Sending an IM to a friend, your tone is far more casual than sending a proposal to a prospective customer. Your tone can also signal how sincere and intelligent or angry and uninformed you appear. Of course, in all your written work, you need to sound professional and knowledgeable. The wrong tone in a letter or a proposal might cost you a customer, as the letter in Figure 4.5 (page 106) demonstrates.

Case Study: A Description of Heparin for Two Different Audiences

In the workplace you will often be faced with the problem of presenting the same information to two completely different audiences. To better understand the impact style and tone can have when solving this problem, read the following two descriptions of heparin, a drug used to prevent blood clots. In both, the message is basically the same. Yet, because the audiences differ, so do the style and the tone.

The first description of heparin appears in a reference work for physicians and other health care providers and is written in a highly technical style with an impersonal tone.

The writer has made the appropriate stylistic choices for the audience, the purpose, and the message. Health care providers understand and expect the jargon and the scientific explanations to prescribe and/or administer heparin correctly. The writer's authoritative, impersonal tone is coldly clinical, which, of course, is also appropriate because the purpose is to convey the accurate, complete scientific facts about this drug, not the writer's or reader's personal opinions or beliefs. The author sounds appropriately both knowledgeable and objective.

> *Heparin Sodium Injection*, USP
> *Sterile Solution*
>
> *Description*: Heparin Sodium Injection, USP is a sterile solution of heparin sodium derived from bovine lung tissue, standardized for anticoagulant activity.
>
> Each ml of the 1,000 and 5,000 USP units per ml preparations contains: heparin sodium 1,000 or 5,000 USP units; 9 mg sodium chloride; 9.45 mg benzyl alcohol added as preservative. Each ml of the 10,000 USP units per ml preparations contains: heparin sodium 10,000 units; 9.45 mg benzyl alcohol added as preservative.
>
> When necessary, the pH of Heparin Sodium Injection, USP was adjusted with hydrochloric acid and/or sodium hydroxide. The pH range is 5.0–7.5.
>
> *Clinical pharmacology*: Heparin inhibits reactions that lead to the clotting of blood and the formation of fibrin clots both *in vitro* and *in vivo*. Heparin acts at multiple sites in the normal coagulation system. Small amounts of heparin in combination with antithrombin III (heparin cofactor) can inhibit thrombosis by inactivating activated Factor X and inhibiting the conversion of prothrombin to thrombin.
>
> *Dosage and administration*: Heparin sodium is not effective by oral administration and should be given by intermittent intravenous injection, intravenous infusion, or deep subcutaneous (intrafrat, i.e., above the iliac crest or abdominal fat layer) injection. **The intramuscular route of administration should be avoided because of the frequent occurrence of hematoma at the injection site.**[1]

The second description of heparin, however, is written in a nontechnical style and with an informal, caring tone. This description is similar to those found on information sheets given to patients about the medications they are receiving in a hospital.

The writer of this patient-centered description has also made appropriate choices for nonspecialists such as patients or their families who do not need elaborate descriptions of the origin and composition of the drug. Using familiar words and adopting a personal, caring tone help to win the patients' confidence and enable them to understand why and how they should take the drug.

> Your doctor has prescribed a drug called *heparin* for you. This drug will prevent any new blood clots from forming in your body. Since heparin cannot be absorbed from your stomach or intestines, you will not receive it in a capsule or tablet. Instead, it will be given into a vein or the fatty tissue of your abdomen. After several days, when the danger of clotting is past, your dosage of heparin will be gradually reduced. Then another medication you can take by mouth will be started.

Characteristics of Job-Related Writing

Job-related writing characteristically serves six basic functions: (1) providing practical information, (2) giving facts rather than impressions, (3) supplying visuals to clarify and condense information, (4) giving accurate measurements, (5) stating responsibilities precisely, and (6) persuading and offering recommendations. These six functions tell you what kind of writing you will produce after you successfully answer the *who? why? what?* and *how?*

1. Providing Practical Information

On-the-job writing requires a practical "Here's what you need to do or to know" approach. One such practical approach is *action oriented*. You instruct the reader to do something—assemble a ceiling fan, test for bacteria, perform an audit, or design a website. Another practical approach of job-related writing is *knowledge oriented*. Here you explain what you want the reader to understand—why a procedure was changed, what caused a problem or solved it, how much progress was made on a job site, or why a new technology should be installed and used.

The following description of an Energy Efficiency Ratio combines both the action-oriented and knowledge-oriented approaches of practical writing.

> Whether you are buying window air-conditioning units or a central air-conditioning system, consider the performance factors and efficiency of the various units on the market. Before you buy, determine the Energy Efficiency Ratio (EER) of the units under consideration. The EER is found by dividing the BTUs (units of heat) that the unit removes from the area to be cooled by the watts (amount of electricity) the unit consumes. The result is usually a number between 5 and 12. The higher the number, the more efficiently the unit will use electricity.[2]

2. Giving Facts, Not Impressions

Occupational writing is concerned with what can be seen, heard, felt, tasted, or smelled. The writer uses *concrete language* and specific details. The emphasis is on facts rather than on the writer's feelings or guesses.

The discussion below, addressed to a group of scientists about the various sources of oil spills and their impact on the environment, is an example of objective, fact-based writing with objectivity. It describes events and causes of oil spills without anger or tears.

> The most critical impact results from the escapement of oil into the ecosystem, both crude oil and refined fuel oils, the latter coming from sources such as marine traffic. Major oil spills occur as a result of accidents such as blowout, pipeline breakage, etc. Technological advances coupled with stringent regulations [can] reduce the chances of such major spills; however, there is [still] a chronic low-level discharge of oil associated with normal drilling and production operations. Waste oils discharged through the river systems and practices

---

[2]Reprinted by permission of New Orleans Public Services, Inc.

Use of a visual to convey information.   Figure 1.5

Using Your Computer Safely

Follow the bulleted guidelines below, and illustrated in the photo to the right, to avoid workplace injuries when using your computer.

- To reduce the possibility of eye damage, maintain a distance of 18 to 24 inches between your eyes and the computer screen and keep your work area well lit.

- To minimize neck strain, position your computer screen so that the top of the screen is at or just below eye level.

- To avoid back and shoulder stress, sit up straight in your chair with your shoulders relaxed and your lower back firmly supported (with a cushion, if necessary).

- To lessen leg and back strain, adjust your chair height so that your upper and lower legs form a 90-degree angle and that your feet are either flat on the floor or on a footrest.

Photo courtesy Ergo Concepts, LLC

associated with tanker transports dump more significant quantities of oils into the ocean, compared to what is introduced by the offshore oil industry. All of this contributes to the chronic low-level discharge of oil into world oceans. The long-range cumulative effect of these discharges is possibly the most significant threat to the ecosystem.[3]

3. Supplying Visuals to Clarify and Condense Information

Visuals are indispensable partners with your words to convey information to your readers. On-the-job writing makes frequent use of visuals such as tables, charts, photographs, flow charts, and drawings to clarify and condense information. Thanks to various software packages, you can easily create and insert visuals into your writing. The use of visuals is discussed in detail in Chapter 6.

Visuals play an important role in the workplace. Note how the photograph in Figure 1.5 can help computer users better understand and avoid workplace injuries. A visual like this, reproduced in an employee handbook or displayed on a website, can significantly reduce employee stress and increase productivity.

---

[3]*Source*: The Offshore Ecology Investigation. Reprinted by permission of Gulf Universities Research Consortium.

4. Giving Accurate Measurements

Much of your work will depend on measurements—acres, bytes, calories, centimeters, degrees, dollars and cents, Euros, grams, percentages, pounds, square feet, units, etc. Numbers are clear and convincing. However, not every culture computes costs in dollars or records temperatures in degrees Fahrenheit.

The following discussion of mixing colored cement for a basement floor would be useless to readers if it did not supply accurate quantities.

> Including permanent color in a basement floor is a good selling point. One way of doing this is by incorporating commercially pure mineral pigments in a topping mixture placed to a 1-inch depth over a normal base slab. The topping mix should range in volume between 1 part portland cement, 1¼ parts sand, and 1¼ parts gravel or crushed stone and 1 part portland cement, 2 parts sand, and 2 parts gravel or crushed stone. Maximum size gravel or crushed stone should be ⅜ inch.[4]

5. Stating Responsibilities Precisely

Because it is directed to a specific audience, job-related writing should make absolutely clear what it expects of, or can do for, that audience. Misunderstandings waste time, cost money and can result in injuries. Directions on order forms, for example, should indicate how and where information is to be listed and how it is to be routed and acted on. The following directions show readers how to perform different tasks and/or explain why.

- Enter agency code numbers in the message box.
- Items 1 through 16 of this form should be completed by the injured employee or by someone acting on his or her behalf, whenever an injury is sustained on the job. The term *injury* includes occupational disease caused by the employment. The form should be given to the employee's official superior within 12–24 hours following the injury. The official superior is that individual having responsible supervision over the employee.

Other kinds of job-related writing deal with the writer's responsibilities rather than the reader's, for example, "Tomorrow I will meet with the district sales manager to discuss (1) July's sales, (2) the necessity of expanding our market, and (3) next fall's production schedule. I will e-mail a report of our discussion by August 3."

6. Persuading and Offering Recommendations

Persuasion is a crucial part of writing on the job. In fact, it is one of the most crucial skills you can learn in the business world. Persuasion means trying to convince your reader(s) to accept your ideas, approve your recommendations, or order your products. Convincing your reader to accept your interpretation or ideas is at the heart of the world of work, whether you are writing to someone outside or inside your company.

---

[4]Reprinted by permission from *Concrete Construction Magazine*, World of Concrete Center, 426 South Westgate, Addison, Illinois 60101.

Writing Persuasively to Clients and Customers

Much of your writing in the business world will promote your company's image by persuading customers and clients (a) to buy a product or service or (b) to adopt a plan of action endorsed by your employer. You will have to convince readers that you (and your company)—your products and services—can save them time and money, increase efficiency, reduce risks, or improve their image, and that you can do this better than your competitors can.

Expect also to be called on to write convincingly about your company's image, as in the case of product recalls or discontinuances (see Figure 3.9, page 89), customer complaints, or damage control after a corporate mistake affecting the environment. You may also have to convince customers around the globe that your company respects cultural diversity and upholds specific ethnic values.

A large part of persuasion, too, is supporting your claims with evidence. You will have to conduct research; provide logical arguments; supply appropriate facts, examples, and statistics; and identify the most relevant information for your particular audience(s). Notice how the advertisement in Figure 1.6 below offers a bulleted list of persuasive reasons—based on cost, time, safety, efficiency, and convenience—to convince correctional officials that they should use General Medical's services rather than those of a hospital or clinic.

An advertisement using arguments based on cost, time, efficiency, safety, and convenience to persuade a potential customer to use a service.   Figure 1.6

GENERAL MEDICAL WILL STOP THE UNNECESSARY TRANSPORTING OF YOUR INMATES.

- We'll bring our X-ray services to your facility, 7 days a week, 24 hours a day.
- We can reduce your X-ray costs by a minimum of 28%. X-ray cost includes radiologist's interpretation and written report.
- Same day service with immediate results telephoned to your facility.
- Save correctional officers' time, thereby saving your facility money.
- Avoid chance of prisoner's escape and possible danger to the public.
- Avoid long waits in overcrowded hospitals.
- Reduce your insurance liabilities.
- Other Services Available: Ultrasound, Two Dimensional Echocardiogram, C.T. Scan, EKG, Blood Lab and Holter Monitor.

General Medical Is Your On-Site Medical Problem Solver

General Medical Services Corp.
A subsidiary of

Federal Medical Industries, Inc. O.T.C.
950 S.W. 12th Avenue, 2nd Floor Suite, Pompano, Florida 33069
(305) 942-1111   FL WATS: 1-800-654-8282

Visual stresses the need for a more efficient way to transport prisoners for medical attention

Bulleted list conviniently and persuasively uses factual data to convince

Writing Persuasively In-House

As much as 60 percent of your writing may be directed to individuals you work with and for. In fact, your very first job-related writing will likely be a persuasive letter of application to obtain a job interview with a potential employer.

On the job, you may have to persuade a manager to buy a new technology or lobby for a change in your office or department. To be successful, you will have to evaluate various products or options by studying, analyzing, and deciding on the most relevant one(s) for your boss. Your reader will expect you to offer clear-cut, logical, and convincing reasons for your choice, backed up with persuasive facts.

As part of your job, too, you will be asked to write convincing memos, e-mails, letters, blogs, and websites to boost the morale of employees, encourage them to be more productive, and compliment them on jobs well done. You can also expect to write persuasively to explain and solve budget, safety, or marketing problems your company faces, as in Figure 3.2 (page 72).

Figure 1.7 is a persuasive e-mail from an employee to a manager reporting a payroll mistake and persuading the reader to correct it. The e-mail contains many of the other characteristics of job-related writing we have discussed. Note how the writer provides factual, not subjective, information; attaches a time sheet (a type of visual); gives accurate details; and identifies her own and her immediate supervisor's responsibilities. The writer's tone is suitably polite yet direct.

Ethical Writing in the Workplace

One of your most important job responsibilities is to ensure that your writing and behavior are ethical. Writing ethically means choosing language that is right and fair, honest and just with your employer, co-workers, and customers. Your reputation and character plus your employer's corporate image will depend on your following an ethical course of action. Many companies are proud of their ethical commitments to the environment, the community, and the global marketplace. Starbucks, for example, tells customers that its "10 percent post-consumer recycled … paper cups helped conserve enough energy to supply 900 homes for a year and save approximately 110,000 trees."

Many of the most significant phrases in the world of business reflect an ethical commitment to honesty and fairness: *accountability, public trust, equal opportunity employer, core values, global citizenship, good faith effort, truth-in-lending, fair play, honest advertising, full disclosure, high professional standards, fair trade, community involvement*, and *corporate responsibility*.

Unethical business dealings, on the other hand, are stigmatized in *cover-ups, dodges, stone walling, shady deals, spin doctors, foul play, bid rigging, employee raiding, misrepresentations, kickbacks, hostile takeovers, planned obsolescence, price gouging, bias*, and *unfair advantage*. Those are the activities that make customers angry and that the Better Business Bureau investigates.

Ethical Requirements on the Job

In the workplace, you will be expected to meet the highest ethical standards by fulfilling the following requirements:

A persuasive e-mail from an employee to a business manager.   Figure 1.7

File Edit View Insert Format Tools Table MathType Window Help Adobe PDF

Reply   Reply to All   Forward

To: <lgriffin@starinstruments.com> (Lee Griffin)

From: <rburke@starinstruments.com> (R. Burke)

Cc:

Subject: Shortage in My October Paycheck

Arial   10   A ▾ B I U

Dear Ms. Griffin,

My paycheck for the two-week period ending October 15 was $75.00 short. For this period I should have been paid $875.00. Instead, my check was for only $800.00. I believe I know why there may have been a discrepancy. The $75.00 additional pay for this period was the result of my having put in five hours of overtime on October 8 and October 12 ($2\frac{1}{2}$ hours each day @ $15.00 per hour). This overtime was not reflected on my current pay stub.

Clearly explains and documents the problem

I have double-checked with my supervisor, Gloria Arrelo, who assured me that she recorded my overtime on the timesheets she sent to your office. She has kindly given me a copy that I have scanned and have attached to this e-mail to verify my hours.

Offers further evidence

Thank you for correcting your records and for crediting me with the additional $75.00 for my overtime.

Closes politely with request

Robbie Burke

- Supplying honest and up-to-date information about yourself in your résumé and job applications. The résumé and your portfolio (see pages 160–179) are key places where you must make ethical decisions with candor and honesty.
- Respecting co-workers, customers, and suppliers in conduct that avoids bullying, discrimination, or any other unfair or unprofessional behavior.
- Refusing to use language that makes false claims or tries to deceive readers with ambiguous words, jargon, or misleading visuals (see pages 230–235).
- Avoiding language that excludes others on the basis of gender, race, national origin, religion, age, physical ability, or sexual orientation (see pages 47–51).
- Maintaining accurate and current records at work. Remember: "If it isn't written, it didn't happen."

- Complying with all local, state, and federal regulations, especially those ensuring a safe, healthy work environment, products, and/or services, for example, following the Occupational Safety and Health Administration guidelines.
- Adhering to your profession's code or standard of ethics, internal audits, licenses, and certificate requirements.
- Following your company's policies and procedures.
- Honoring guarantees and warranties and meeting customers' needs impartially.
- Cooperating fairly and on a timely basis with your collaborative team.
- Respecting all copyright obligations and privileges.

Following these guidelines is not only an ethical requirement; it could also be a legal one. For example, doing personal (or outside consulting) work on company time is unethical and illegal. It would also be neglectful and unethical to allow an unsafe product to stay on the market just to spare your company the expense and embarrassment of a product recall.

Computer Ethics

Computer ethics are essential in the e-commerce world of work. A good rule to follow is never to do anything online that you wouldn't do offline. Never use a company computer for any activity not directly related to your job. It would be grossly unethical to erase a computer program intentionally, violate a software licensing agreement, or misrepresent (by fabrication or exaggeration) the scope of a database. Also, posting anything that attacks a competitor, colleagues, your boss, or your company is considered unethical. Follow the Ten Commandments of Computer Ethics prepared by the Computer Ethics Institute listed in Figure 1.8.

Figure 1.8   The Ten Commandments of Computer Ethics.

1. Thou shalt not use a computer to harm other people.
2. Thou shalt not interfere with other people's computer work.
3. Thou shalt not snoop around in other people's computer files.
4. Thou shalt not use a computer to steal.
5. Thou shalt not use a computer to bear false witness.
6. Thou shalt not copy or use proprietary software for which you have not paid.
7. Thou shalt not use other people's computer resources without authorization or proper compensation.
8. Thou shalt not appropriate other people's intellectual output.
9. Thou shalt think about the social consequences of the program you are writing or the system you are designing.
10. Thou shalt always use a computer in ways that ensure consideration and respect for your fellow human beings.

Source: Computer Ethics Institute, London.

You are also ethically bound to protect your computer at work from security risks and possible system malfunctions. Never be afraid to ask for advice from a co-worker or someone in your firm's information technology (IT) department who knows what to do and how if there is a computer emergency.

Here are some other specific guidelines to follow about using your computer at work:

- Protect your password to access your company's documents as well as its proprietary databases, templates, and other customized applications. Do not share your password, and never use a password belonging to someone else.
- Back up your files. If your computer crashes or your server goes down, you will save yourself, your co-workers, your boss, and your clients time and stress by backing up essential files regularly so that you do not risk losing important information.
- Always save sensitive e-mails, IMs, blogs, memos, letters, etc. that you or your employer may need to document actions or decisions.
- Protect your computer from viruses, worms, phishing (see page 83), etc. by checking with your IT department to make sure the most recent updates to your antivirus, spyware, and malware are installed on your computer.
- Be especially careful in opening attachments or anything you suspect may be infected, such as spam. Never forward a document you think may have a virus.
- Do not use your work e-mail account for personal e-mails (see pages 81–83). Instead, use an alternate e-mail address; you can sign up for a free e-mail account through Hotmail, Yahoo!, Gmail, or similar systems. If you cannot access your e-mail on the job because of a computer emergency, you can use this alternate e-mail address until the problem is solved.

Employers Insist On and Monitor Ethical Behavior

Ethical behavior is crucial to your success in the workplace. Your employer will insist that you are honest, show integrity, and exhibit loyalty in your professional relationships with clients, co-workers, supervisors, inspectors, and vendors.

To make sure you are acting responsibly on the job, employers can legally monitor your work—electronically, through cameras, or by personal visits. Some of these visits are not announced (such as the secret shopper who reports on the customer service he/she received at a store). How many times have you made a call to an organization and heard, "This call may be monitored for quality assurance"? According to a survey conducted by the American Management Association, monitoring employees has risen 45 percent in the past few years and extends to voice mail, e-mail, IMs, and social network sites, like Facebook and Twitter.

Employers can monitor the behavior of their employees for several reasons:

- to determine if a worker is doing his/her job correctly
- to identify wrongdoing on the employee's part
- to improve service, production, communication, transportation
- to ensure legal compliance with state and federal codes
- to heighten security measures

Monitoring gives management solid facts about employee training, performance reviews, and adhering to company policy. But working with integrity means doing the right thing—even when no one is watching.

Some Guidelines to Help You Reach Ethical Decisions

The workplace presents conflicts over who is right and who is wrong, what is best for the company and what is not, and whether a service or product should be changed and why. You will be asked to make a decision and justify it. While this book cannot cover all kinds of ethical problems, here are a few guidelines to help you respond ethically on the job.

1. **Follow your conscience and "to thine own self be true."** You cannot authorize something that you believe is wrong, dangerous, unfair, contradictory, or incomplete. But don't be hasty. Leave plenty of room for diplomacy and for careful questioning and researching. Don't blow a small matter out of proportion.

2. **Be suspicious of convenient (and false) appeals that go against your beliefs.** Watch out for red flags that anyone places in the way of your conscience: "No one will ever know." "It's OK to cut corners every once in a while." "We got away with it last time." "Don't rock the boat." "No one's looking." "As long as the company makes money, who cares?" These rationalizations are traps you must avoid.

3. **Meet your obligations to your employer, your co-workers, your customers, and the global community.** Keeping information from a co-worker who needs it, omitting a fact, justifying unnecessary expenses, concealing something risky about a product or service from an international customer that you otherwise would disclose to a U.S. consumer—all of these are unethical acts.

4. **Take responsibility for your actions.** Saying "I do not know" when you do know can constitute a serious ethical violation. Keep your records up-to-date and accurate, sign and date your work, and never backdate a document to delete information or to fix an error that you committed. Failing to test a set of instructions thoroughly, for example, might endanger readers around the globe.

5. **Honor confidentiality at work.** Never share sensitive/confidential information with individuals who are *not* entitled to see or hear it. You violate corporate trust by telling others about your company's marketing strategies, sales records, personnel decisions, or customer/client interactions. You also have to respect an individual's right to privacy laws. For example, according to Heath Insurance Portability and Accountability Act (HIPPA) guidelines, heath care professionals are not allowed to share a patient's records with unauthorized individuals. It is equally unethical to divulge personal information that a co-worker or supervisor has asked you to keep confidential. In fact, your employer may rightfuly insist that you sign a confidentiality agreement when you are hired.

6. **Document your work carefully and honestly.** Rely on hard evidence: documentation, tests, testimony, valid precedents. Do your homework by studying codes, specifications, books and agency handbooks; confer with a customer or a

co-worker when you are in doubt about an issue. Familiarize yourself with your company's protocols, methods, and materials. Make sure your documents are accurate and comply with appropriate city, state, federal, and international regulations.

7. Keep others in the loop. Confer regularly with your collaborative writing team (see Chapter 2, pages 51–60) and report to your boss as often as you are instructed to give progress reports and to alert him or her about problems. Don't wait to tell your supervisor and/or co-workers about a problem until it gets worse. Prompt and honest notifications are essential to the safety, security, progress, and success of a company.

8. Treat company property respectfully. Use company supplies, networks/computers, equipment, technology, and vehicles responsibly and only for work-related business. Taking supplies home, charging non–work-related expenses (meals, clothes, travel) on a company credit card, surfing the Internet when you are at work—these are just a few instances of unethical behavior.

9. Think green in the workplace. Closely related to number 8 above is being respectful of the environment—whether at the office, at a work site, or in the communities your company serves. Many firms have adopted a green philosophy. To follow this policy at your office, conserve energy by turning off all computers, copiers, and other office machines when you leave work; replace old light bulbs with longer-life ones; recycle paper; copy and print on both sides of paper; view documents on your computer screen instead of printing them; adjust thermostats when you are gone for the day or weekend; and carpool or vanpool. You can also reduce toxic chemicals in the atmosphere by using soy-based ink and inspecting vehicles regularly and reporting any pollution.

10. Weigh all sides before you commit to a conclusion. Research what you write and communicate orally. Do your homework by conferring with co-workers, checking the history of a transaction or other corporate decision, and familiarizing yourself with company policies. Don't rely on office gossip or create problems where there are none. Give people the benefit of the doubt until you have hard evidence to the contrary.

Ethical Dilemmas: Some Scenarios

Sometimes in the workplace you will face situations where there is no clear-cut right or wrong choice. Here are a few scenarios, similar to ones in which you may find yourself, that are gray areas, ethically speaking, along with some possible solutions.

- You work with an office bully who often intimidates co-workers, including you, by talking down to them, interrupting them, or insulting them for their suggestions. At times, this bully has even sent sarcastic e-mails and IMs. You are upset that this behavior has not been reported to management. But you are concerned that if the bully finds out that you have reported the situation the entire office may suffer. How should you handle the problem?

You cannot allow such insulting behavior to go unreported. But first you need to provide documentation about where, when, and how often the bullying has occurred. If you feel uncomfortable talking to the bully, go to your boss and report how this person's actions have threatened the workplace, and ask for assistance from your company's employee assistance program or human resources department. State and federal laws insist that companies provide a safe work "environment", free from intimidation, harassment, or threats of dismissal for reporting bullying.

■ You see an opening for a job in your area but the employer wants someone with a minimum of two years of field experience. You have just completed an internship and had one summer's (12 weeks) experience, which together total almost seven months. Should you apply for the job, describing yourself as "experienced"?

Yes, but honestly state the type and the extent of your field experience and the conditions under which you obtained it.

■ You work for a company that usually assigns commissions to the salesperson a customer requests. One afternoon a customer asks for a salesperson who happens to have the day off. You assist the customer all afternoon and even arrange to have an item shipped overnight so that the customer can have it in the morning. When you ring up the sale, should you list your employee number for the commission or the off-duty employee's?

You probably should defer crediting the sale to either number until you speak to the absent employee and suggest a compromise — possibly splitting the commission.

■ A piece of computer equipment, scheduled for delivery to your customer the next day, arrives with a damaged part. You decide to replace it at your store before the customer receives it. Should you inform the customer?

Yes, but assure the customer that the equipment is still under the same warranty and that the replacement part is new and also under the same warranty. If the customer protests, agree to let him or her use the computer until a new unit arrives.

As these brief scenarios suggest, sometimes you have to make concessions and compromises to be ethical in the world of work.

Writing Ethically on the Job

Your writing as well as your behavior must be ethical. Words, like actions, have implications and consequences. If you slant your words to conceal the truth or gain an unfair advantage, you are not being ethical. False reporting and advertising are false writing. Bias and omission of facts are wrong. Strive to be fair, reliable, and accurate in reporting events, statistics, and trends.

Unethical writing is usually guilty of one or more of the following faults, which can conveniently be listed as the three *M*'s: *m*isquotation, *m*isrepresentation, and *m*anipulation. Here are eight examples:

1. **Plagiarism** is stealing someone else's words and claiming them as your own without documenting the source. Do not think that by changing a few words of someone else's writing here and there you are not plagiarizing. Give proper credit to your source, whether in print, in person (through an interview), or online. The penalties for plagiarism are severe—a reprimand or even the loss of your job. See pages 328–329 for further advice on how to avoid plagiarism.

2. **Selective misquoting** deliberately omits damaging or unflattering comments to paint a better (but untruthful) picture of you or your company. By picking and choosing only a few words from a quotation, you unethically misrepresent what the speaker or writer originally intended.

Full Quotation:	I've enjoyed at times our firm's association with Graphics Inc., although I was troubled by the uneven quality of their service. At times, it was excellent while at others it was far less so.
Selective Misquotation:	I've enjoyed … our firm's association with Graphics Inc. The quality of their service was … excellent.

The dots, called *ellipses*, unethically suggest that only extraneous or unimportant details were omitted.

3. **Arbitrary embellishment of numbers** unethically misrepresents by increasing or decreasing percentages or statistical and other numerical information. It is unethical to stretch the differences between competing plans or proposals to gain an unfair advantage or to express accurate figures in an inaccurate way.

Embellishment:	An overwhelming majority of residents voted for the new plan.
Ethical:	The new plan was passed by a vote of 53 to 49.

Embellishment:	Our competitor's sales volume increased by only 10 percent in the preceding year while ours doubled.
Ethical:	Our competitor controls 90 percent of the market, yet we increased our share of that market from 5 percent to 10 percent last year.

4. **Manipulation of information or context.** Closely related to number 3, is the misrepresentation of events, usually to "put a good face" on a bad situation. The writer here unethically uses slanted language and intentionally misleading euphemisms to misinterpret events for readers.

Manipulation:	Looking ahead to 2012, the United Funds Group is exceptionally optimistic about its long-term prospects in an expanding global market. We are happy to report steady to moderate activity in an expanding sales environment last year. The United Funds Group seeks to build on sustaining investment opportunities beneficial to all subscribers.
Ethical:	Looking ahead to 2012, the United Funds Group is optimistic about its long-term prospects in an expanding global market. Though the market suffered from the recession this year, the United Funds Group hopes to recoup its losses in the year ahead.

The writer minimizes the negative effects of the recession by calling it "an expanding sales environment."

5. Using fictitious benefits to promote a product or service seemingly promises customers advantages but delivers none.

False Benefit: Our bottled water is naturally hydrogenated from clear underground springs.

Truth: All water is hydrogenated because it contains hydrogen.

6. Unfairly characterizing (by exaggerating or minimizing) hiring or firing conditions is unethical.

Unethical: One of the benefits of working for Spelco is the double pay you earn for overtime.

Truth: Overtime at Spelco is assigned on the basis of seniority.

Unethical: Our corporate restructuring will create a more efficient and streamlined company, benefiting management and workers alike.

Truth: Downsizing has led to 250 layoffs this quarter.

Companies faced with laying off employees want to protect their corporate image and maintain their stockholders' good faith to do so, so they often "put the best face" on such an action as the unethical example above shows.

7. Manipulating international readers by adopting a condescending view of their culture and economy is unethical.

Unethical: Since our product has appealed to U.S. customers for the last 16 months, there's no doubt that it will be popular in your country as well.

Fair: Please let us know if any changes in product design or construction may be necessary for potential customers in your country.

8. Misrepresenting through distortion or slanted visuals is one of the most common types of unethical communication. Making a visual appear bigger, smaller, or more or less favorable is all too easy with graphics software packages. See pages 230–235 of Chapter 6 for guidelines on how to prepare ethical visuals.

In sum ethical writing is clear, accurate, fair, and honest. These are among the most important goals of any workplace communication. Because ethics is such an important topic in writing for the business world, it will be emphasized throughout this book.

Successful Employees Are Successful Writers

As this chapter has stressed, being a successful employee means being a successful writer at work. The following guidelines, which summarize the key points of this chapter, will help you to be both.

1. Know your job—assignments, roles, responsibilities, goals, what you need to write, and what you shouldn't, and to whom.
2. Analyze your audience's needs and what they will expect to find in your written work.
3. Write clearly and appropriately for your international readers.
4. Respect the cultural diversity of co-workers, managers, customers, and vendors whether they live in the United States or in another country.

5. Make sure your written work is accurate, relevant, and practical, and include culturally appropriate visuals to help readers understand your message.
6. Document, document, document. Submit everything you write with clear-cut evidence based on factual details and persuasive, logical interpretations.
7. Keep your computer safe from viruses and other computer emergencies.
8. Be prepared to give and to receive feedback from co-workers, managers, and customers to meet all deadlines.
9. Follow your company's policy, and promote your company's image and traditions.
10. Be ethical in what you say, write, illustrate, and do.

 ## Revision Checklist

At the end of each chapter of this book you will find a checklist you should review before you submit the final copy of your work, either to your instructor or to your boss. These checklists specify the types of research, planning, drafting, editing, and revising you should do to ensure the success of your work. Regard each checklist as a summary of the main ideas in the chapter as well as a handy guide to quality control. You may find it helpful to check each box as you verify that you have performed the necessary revision/review. Keep in mind that effective writers are also careful editors.

- [ ] Displayed respect for and appropriately shaped my message for the global marketplace with international readers.
- [ ] Identified my audience—their background, knowledge of English, reason for reading my work, and likely response to my work and me.
- [ ] Tailored my message to my audience's needs and background, giving them neither too little nor too much information.
- [ ] Pushed to the main point right away; did not waste my reader's time.
- [ ] Selected the most appropriate language, technical level, tone, and level of formality.
- [ ] Did not waste my audience's time with unsupported generalizations or opinions; instead gave them accurate measurements, facts, and carefully researched material.
- [ ] Included appropriate visuals to make my work easier for my audience to follow.
- [ ] Used persuasive reasons and data to convince my reader to accept my plan or work.
- [ ] Ensured that my writing and visuals are ethical—accurate, fair, honest, a true reflection of the situation or condition I am explaining or describing, for U.S. as well as global audiences.
- [ ] Followed the "Ten Commandments of Computer Ethics".
- [ ] Adhered to the ethical codes of my profession as well as those policies and regulations set down by my employer.
- [ ] Gave full and complete credit to any sources I used, including resource people.
- [ ] Avoided plagiarism and the unfair or dishonest use of copyrighted materials, both written and visual, including all electronic media.

Exercises

1. Write a memo (see pages 70–77 for format) addressed to a prospective supervisor to introduce yourself. Your memo should have four headings: **Education**—including goals and accomplishments; **Job Information**—where you have worked and your responsibilities; **Community Service**—volunteer work, church work, youth groups; and **Writing Experience**—your strengths and the audiences for whom you have written.

2. Bring to class a set of printed instructions, a memo, a sales letter, a brochure, or a homepage. Comment on how well the example answers the following questions.
   a. Who is the audience?
   b. Why was the material written?
   c. What is the message?
   d. Are the style and tone appropriate for the audience, the purpose, and the message? Why? Why not?
   e. Discuss the use of any visuals and color in the document. For instance how does color (or the lack of it) affect an audience's response to the message?

3. Find an advertisement that contains a drawing or photograph. Bring it to class together with a paragraph of your own (75–100 words) describing how the message of the ad is directed to a particular audience and commenting on why the illustration was selected for that audience.

4. Pick one of the following topics, and write two descriptions of it. In the first description, use technical vocabulary. In the second, use language suitable for the general public.
   a. spark plug
   b. blood pressure cuff
   c. flash drive
   d. computer chip
   e. bluetooth
   f. legal contract
   g. electric sander
   h. cyberspace
   i. muscle
   j. protein
   k. BlackBerry
   l. social networking
   m. bread
   n. money
   o. iPod
   p. soap
   q. blogging
   r. computer virus
   s. swine flu
   t. thermostat
   u. trees
   v. food processor
   w. earthquake
   x. recycling

5. Redo Exercise 4 as a collaborative writing project.

6. Select one article from a newspaper and one article either from a professional journal in your major field or from one of the following journals: *Advertising Age, American Journal of Nursing, Business Marketing, Business Week, Computer, Computer Design, Construction Equipment, Criminal Justice Review, E-Commerce, Food Service Marketing, Journal of Forestry, Journal of Soil and Water Conservation, National Safety News, Nutrition Action, Office Machines, Park Maintenance, Scientific American*. State how the two articles you selected differ in terms of audience, purpose, message, style, and tone.

7. Assume that you work for Appliance Rentals, Inc., a company that rents TVs, microwave ovens, stereo components, and the like. Write a persuasive letter

to the members of a campus organization or civic club urging them to rent an appropriate appliance or appliances. Include details in your letter that might have special relevance to members of this specific organization.

8. How do the visuals and the text of the Sodexho advertisement on page 30 stress to current (and potential) employees, customers, and stockholders that the company is committed to diversity in the workplace? Also explain how the ad illustrates the functions of on-the-job writing as defined on pages 14–16.

9. Write a letter to an Internet service provider that has mistakenly billed you for caller ID equipment that you never ordered, received, or needed.

10. The following statements contain embellishments, selected misquotations, false benefits, and other types of unethical tactics. Revise each statement to eliminate the unethical aspects.
    a. Storm damage done to water filtration plant #3 was minimal. While we had to shut down temporarily, service resumed to meet residents' needs.
    b. All customers qualify for the maximum discount available.
    c. The service contract … on the whole … applied to upgrades.
    d. We followed the protocols precisely with test results yielding further opportunities for experimentation.
    e. All our costs were within fair-use guidelines.
    f. Customers' complaints have been held to a minimum.
    g. All the lots we are selling offer a relatively easy access to the lake.

11. You work for a large international company and a co-worker tells you that he has no plans to return to his job after he takes his annual two-week vacation. You know that your department cannot meet its deadlines short-handed and that your company will need at least two or three weeks to recruit and hire a qualified replacement. You also know that it is your company's policy not to give paid vacations to employees who do not agree to work for at least three months following their return. What should you do? What points would you make in a confidential e-mail to your boss? What points would you raise to your co-worker?

12. Your company is regulated and inspected by the Environmental Protection Agency (EPA). In 90 days, the EPA will relax a particular regulation about dumping occupational waste. Your company's management is considering cutting costs by relaxing the standard now, before the new, easier regulation is in place. You know that the EPA inspector probably will not return before the 90-day period elapses. What do you recommend to management?

13. You and your co-workers have been intimidated by an office bully, a 12-year employee who has seniority. As a collaborative writing project (see pages 51–60), draft a letter to the head of your human resources department documenting instances of the bully's actions and specific assistance and asking for advice on how to proceed.

14. A co-worker takes extended lunch breaks and often comes in late and leaves early. Your department is under minimal supervision from an office manager and so there is no boss looking over your colleague's shoulder day by day. You have covered for him several times because you do not want him to be reprimanded or,

Advertisement for Exercise 8

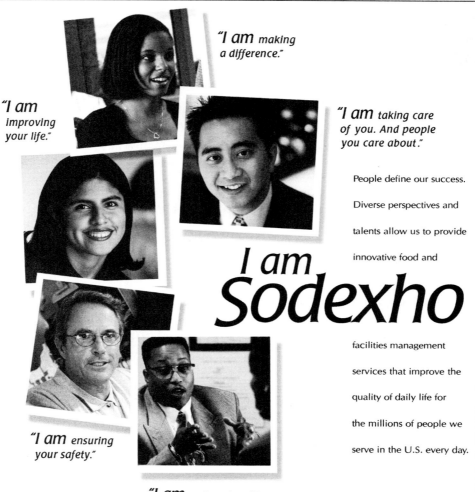

Courtesy of Sodexo, Inc. (Sodexho, Inc. became Sodexo, Inc. in 2008)

worse yet, fired. He is taking advantage of your friendship and now unfairly wants you to cover for him routinely. Write an appropriate memo to your co-worker and to the office managers.

2

The Writing Process and Collaboration at Work

In Chapter 1 you learned about the different functions of writing for the world of work and also explored some basic concepts all writers must master. To be a successful writer, you need to

- identify your audience's needs
- determine your purpose in writing to that audience
- make sure your message meets your audience's needs
- use the most appropriate style and tone for your message
- format your work so that it clearly reflects your message

Visit www. cengage.com/ english/kolin/ writingatwork-concise3e for this chapter's online exercises, ACE quizzes, and Web links.

Just as significant to your success is knowing how effective writers in the business world actually create their work for their audiences. This chapter gives you some practical information about the strategies and techniques careful writers use when they work. These procedures are a vital part of what is known as the *writing process*. This process involves such matters as how writers gather information, how they transform their ideas into written documents, and how they organize and revise what they have written to make it suitable for their audiences. This chapter will also show you how writers in the world of work collaborate to create a document.

What Writing Is and Is Not

As you begin your study of writing for the world of work, it might be helpful to identify some notions about what writing is and what it is not.

What Writing Is

- **Writing is a dynamic process; it is not static.** It enables you to discover and evaluate your thoughts as you carefully draft and revise.
- **A piece of writing changes as your thoughts and information change** and as your view of the material changes.

- **Writing takes time.** Some people think that revising and polishing are too time consuming. But poor writing actually takes more time and costs more money in the end. It can lead to misunderstandings, lost sales, product recalls, low morale, and even damage to your reputation and that of your company.
- **Writing means making a number of judgment calls.** You have choices to make to meet an audience's needs (see pages 6–10).
- **Writing grows sometimes in bits and pieces and sometimes in larger spurts.** It needs many revisions; an early draft is never a final copy.

What Writing Is Not

- **Writing is not a mysterious process, known only to a few.** Even if you have not done much writing before, you can learn to do it effectively.
- **Writing is not simply following a magical formula.** Successful writing requires hard work and thoughtful effort, not simply following a formula, as if you were painting by numbers. Writing does not proceed in some predictable way where introductions are always written first and conclusions last.
- **Writing is not completed in a first attempt.** Just because you put something down on paper or on a computer screen does not mean it is permanent and unchangeable. Writing means *re*writing, *re*vising, and *re*thinking. The better a piece of writing is, the more a writer or group of writers have reworked it.

Researching

Before you start to compose any e-mail, memo, letter, blog, proposal, or report, you'll need to do some research. Research is crucial to obtain the right information for your audience. Information must be factually correct, timely, intellectually significant, and appropriate. The world of work is based on conveying information—the logical presentation and sensible interpretation of facts.

Don't ever think you are wasting time by not starting to write your report or letter immediately. Actually, you will waste time and risk doing a poor job if you do not find out as much as possible about your topic (and your audience's interest in it). Do some necessary research.

First, find out as much as you can about your assignment and your readers, and what they expect from your written work. Next, determine the exact kind of research you must do to gather and interpret the information your audience needs. Your research can include

- conferring with people inside and outside your company
- consulting notes from conferences and meetings
- collaborating in person, by e-mail, or by instant messaging (IM)
- getting briefings from sales or technical staff

- doing Internet searches
- locating and evaluating websites
- reading current periodicals, reports, and other relevant documents
- evaluating reports, products, and services
- contacting vendors, customers, and inspectors
- comparing/contrasting products, services, locations
- surveying customers' views/opinions

Keep in mind that research is not confined to just the beginning of the writing process; it goes on throughout the various stages of creating a document.

Planning

At this stage in the writing process, your goal is to get something—anything—down on paper or on your computer screen. For most writers, getting started is the hardest part of the job. But you will feel more comfortable and confident once you begin to see your ideas written down before your eyes. It is always easier to clarify and criticize something you can see.

Getting started is also easier if you have researched your topic, because you have something concrete to say and to build on. Each part of the process relates to and supports the next. Careful research prepares you to begin writing.

Still, getting started is not easy. Take advantage of a number of widely used strategies that can help you to develop, organize, and tailor the right information for your audience. Use any one of the following techniques, alone or in combination.

1. Clustering. In the middle of a sheet of paper, write the word or phrase that best describes your topic, and then start writing other words or phrases that come to mind. (It is also possible to do this on your computer screen using a "mind mapping" software program such as MindGenius or MindMapper to create clusters.) As you write, circle each word or phrase and connect it to the word from which it sprang. Note the clustered grouping in Figure 2.1 for a report encouraging a manager to switch to flextime—a system in which employees can work on a flexible time schedule within certain limits. The resulting diagram gives the writer a rough sense of some of the major divisions of the topic and where they may belong in the report.

2. Brainstorming. At the top of a sheet of paper or your computer screen, describe your topic in a word or phrase and then list any information you know or found out about that topic—in any order and as quickly as you can. Brainstorming is like thinking aloud except that you are recording your thoughts.

- Don't stop to delete, rearrange, or rewrite anything, and don't dwell on any one item.
- Don't worry about spelling, punctuation, grammar, or whether you are using words and phrases instead of complete sentences.

Figure 2.1   **Clustering of ideas to prepare a report on flextime.**

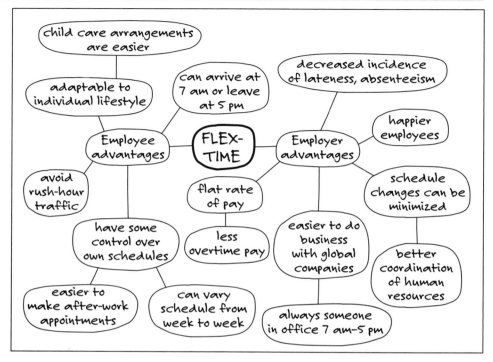

- Keep the ideas flowing. The result may well be an odd assortment of details, comments, and opinions.
- After stepping away from the list for a few minutes (or hours) and returning with fresh eyes, expect to add and delete some ideas or combine or rearrange others as you start to develop your topic in more detail.

Figure 2.2 (p. 35) shows Marcus Weekley's initial brainstormed list for a report to his boss on purchasing a new color laser all-in-one printer. After he began to revise it, he realized that some items were not relevant for his audience (6, 8, and 13). He also recognized that some items were repetitious (1, 2, and 11). Further investigation revealed that his company could purchase a printer for far less than his initial high guess (17). As Weekley continued to work on his list and the overall topic became clearer to him, he added and deleted points.

3. Outlining. For most writers, outlining may be the easiest and most comfortable way to begin or to continue planning their report or letter. Outlines typically go through several revisions and are for no one's eyes but your own, so don't worry if your first attempt is brief, messy, informal, or incomplete. Clarify and tighten your outline as you revise.

You can make easy-to-read, easy-to-change outlines on your computer. A word processing program can help you create an outline with main and subpoints

Marcus Weekley's initial, unrevised brainstormed list.   Figure 2.2

1. combines four separate pieces of equip—printer, copier, fax, and scanner

2. more comprehensive than our current configuration of four pieces of equip

3. would coordinate with office furniture

4. energy efficiency increased due to fewer machines being used

5. more scalable fonts

6. one machine interfaces with all others in same-case housing

7. scanner makes photographic-quality pictures

8. print capabilities are a real contribution to technology

9. increase communication abilities through fax machine

10. new scanner picture quality better than current scanner

11. only have to buy one machine as opposed to four

12. speed of fax allows quick response time

13. stock is doing better on Wall Street compared to other equities

14. reducing our advertising costs through use of color printer

15. increase work area available

16. would help us do our work better

17. top-of-line models can be bought for $4,500

List is not organized but simply reflects writer's ideas about possible topics

1, 2, and 11 are repetitious

6, 8, and 13 are not relevant for audience or purpose

4, 9, 12, and 14 are of special interest to decision makers concerned about costs

Research will show price is too high

to help you organize your ideas when you start to draft. You can also add notes to remind yourself where you may have to fill in gaps or insert a transition from one idea to another. Note how Marcus Weekley organized his revised brainstormed list into an outline in Figure 2.3 on page 36.

Figure 2.3 **Marcus Weekley's early outline after revising his brainstormed list.**

Outline form helps writer group ideas/topics and go to next step in revision

Headings correspond to major sections of report

Outline reflects scope and details of writer's research

Last section of outline also functions as a conclusion

I. Convenience/Capabilities of all-in-one in laser printer

A. Would reduce number of machines having to be serviced

B. Can be configured easily for our network system

C. Easy to install and to operate

D. 33.6 Kbps fax machine would increase our communication

E. 2400 × 1200 dpi copier means better copy quality

F. 4800 × 4800 dpi scanner means higher quality pictures than current scanner provides

II. Time/Efficiency

A. 50 ppm printer is nearly twice as fast as current printer

B. 33.6 Kbps speed fax allows quick response time

C. Greater graphics capability—130 scalable fonts

D. Scanner compatible with our current PhotoEdit imaging/graphics software

E. 100,000-page monthly duty cycle means less maintenance

III. Money

A. Costs less overall for multitasking printer than combined four machines

B. Reduced monthly power bill by using one machine rather than four

C. Included ScanText and WordPort software means not buying new software to network office computers

D. Save on service costs

E. Reduced advertising costs through printer 50–400% enlargement/reduction options, which allow for more in-house advertising

Drafting

If you have done your planning carefully, you will find it easier to start your first draft. When you draft, you convert the words and phrases from your outlines, brainstormed lists, or clustered groups into paragraphs. During drafting, as elsewhere in the writing process, you will see some overlap as you look back over your lists or outlines to shape your document.

Don't expect to wind up with a polished, complete version of your letter, report, or proposal after working on only one draft. In most cases, you will have to work through many drafts, but each draft should be less rough and more targeted than the preceding one.

Key Questions to Ask as You Draft

As you work on your drafts, ask yourself the following questions about your content and organization.

- Am I giving my readers too much or too little information?
- Does this point belong where I have it, or would it more logically follow or precede something else?
- Is this point necessary and relevant?
- Am I repeating or contradicting myself?
- Have I ended appropriately for my audience?

To answer the questions successfully, you may have to continue researching your topic and reexamining your audience's needs. But, in the process, new and even better ideas may come to you, and the ideas you originally thought were essential may in time appear to be unworkable and unnecessary.

Guidelines for Successful Drafting

Following are some suggestions to help your drafting go more smoothly and efficiently.

- In an early draft, write the easiest part first. Some writers feel more comfortable drafting the body (or middle) of their work first. See the differences between the short report in Figures 2.4 and 2.5, pages 39–40.
- As you work on a later draft, write straight through. Do not worry about spelling, punctuation, or the way a word or sentence sounds. Save those concerns for later stages.
- Allow enough time between drafts so that you can evaluate your work with fresh eyes and a clear mind.
- Get frequent outside opinions. Show, e-mail, or fax a draft to a co-worker or a supervisor for comment. A new pair of eyes will see things you missed. As we'll see, collaboration is essential in the workplace (see pages 51–58).
- Start considering if visuals would enhance the quality of your work and, if so, decide on what types and where best to insert them.

Revising

Revision is an essential stage in the writing process. It requires more than giving your work another quick glance. Do not be tempted to skip the revision stage just because you have written the required number of words or sections or because you think you have put in too much time already. Revision is done *after* you produce a draft that you think conveys the appropriate message for your audience. The quality of your memo, letter, or report depends on the revisions you make now.

Allow Enough Time to Revise

Like planning or drafting, revision is not done well in one big push. It evolves over a period of time. Make sure you budget enough time to do it carefully.

- Avoid drafting and revising in one sitting. If possible, wait at least a day before you start to revise. (In the busy work world, waiting a couple of hours may have to suffice.)
- Ask a co-worker or friend familiar with your topic to comment on your work.
- Plan to read your revised work more than once.

Revision Is Rethinking

When you revise, you *resee*, *rethink*, and *reconsider* your entire document. You ask questions about the major issues of content, organization, tone, and format (see pages 198–207). Revision involves going back and repeating earlier steps in the writing process.

Revision means asking again the questions you have already asked and answered during the planning and drafting stages. During the process, you will discover gaps to fill, points to change, and errors to correct in your draft. Revision gives you a second (or third or fourth) chance to get things right for your audience. Take advantage of the document tracking options (see pages 58–59) such as Track Changes and Edit that allow you to see your additions, cuts, and moves in a different color.

Key Questions to Ask as You Revise

By asking and successfully answering the following questions as you revise, you can discover gaps or omissions, points to change, and errors to correct in your draft.

Content

1. Is it accurate? Are my facts (figures, names, addresses, dates, costs, references, warranty terms, statistics) correct?
2. Is it relevant for my audience and purpose? Have I included information that is unnecessary, too technical, concrete or inappropriate?
3. Have I given enough concrete evidence to explain things adequately and to persuade my readers? (Too little information will make readers skeptical about what you are describing or proposing.) Have I left anything out?

Organization

1. Have I clearly identified my main points and shown readers why those points are important?
2. Is everything in the right, most effective order? Should anything be switched or moved closer to the beginning or the end of my document?
3. Have I spent too much (or too little) effort on one section? Do I repeat myself? What can be cut? Where and why?
4. Have I grouped related items in the same part of my report or letter, or have I scattered details that need to appear in one paragraph or section?

Tone

1. How do I sound to my readers—professional and sincere, or arrogant and unreliable? What attitude/tone do my words or expressions convey?
2. How will my readers, native speakers as well as an international audience, think I perceive them—honest and intelligent or unprofessional and uncooperative?

Case Study: A "Before" and "After" Revision

Mary Fonseca, a staff member at Seacoast Labs, was asked by her supervisor to prepare a short report for the general public on the Lab's most recent experiments. Figures 2.4 and 2.5 show her "before" and "after" revisions.

Unorganized opening paragraphs of Mary Fonseca's "before" draft.   Figure 2.4

Drag is an important concept in the world of science and technology. It has many implications. Drag occurs when a ship moves through the water and eddies build up. Ships on the high seas have to fight the eddies, which results in drag. In the same way, an airplane has to fight the winds at various altitudes at which it flies; these winds are very forceful, moving at many knots per hour. All these forces of nature are around us. Sometimes we can feel them, too. We get tired walking against a strong wind. The eddies around a ship are the same thing. These eddies form various barriers around the ship's hull. They come from a combination of different molecules around the ship's hull and exert quite a force. Both types of molecules pull against the ship. This is where the eddies come in.

Information hard to follow and not relevant for audience

Does not explain process very well

Scientists at Seacoast Labs are concerned about drag. Dr. Karen Runnels, who joined Seacoast about three years ago, is the chief investigator. She and her team of highly qualified experts have constructed some fascinating multilevel water tunnels. These tunnels should be useful to ship owners. Drag wastes a ship's fuel.

Important point not developed

Figure 2.5   The "after" draft—a revision of Mary Fonseca's "before" draft in Figure 2.4.

Effective use of definition and headings

What is Drag?

We cannot see or hear many of the forces around us, but we can certainly detect their presence. Walking or running into a strong wind, for example, requires a great deal of effort and often quickly leaves us feeling tired. When a ship sails through the water, it also experiences these opposing forces known as **drag**. Overcoming drag causes a ship to reduce its energy efficiency, which leads to higher fuel costs.

Describes cause and effect of drag

How Drag Works

It is not easy for a ship to fight drag. As the ship moves through the water, it drags the water molecules around its hull at the same rate the ship is moving. Because of the cohesive force of those molecules, other water molecules immediately outside the ship's path get pulled into its way. All the molecules become tangled rather than simply sliding past each other. The result is an eddy, or small circling burst of water around the ship's hull, which intensifies the drag. Dr. Jorge Fröes, a highly respected structural engineer, explains the process using this analogy: "When you put a spoon in honey and pull it out, half the honey comes out with the spoon. That's what is happening to ships. The ship is moving and at the same time dragging the ocean with it."

Uses an easy-to-follow analogy

Reducing Drag

At Seacoast Labs, scientists are working to find ways to reduce drag on ships. Dr. Karen Runnels, the principal investigator, and a team of researchers have constructed water tunnels to simulate the movement of ships at sea. The drag a ship encounters is measured from the tiny air bubbles emitted in the water tunnel. Dr. Runnels's team has also developed the use of polymers, or long carbon chain molecules, to reduce drag. The polymers act like a slimy coating for the ship's hull to help it glide through the water more easily. When asbestos fibers were added to the polymer solutions, the investigators measured a 90 percent reduction in drag. The team has also experimented with an external pump attached to the hull of a ship, which pushes the water away from a ship's path, saving fuel and time.

Clearly explains the research and its importance for readers

When starting to revise Figure 2.4, Fonseca realized that it lacked focus. It jumped back and forth between drag on ships and drag on airplanes. Because Seacoast Labs did not work on planes, she wisely decided to drop that idea. She also realized that the information on the effects of drag, something Seacoast was work-ing on, was so important it deserved a separate paragraph. In light of this key idea, she realized that her explanation of molecules, eddies, and drag needed to be made more reader-friendly. Researching further, she decided to add a new paragraph on the causes and effects of drag, which became paragraph 2 in Figure 2.5.

Yet by pulling ideas about drag and its effects from her longish first paragraph in Figure 2.4, Fonseca was left with the job of finding an opening for her report. Buried in her original opening paragraph was the idea that we cannot always see the forces of nature but "we can feel them." She thought this analogy of walking against the wind and drag would work better for her audience of nonspecialists than the original wooden remarks she had started with.

Although her organization and ideas were far better in her second draft than those in her "before" draft in Figure 2.4, she realized she had said very little about her employer, Seacoast Labs. Doing more research, she found information about Seacoast's experiments and why they were so important in saving money. This information was far more significant and relevant than saying Dr. Runnels had been at Seacoast for three years.

Through revision and further research, then, Mary Fonseca transformed two poorly organized and incomplete paragraphs into three separate yet logically connected ones that highlighted her employer's work. Thanks to her revision (Figure 2.5), she came up with two very helpful headings—"What is Drag?" and "How Drag Works"—for her nonspecialist readers.

Editing

Editing is quality control for your reader. This last stage in the writing proc-ess might be compared to detailing an automobile—the preparation a dealer goes through to ready a new car for prospective buyers. Editing is done only after you are completely satisfied that you have made all the big decisions about content and organization—that you have said what you wanted to, where and how you intended, for your audience.

When you edit, you will check your work for

- sentences
- word choices
- punctuation and spelling
- grammar and usage
- clarity

As with revising, don't skip or rush through the editing process, thinking that once your ideas are down, your work is done. Your style, punctuation, spelling, and grammar matter a great deal to readers. If your work is hard to read or contains mistakes in spelling or punctuation, readers will think that your ideas and your research are also faulty. Also edit for sexist language (see pages 47–50).

The following sections give you some basic guidelines about what to look for when you edit your sentences and words. The appendix, "A Writer's Brief Guide to Paragraphs, Sentences, and Words" (pages 376–394), will give you helpful suggestions on correct spelling and punctuation.

Editing Guidelines for Writing Lean and Clear Sentences

Here are four of the most frequent complaints readers voice about poorly edited writing in the world of work:

- The sentences are too long. I could not follow the writer's meaning.
- The sentences are too complex. I could not understand what the writer meant the first time I read the work; I had to reread it several times.
- The sentences are unclear. Even after I reread them, I am not sure I understood the writer's message.
- The sentences are too short and simplistic. The writing felt "dumbed down."

Writing clear, concise, readable sentences is not always easy. It takes effort, but the time you spend editing will pay off in rich dividends for you and your readers. The seven guidelines that follow should help with the editing phase of your work.

1. Avoid needlessly complex or lengthy sentences. Do not pile words on top of words. Instead, edit one overly long sentence into two or even three more manageable ones.

Too long:	The planning committee decided that the awards banquet should be held on March 15 at 6:30, since the other two dates (March 7 and March 22) suggested by the hospitality committee conflict with local sports events, even though one of those events could be changed to fit our needs.
Edited for easier reading:	The planning committee has decided to hold the awards banquet on March 15 at 6:30. The other dates suggested by the hospitality committee—March 7 and March 22—conflict with two local sports events. Although the date of one of those sports events could be changed, the planning committee still believes that March 15 is our best choice.

2. Combine short, choppy sentences. Don't shorten long, complex sentences, only to turn them into choppy, simplistic ones. A memo, e-mail, letter, or report written exclusively in short, staccato sentences sounds immature.

When you find yourself looking at a series of short, blunt sentences, as in the following example, combine them where possible and use connective words similar to those italicized in the edited version.

Choppy:	Medical transcriptionists have many responsibilities. Their responsibilities are important. They must be familiar with medical terminology. They must listen to dictation. Sometimes physicians talk very fast. Then the transcriptionist must be quick to transcribe what is heard. Words could be missed. Transcriptionists must forward reports. These reports have to be approved. This will take a great deal of time and concentration. These final reports are copied and stored properly for reference.
Edited:	Medical transcriptionists have many important responsibilities. *These* include transcribing physicians' orders using correct medical terminology. *When* physicians talk rapidly, transcriptionists have to keyboard accurately

so that no words are omitted. *Among their most demanding* duties are keyboarding and forwarding transcriptions *and then*, after approval, storing copies properly for future reference.

3. Edit sentences to tell who does what to whom or what. The clearest sentence pattern in English is the subject-verb-object (s-v-o) pattern.

> s   v   o
> Sue mowed the grass.

>   s   v   o
> Our website contains a link to key training software programs.

Readers find this pattern easiest to understand because it provides direct and specific information about the action. Hard-to-read sentences obscure or scramble information about the subject, the verb, or the object. In the following unedited sentence, subjects are hidden in the middle rather than being placed in the most crucial subject position.

> Unclear: The control of the ceiling limits of glycidyl ethers on the part of the employers for the optimal safety of workers in the workplace is necessary. (Who is responsible for taking action? What action must they take? For whom is such action taken?)
>
> Edited: Employers must control the ceiling limits of glycidyl ethers for their workers' safety.

4. Use strong, active verbs rather than verb phrases. In trying to sound important, many bureaucratic writers avoid using simple, graphic verbs. Instead, these writers use a weak verb phrase (for example, *provide maintenance of* instead of *maintain*, *work in cooperation with* instead of *cooperate*). Such verb phrases imprison the active verb inside a noun format and slow a reader down. Note how the edited version here rewrites the weak verb phrase.

> Weak: The city provided the employment of two work crews to assist the strengthening of the dam.
>
> Strong: The city employed two work crews to strengthen the dam.

5. Avoid piling modifiers in front of nouns. Putting too many modifiers (words used as adjectives) in the reader's path to the noun is confusing for readers, who will have trouble deciphering how one modifier relates to another modifier or to the noun. To avoid that problem, edit the sentence to place some of the modifiers after or before the nouns they modify.

> Crowded: The vibration noise control heat pump condenser quieter can make your customers happier.
>
> Readable: The quieter on the condenser for the heat pump will make your customers happier by controlling noise and vibrations.

6. Replace wordy phrases or clauses with one- or two-word synonyms.

> Wordy: The college has parking zones for different areas for people living on campus as well as for those who do not live on campus and who commute to school.

Edited: The college has different parking zones for resident and commuter students. (Twenty words of the original sentence—everything after "areas for"—have been reduced to four words: "resident and commuter students.")

7. Combine sentences beginning with the same subject or ending with an object that becomes the subject of the next sentence.

Wordy: I asked the inspector if she were going to visit the plant this afternoon. I also asked her if she would come alone.

Edited: I asked the inspector if she were going to visit the plant alone this afternoon.

Wordy: Homeowners want to buy low-maintenance bushes. These low-maintenance bushes include the ever-popular holly and boxwood varieties. These bushes are also inexpensive.

Edited: Homeowners want to buy low-maintenance and inexpensive bushes such as holly and boxwood. (This revision combines three sentences into one, condenses 24 words into 14, and joins three related thoughts.)

Editing Guidelines for Eliminating Unnecessary Words

Too many people in business think the more words, the better. Nothing could be more self-defeating. Your readers are busy; unnecessary words slow them down. Make every word work. Cut out any words you can from your sentences. If the sentence still makes sense and reads correctly, you have eliminated wordiness.

1. Replace wordy phrases with precise ones. For example, in the following list, the wordy phrases on the left should be replaced with the precise words on the right.

Wordy	Concisely Edited
at a slow rate	slowly
at this point in time	now
be in agreement with	agree
bring to a conclusion	conclude, end
come to terms with	agree, accept
due to the fact that	because
express an opinion that	affirm
for the period of	for
in such a manner that	so that
in the area /case /field of	in
in the neighborhood of	approximately
long period of time	long
look something like	resemble
show a tendency to	tend
with a view to	to
with reference to	regarding, about
with the result that	so

2. Use concise, not redundant, phrases. Another kind of wordiness comes from using redundant expressions—saying the same thing a second time, only in different words. "Fellow colleague," "component parts," "corrosive acid," and "free gift" are phrases that contain this kind of double speech; a fellow *is* a colleague, a component *is* a part, acid *is* corrosive, and a gift *is* free. In the examples below, the suggested changes on the right are preferable to the redundant phrases on the left.

Redundant	Concise
absolutely essential	essential
advance reservations	reservations
basic necessities	necessities, needs
close proximity	proximity, nearness
each and every	each
end result	result
final conclusions/final outcome	conclusions/outcome
first and foremost	first
full and complete	full, complete
future plans	plans
personal opinion	opinion
present status	status
refer back	refer
tried and true	tried, proven

3. Watch for repetitive words, phrases, or clauses within a sentence. Sometimes one sentence or one part of a sentence needlessly duplicates another.

Redundant: To provide more room for employees' cars, the security department is studying ways to expand the employees' parking lot.

Edited: The security department is studying ways to expand the employees' parking lot. (Because the first phrase says nothing that the reader does not know from the independent clause, cut it.)

4. Avoid unnecessary prepositional phrases. Adding a prepositional phrase can sometimes contribute to redundancy. The italicized words below are unnecessary. Be on the lookout for these phrases and delete them.

audible *to the ear*	second *in sequence*	soft *in texture*
bitter *in taste*	short *in duration*	tall *in height*
fly *through the air*	hard *to the touch*	twenty *in number*
orange *in color*	honest *in character*	visible *to the eye*
rectangular *in shape*	light *in weight*	loud *in volume*

Figure 2.6 shows an e-mail that Trudy Wallace wants to send to her boss, Lee Chadwick, about issuing smartphones to the sales force. Her unedited work is bloated with unnecessary words, expendable phrases, and repetitive ideas.

Figure 2.6   Wordy, unedited e-mail.

Wordy and unfocused subject

One long, unbroken paragraph is hard to follow

Repeats same idea in two or three sentences

Uses awkward and wordy sentences

Does not say what writer will do about problem

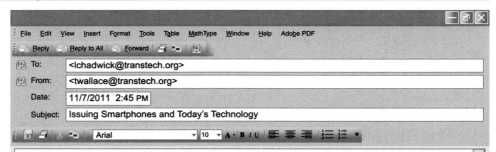

To: <lchadwick@transtech.org>

From: <twallace@transtech.org>

Date: 11/7/2011 2:45 PM

Subject: Issuing Smartphones and Today's Technology

Dear Lee,

Due to the inescapable reliance on technology, specifically on e-mail and Internet communications, within our company, I believe it would be beneficial to look into the possibility of issuing smartphones such as BlackBerrys, iPhones, or Palm Pixis to our employees. Issuing smartphones would have a variety of positive implications for efficiency of our company. Unlike normal cell phones, these smartphones have many new features that will help our employees in their daily work, since they combine cellular phone technology with e-mail and document transfer capabilities, as well as many other important features. The employees could increase their efficiency due to the fact that they could constantly keep track of appointments on their schedule for each day. The employees would also benefit from smartphones by having access to their contacts even when they are out on the road traveling, whether they are at local office meetings or on cross-country business trips. By means of smartphones I feel quite certain that our company's correspondence would be dealt with much more speedily, since these devices will allow our employees to access their e-mail at all times. I think it would be absolutely essential for the satisfaction of our customers and to the ongoing operation of our company's business today to respond fully and completely to the possibility such a proposal affords us. It would therefore appear safe to conclude that with reference to the issue of smartphones that every means at our disposal would be brought to bear on issuing such smartphones to our employees.

Thanks,

Trudy

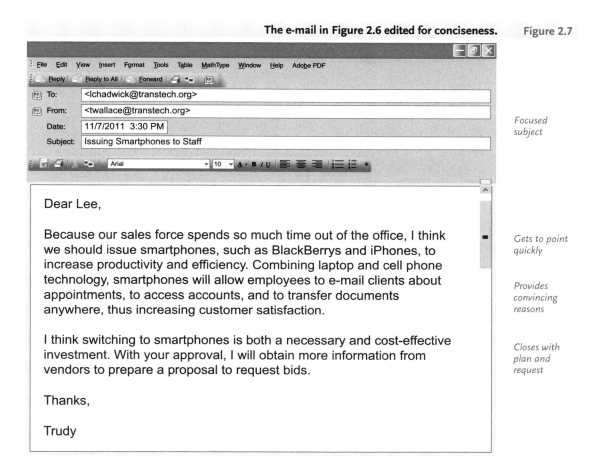

The e-mail in Figure 2.6 edited for conciseness.   Figure 2.7

Focused subject

Gets to point quickly

Provides convincing reasons

Closes with plan and request

After careful editing, see how Trudy Wallace streamlined her e-mail to her boss, Lee Chadwick in Figure 2.7. She pruned wordy expressions and combined sentences to cut out duplication. The revised version is only 86 words, as opposed to the 253 words in the draft. Not only has Wallace shortened her message, but she has also made it easier to read.

Editing Guidelines to Eliminate Sexist Language

Editing involves far more than just making sure that your sentences are readable. It also reflects your professional style—how you see and characterize the world of work and the individuals in it, not to mention how you want your readers to see you. Your words should reflect a high degree of ethics and honesty, free from bias, offense, and stereotypes. They need to be sensitive to the needs of your international audience as well (see pages 134–139).

Sexist language in particular offers a distorted view of a job force and discriminates in favor of one sex at the expense of another, usually women. Using sexist language offends and demeans female readers by depriving them of their equal rights. It may cost your company business as well. You can avoid gender bias by using inclusive language for women and men alike, treating them equally and fairly.

Sexist language is often based on stereotypes that depict men as superior to women. For example, calling politicians *city fathers* or *favorite sons* follows the stereotypical picture of seeing politicians as male. Such phrases discriminate against women who do or could hold public office at all levels of government. Similarly, do not use such sexist designations as *male nurse*, *male models*, or *male secretary*. Never assume or imply a person's gender based on his or her profession.

As the examples above show, such language prejudiciously labels some professions as masculine and others as feminine. Keep in mind that sexist phrases assume engineers, physicians, and pilots are male (*he, his,* and *him* are often linked with these professions in descriptions) while social workers, nurses, and secretaries are female (*she, her*), although members of both sexes work in all those professions. Sexist language also wrongly points out gender identities when such roles do not seem to follow biased expectations—*lady lawyer, male secretary, female surgeon,* or *female astronaut*. Such offensive distinctions reflect prejudiced attitudes that you should eliminate from your writing.

Always prune the following sexist phrases: *every man for himself, gal Friday, little woman, lady of the house, old maid, the best man for the job, to man a desk* or *post, the weaker sex, woman's work, working wives, a manly thing to do,* and *young man on the way up.*

Finally, don't assume all employees are male. Instead of writing, "All staff members and their wives are invited to attend," simply say, "All staff members and their guests are invited to attend."

Ways to Avoid Sexist Language

1. **Replace sexist words with neutral ones.** Neutral words do *not* refer to a specific sex; they are genderless. The sexist words on the left in the following list can be replaced by the neutral nonsexist substitutes on the right.

Sexist	Neutral
actress	actor
alderman; assemblyman	representative
businessman	businessperson
cameraman	photographer
chairman	chair, chairperson
congressman	representative
craftsman	skilled worker
divorcée	divorced person
fireman	firefighter
foreman	supervisor, manager
housewife	homemaker
janitress	cleaning person
landlord, landlady	owner

Sexist	Neutral
maiden name	family name
mailman/postman	mail carrier
man hours	work hours, staff hours
mankind	humanity, human beings
manmade	synthetic, artificial
manpower	strength, power
man to man	candidly
men	human beings, people
policeman	police officer
repairman	repair person
salesman	salesperson, clerk
spokesman	spokesperson
waitress	server
weatherman	meteorologist
woman's intuition	intuition
workman	worker

2. Watch masculine pronouns. Avoid using the masculine pronouns (*he, his, him*) when referring to a group that includes both men and women.

Every worker must submit his travel expenses by Monday.

Workers may include women as well as men, and to assume that all workers are men is misleading and unfair to women. You can edit such sexist language in several ways.

a. Make the subject of your sentence plural and thus neutral.
Workers must submit their travel expenses by Monday.

b. Replace the pronoun *his* with *the* or *a* or drop it altogether.
Every employee is to submit a travel expense report by Monday. **or**
Every worker must submit travel expenses by Monday.

c. Use *his or her* instead of *his*.
Every worker must submit his or her travel expenses by Monday.

d. Reword the sentence using the passive voice.
All travel expenses must be submitted by Monday.

Moreover, in some contexts exclusive use of the masculine pronoun might invite a lawsuit. For example, you would be violating federal employment laws prohibiting discrimination on the basis of sex if you wrote the following in a help-wanted ad for your company.

Each applicant must submit his transcript with his application. He must also supply three letters of recommendation from individuals familiar with his work.

The language of such a notice implies that only men can apply for the position.

Keep in mind that international readers may find these guidelines on using *him/her/his* confusing since many languages (e.g., French, Spanish) follow grammatical instead of natural gender. In French, the word for *doctor* is masculine.

3. Avoid using sexist words that end in -ess or -ette: *stewardess* (use *flight attendant*); *poetess* (for *poet*); *usherette* (for *usher*); or *drum majorette* (for *drummer*).

4. Eliminate sexist salutations. Never use the following salutations when you are unsure of who your readers are:

- Dear Sir
- Gentlemen
- Dear Madam

Any woman in the audience will surely be offended by the first two greetings and may also be unhappy with the pompous and obsolete *madam*. It is usually best to write to a specific individual, but if you cannot do that, direct your letter to a particular department or office: *Dear Warranty Department* or *Dear Selection Committee.*

Be careful, too, about using the titles *Miss, Mr.,* and *Mrs.* Sexist distinctions are unjust and insulting. In the business world Ms. is widely used to refer to a woman. Hence, it is be preferable to write *Dear Ms. McCarty* rather than *Dear Miss or Mrs. McCarty* unless a woman expressly asked to be addressed as Miss or Mrs. A woman's marital status should not be an issue. When you are in doubt about how to address a correspondent, write the person's first and last name as in *Dear Indira Kumar.* Chapter 4 shows you acceptable salutations to use in your letters (see page 101).

5. Never single out a person's physical appearance. Sexist physical references negatively draw attention to a woman's gender, as the example below illustrates

The manager is a tall blonde who received training at Mason Technical Institute.

Avoiding Other Types of Stereotypical Language

In addition to sexist language, avoid any references that stereotype an individual because of race, national origin, age, disability or sexual orientation. Not only are such references almost always irrelevant in the workplace (except for Equal Employment Opportunity Commission reports or health care documentation), they are discriminatory, culturally insensitive, and ethically wrong.

To eliminate biased language in your workplace writing, follow the guidelines below.

1. Do not single out an individual because of race or national origin or stereotype him or her because of it. Be especially sensitive when referring to someone's ethnic identity.

Wrong: Bill, who is African American, is one of the company's top sales reps.
Right: Bill is one of the company's top sales reps.

2. **Identify members of an international community accurately**. Not every native Spanish speaker is Latin American or Hispanic. There are significant cultural differences you need to be sensitive to, e.g., Cuban American, Mexican American.

3. **Avoid words or phrases that discriminate against an individual because of age**. For example, do not use *middle aged*, *elderly*, *up in years*, *old timer*, *over the hill*, *senior moment*, or the adjectives *spry* or *frail* when they are applied to someone's age: "a spry 71." Making someone's age an issue is unfair, whatever it may be.

> Wrong: Jerry Fox, who will be 57 next month, comes up with obsolete plans from time to time.
> Right: Some of Jerry Fox's plans have not been adopted.

4. **Respect individuals who may have a disability**. Avoid derogatory words such as *amputee, crippled, handicapped, impaired*, or *lame* (physical disabilities) or *retarded* or *slow* (mental disabilities). Emphasize the individual instead of the physical or mental condition as if it solely determined the person's abilities.

> Wrong: Tom suffers from MS.
> Right: Tom is a person living with MS.

> Wrong: Sarah, who is crippled, still does an excellent job of keyboarding.
> Right: Sarah's disability does not prevent her from keyboarding.

Keep in mind that the Americans with Disabilities Act prohibits employers from asking if a job applicant has a disability.

Also avoid using discriminatory expressions such as a *crippled economy*, *lame excuse*, *mentally challenged*, *mental midget*, or *crazy scheme*.

5. **Don't stereotype based on sexual orientation**. It is unfair to label a person by sexual orientation as if that is the only significant aspect of that individual's life.

> Wrong: Paula Smith, a lesbian, hosts a successful daytime talk show.
> Right: Paula Smith hosts a successful daytime talk show.

Collaboration Is Crucial to the Writing Process

In the world of work, writing skills, such as researching, planning, drafting, revising, and editing—just discussed in this chapter—are vital for your success. But you will often communicate as part of a team (including managers and co-workers) to write a report, a proposal, or even a letter successfully. One survey estimates that 90 percent of all businesspeople spend some time writing as part of a collaborative team as depicted in Figure 2.8. Being a team player is one of the most prized skills you can possess in the world of work.

Over the next few sections of this chapter, we'll explore the advantages of the collaborative writing process, guidelines for effective group writing, and ways

Figure 2.8   A collaborative writing team at work

© Golden Pixels LLC/Alamy

to help resolve conflicts in the group-writing process. Because an overwhelming majority of workplace collaborative writing takes place online, we'll also investigate how technology enhances the collaborative writing process and see how a document can be collaboratively created online.

Advantages of Collaborative Writing

Collaborative writing teams benefit both employers and employees. Collaboration gets writing done more easily and efficiently in the work place. Here are some of the advantages of writing as part of a team.

1. **Builds on collective talents**. Many heads are better than one. A writing team profits from the diverse backgrounds, skills, and research of its individual members. Collaboration joins individuals from diverse disciplines in the corporate world—IT staff, accountants, lawyers, media specialists, and security experts, among others.

2. **Allows for productive feedback and critique**. Collaborative writing encourages all members of the team to get involved, offering both their agreement as well as their critical viewpoints to produce the highest-quality document possible.

3. **Increases productivity and saves time**. When a group has planned carefully, collaboration actually cuts down on the number of meetings and conferences, saving a company time and employees travel.

4. Ensures overall writing effectiveness. The more people involved in developing a document, the greater the chances are for thoroughness and cohesion. Team effort helps to uncover inconsistencies, find omissions, and correct problems.

5. Accelerates decision-making time. A group investigating problems and offering pertinent solutions can considerably reduce the time it takes to prepare a document.

6. Boosts employee morale and confidence while decreasing stress. Working as part of a team can relieve individuals of some job stress by not making them solely responsible for planning, drafting, and revising a document. Members can discuss ideas with each other and help one another in drafting, organizing, and editing a document. A team, therefore, provides a safety net by assuring individual members that they can always talk over problems and they will have help in meeting deadlines.

7. Contributes to customer service and satisfaction. By pooling their knowledge of an audience's needs, a collaborative writing team is much more likely — through discussions, interactions, even disagreements — to anticipate a customer's, or an agency's, requests and complaints.

8. Affords a greater opportunity to understand global perspectives. By working with a multinational work force, such as that described in the long report in Chapter 9 (pages 338–354), individuals can develop greater sensitivity to and appreciation of the needs and problems of an international audience.

Seven Guidelines for Successful Group Writing

To be successful, a collaborative writing team should observe the following seven guidelines helpful.

1. Understand and agree on the purpose, audience, scope, organization, and deadlines for the report. Everyone needs to be on the "same page" from start to finish.

2. Establish group rules early on and stick to them. Decide when and where the group will meet, how and when members are to communicate with each other (face-to-face, telephone, e-mail, other online technologies). Also establish guidelines on how digital and hard-copy information is to be shared and with whom and how often, when various tasks are to be completed, and what "fail-safe" mechanisms are in place if and when problems arise.

3. Put the good of the group ahead of individual egos. Group harmony and productivity are essential if the report or proposal is to get done on time. Individuals need to be active participants and keen listeners. Getting one's own way can slow down the overall success of the document. In the business world, sometimes a report bears only the boss's name, not the name of the group that prepared it.

4. Agree on the group's organization. The group can appoint a leader who keeps the team on task by being a coordinator, cheerleader, a scheduler, an evaluator, and a peacemaker who can resolve conflicts quickly, and also a referee who knows when to call time-out.[1] The leader must be skilled in interpersonal interactions. In many organizations, the leader serves as the liaison with management and submits the final document. A group secretary or recorder may also be appointed.

5. Identify each member's responsibilities precisely. There should be a fair distribution of labor so that each member can use his or her particular and proven skills. One member of the group may be responsible for document design and visuals, another for research and documentation, another for writing drafts, and still another for making a presentation. The entire group, however, needs to share responsibility for the overall preparation, design, writing, and proofing of the report.

6. Provide clear and positive feedback at each meeting and for each part of the report the group prepares. Members need to come to meetings prepared, raise important questions, and make thoughtful recommendations. They should also back up their ideas with facts, dates, costs, model numbers, etc., not vague unsupported generalizations. Not showing up, being unprepared or failing to turn assignments in on time jeopardizes the entire project.

7. Follow an agreed-on timetable, but leave room for flexibility. The group should estimate a realistic time frame necessary to complete the various stages of their work—when drafts and revisions are due or when editing must be concluded, for example. A project schedule based on that estimate should then be prepared. **But remember: Projects always take longer than initially planned.** New information may surface or you may need to do additional research. Prepare for a possible delay at any one stage. Moreover, consider the needs of international colleagues, who live in different time zones. The group may have to submit progress reports (see pages 288–292) to its members as well as to management.

Sources of Conflict in Group Dynamics and How to Solve Them

The success of collaborative writing depends on how well the team interacts. They have to meet and plan before they can even begin researching, set ground roles, decide on responsibilities, and work together on solving problems and come up with solutions. Discussion and criticism are essential to the process of creating any successful document—report, proposal, etc. "Conflict" in the sense of conflicting opinions—a healthy give-and-take—can be positive if it alerts the group to problems (inconsistencies, redundancies, incompleteness) and provides ways to resolve

---

[1]Adapted from Hendrie Wesigner, *Emotional Intelligence at Work* (New York: Jossey-Bass, 1997).

them. A conflict can even help the group generate and refine ideas, leading to a better organized and more carefully written document.

But when conflict translates into ego tripping and personal attacks, nothing productive emerges. Everyone in the group must agree beforehand on three iron-clad working policies of group dynamics: (1) individuals must seek and adhere to group consensus; (2) compromise may be advisable, even necessary, to meet a deadline; and (3) if the group decides to accept compromise, the group leader's final decision on resolving conflicts must be accepted.

Common Problems, Practical Solutions

Following are some common problems in group dynamics, with suggestions on how to avoid or solve them.

1. Resisting constructive criticism. No one likes to be criticized, yet criticism can be vital to the group effort. Be open to suggestions. Individuals who insist on "their way or no way" can become hostile to any change or revision, no matter how small.

Solution: When emotions become too heated, the group leader may wisely move the discussion to another section of the document or to another issue and allow some cooling-off time. Negotiation is an essential job skill.

2. Giving only negative criticism. Do not saturate a meeting with nothing but negatives. You will block communication if you start criticizing the group's efforts with words such as "Why don't you try ...," "What you need is ...," "Don't you realize that ...," or "If you don't...."

Solution: When you criticize an idea, diplomatically remind the individual of the team's goals and point to ways in which revision (criticism) furthers them. Explain the problem, and offer a helpful, relevant revision. Never attack a group member. Mutual respect is everyone's right and obligation. Be objective, constructive, and cooperative.

3. Dominating a meeting. The group process is about sharing and responding to ideas, not about taking over. When one group member dominates the discussion and becomes aggressive and territorial, the group process suffers.

Solution: The leader should let the group know that the participation of all members is valued, but then say "we need to hear from the rest of the group." Some groups follow a three-minute rule—each member has three minute to make comments and does not get the floor again until everyone has had a chance to speak. If one group member still continues to dominate, the leader may (in private) have to speak to him/her.

4. Refusing to participate. Withholding your opinions hurts the group efforts; identify what you believe are major problems and give the group a chance to consider them.

Solution: If you don't feel sure of yourself or your points, talk to another member of the group, a listening partner, before a meeting to "test" your ideas or to write down your suggestions before a meeting to share them with the group.

5. Interrupting with incessant questions. Some people interrupt a meeting so many times with questions that all group work stops. The individual may simply be unprepared or may be trying to exercise his or her control of the group.

Solution: When that happens, a group leader can remark, "We appreciate your interest, but would you try an experiment, please, and attempt to answer your own questions?"or if the person claims not to know, the leader might then say, "Why don't you think about it for a while and then get back to us?"

6. Inflating small details out of proportion. Some individuals waste valuable discussion and revision time by dwelling on relatively insignificant points—the choice of a single word, e.g., an optional comma, or steer the group away from larger, more important issues (e.g., costs, schedules, etc.). Nitpickers can derail the group.

Solution: If there is consensus about a matter, leave it alone and turn to more important issues. The leader should remind the group (without singling anyone out) about the bigger picture and caution them to stay on track.

7. Being overly deferential to avoid conflict. This problem is the opposite of that described in #1. You will not help your group by being a "yes person" simply to appease a strong-willed member of the group.

Solution: Feel free to express your opinions politely; if tempers begin to flare, call in the group leader or seek the opinions of others on the team. The leader needs to promote and protect meetings as a safe place to express ideas.

8. Not respecting cultural differences. You may have individuals from different countries or cultures on your team. Disregarding or misjudging the way they interact with the group can seriously threaten group success and harmony. Moreover, discrimination of any type—based on race, age, religious beliefs, nationality—is unacceptable in a collaborative group or workplace.

Solution: Some companies offer employees seminars on cultural sensitivity in the workplace. But a group leader must also ensure that diversity is honored and, if anyone does not, he/she may write that person up.

9. Violating confidentiality. Leaking confidential information about a personal issue, product, procedure, or operation is a serious violation in the world of work. But realize that sometimes meetings can be closed to everyone in the company except for those who are a part of the group.

Solution: Violating confidentiality is often grounds for dismissal. At a preliminary meeting, and periodically during the writing process, the leader should emphasize the whys, hows, and whens of confidentiality.

10. Not finishing on time or submitting an incomplete document. Meeting established deadlines is the group's most important obligation to one another and to their company. When some members are not involved in the planning stages or when they skip meetings or ignore group communications, deadlines are invariably missed. If you miss a meeting, get briefed by an individual who was there.

Solution: The group leader can institute networking through e-mail to announce meetings, keep members updated, or provide for ongoing communication and questions. See page 58.

Computer-Supported Collaboration

To be successful writers, employees must be proficient in using multiple types of collaborative software systems, otherwise know as **groupware**. You will be expected to know—or at least be adaptable to learning—not only how to use e-mail for collaborative coomunications but also how to navigate document tracking systems (such as Microsoft Word's Track Changes feature or Adobe Acrobat) as well as Web-based collaboration systems (wikis).

Groupware does not, of course, completely eliminate the need for a group to meet in person to discuss priorities, clarify issues, or build team spirit. But face-to-face communications, although sometimes essential, are frequently accompanied by computer-supported collaboration via groupware.

Advantages of Computer-Supported Collaboration

Collaboration online offers significant advantages to both employers and workers. Here are some of those benefits:

1. **Increased opportunities to "meet."** Today's meeting-heavy and travel-filled work environments often make collaborating in person difficult and costly. Various types of groupware provide a virtual shared work place where team members can ask questions, share information, make suggestions and revisions, and troubleshoot.

2. **Reduced stress in updating new group members**. Groupware can efficiently and quickly bring new team members into the ongoing conversation, saving time and effort and reducing the stress of face-to-face meetings.

3. **Expanded options to communicate worldwide**. Groupware ensures the flexibility to contact group members anywhere—at a home office (telecommuters, freelancers), on the road, at another office, or in another country.

4. **Improved feedback and accountability**. Because it is accessible and easy to use, groupware helps team members participate actively in the collaboration process. Groupware encourages team members to get to the point and make sure other members understand their message clearly and quickly. Because each member's contributions are documented, everyone's participation becomes a matter of record.

5. **Enhanced possibility of complete and clear information**. Groupware allows everyone involved in the team to truly be "on the same page." It provides a running record, thus saving all information and preventing misinterpretation.

Types of Groupware

There are essentially three types of groupware commonly used to produce collaboratively written documents in today's workplace: e-mail, document tracking systems, and wikis.

E-Mail

E-mail is used for many jobs in the world of work, as you will see in Chapter 3. It has an important role to play in collaborative writing online as well. While e-mail is not the place to create or revise a collaboratively-written document (because edits and other changes are hard to incorporate and keep track of), e-mail nevertheless makes online collaboration possible for the following reasons:

1. E-mail is used to send collaboratively written documents as attachments or PDF files. Team members from around the globe can then access the same document and share their feedback.

2. E-mail helps to conduct group business—setting up meeting times, notifying members about the status of a report, alerting members to a problem, etc.

3. E-mail makes sure every member of the team is working on the same document by identifying each document by name and number (*Report on Parking, Rev. 4* or *Proposal on Recycling, Draft 2*).

4. E-mail saves the group time by decreasing the number of face-to-face meetings it must have, but note that e-mail cannot take the place of a face-to-face exchange.

Document Tracking Software

Document tracking software, such as Microsoft Word's Track Changes feature or Adobe Acrobat, provide another way for collaborative writing teams to share, comment, and revise their work online. Figure 2.9 shows an example of a collaboratively written document using Microsoft Word—the first draft of a section of a report on increasing parking spaces at a hospital—and how it has been revised and edited by several team members using the Track Changes feature. (Keep in mind, though, that depending on the document tracking system your company is using, your tracked documents may look somewhat different.)

Sent as an e-mail attachment to every team member, Figure 2.9 preserves all the original text of Draft 1 while automatically showing and identifying comments from each individual. Notice that each of the three reader's initials and his or her changes and comments are "tracked" in the markup balloons or boxes on the right. (Microsoft Word automatically assigns a different color to each team member so their edits and comments are easily distinguishable from each other.) This tracked document is then shared by other team members who can tag areas of the document that they would like to comment on. Team members can insert headings, clarify and verify factual data, call for visuals, and ask for and even supply new text, as MM does with her newly inserted text in Figure 2.9. Team members can also edit sentences in the document, and Word will automatically track these changes.

The team leader or group member responsible for working on revising the next version of this section of the report can accept or reject the tracked changes but only after they are agreed on by the group. This is where the dynamics of collaborative writing must work for the good of the entire group and to produce the document on time. After deciding on the changes made about Draft 1, the group will then receive a clean document of Draft 2 (free of tracked changes/comments) and can continue with the writing process leading toward a final draft.

Collaborative editing using a document tracking system.   Figure 2.9

Draft 1 — Proposal to Expand Hospital Parking Facilities ~~How We Can Expand the Hospital's Parking~~

~~CGH~~ Community General Hospital needs to expand its parking facilities. Right now there is just too little room for visitors, staff, and patients. These inadequate parking facilities are a ~~detrime~~ detriment to the overall growth of pt. care. They have been an important talking point since the inception of this committee. Maybe they were ok when the hospital opened its doors in 1973, but not today. ~~CGH is a good place to work and t~~The new parking facilities would definitely benefit the present staff and visitors from walking ~~long distances~~ two blocks in the rain and ice.

The exact number of new spots is hard to estimate now but I am thinking around 500 might be just right. The problem is the traffic flow around the hospital. While the new parking facilities would alleviate it, it also raises a central question about how to get it done. Perhaps Wentworth Avenue, East to West, might be turned into a one-way street. That way we could add up to 11 new spots in the front of the ER and thus resolve the congestion that has hampered easy ~~egress~~ entrance and ~~ingress~~ exit. Another possibility worth considering is changing Taylor Street—right now it is a two-way street and we could make it one-way West to East.

At any rate, the traffic flow is a key issue the hospital needs to solve if it is to expand its parking facilities. But there are other important engineering problems that must be solved. Eleanor Yi, the hospital engineer, has studied the stress points, pre-cast concrete, and the slope of vehicular access ramps that would accommodate increased traffic flow. As you can see, she believes that the hospital does not have the space to locate all the new parking spots in one plane area. She recommends a two-story structure and believes the North side of the ER might be the best place.

LB 5/5/11 10:13 AM
Comment: Centered and rephrased the title to matchhouse style for internal proposals.

KT 5/5/11 8:58 AM
Comment: Let's use the full name of the hospital here.

MM 5/5/11 11:28 AM
Comment: We should cite source and statistics. See hospital report CGH-GR-2010

KT 5/5/11 9:00 AM
Comment: We need to take out abbreviations in the final copy.

LB 5/5/11 10:15 AM
Comment: We must be precise. It was 417, according to theengineering proposal.

MM 5/5/11 11:38 AM
Comment: I think it might be best to start a new section here and title it "Increased Traffic Flow."

MM 5/5/11 11:39 AM
Comment: We want to include an alternate plan. I've inserted text to address this.

LB 5/5/11 10:17 AM
Comment: Let's put her documentation in an appendix.

MM 5/5/11 1:40 AM
Comment: That's a good idea; her solution appears workable and falls within budgetary limits.

KT 5/5/11 9:04 AM
Comment: I don't understand this terminology. Should we use a different phrase here?

LB 5/5/11 10:19 AM
Comment: We have to provide further commentary in terms of how many ramps and perhaps confirmation from engineering consulting firm to corroborate Yi's findings.

Visit www.
cengage.com/
english/kolin/
writingatwork
concise3e
for an online
exercise,
"Collaborating
Using a
Document
Tracking
System."

Wikis

Wikis are similar to document tracking systems, but they have a few crucially different characteristics. First, wikis are not systems found within a software package such as Microsoft Office. They are websites for which team members are given access passwords, enabling them to check documents in and out of the site. Second, wikis typically do not show tracked changes right on the document. Instead, when an edited document is uploaded back to the website, each version is assigned a new version number. Versions of the document can be compared and the differences between versions can be viewed, but these differences will not show up within a single version.

The advantage of wikis over tracked documents is that subsequent versions of a document are as easy to read from start to finish as the first draft. Each wiki version is a clean document free of complicated tracked edits. Therefore, when team members are revising, they will need to proofread only the final draft, rather than go through the laborious process of accepting/rejecting changes, deleting comments, and troubleshooting lingering inconsistencies.

The disadvantage of wikis, when compared with tracked documents, is that wikis may result in a lack of quality control. Because each group member's changes are not clearly tracked, it may be difficult for group members to keep up with the succession of changes. When using wikis, teams should set up a clear protocol outlining who may make changes to the document and when.

Avoiding Problems with Online Collaboration: A Summary

Regardless of the online collaborative method your team uses, it must establish ground rules by which documents are created, posted, shared, protected, and submitted. By following the guidelines listed below, your team can avoid common problems in its online collaboration:

1. **Be sure that all team members have access and authorization**. Team members cannot give input if they cannot open, edit, save, or share a document. To do this, every team member must use the same software.
2. **Everyone in the group must be "in the loop."** Verify that you and your team members are all working on the same (and correct) version of the document at the same time. Problems result when a team member wastes time and delays a deadline by editing an earlier-discarded or otherwise incorrect version.
3. **Save the original draft and subsequent ones in separate files** in case the team needs to return to these earlier copies to verify that changes have been made.
4. **Link each revision with the individual who made it**. Track each contributor's changes by having an identifying color code for each one's suggested revisions and/or by using initials to identify who has made any changes, as in Figure 2.9.
5. **Maintain confidentiality to protect the document from unauthorized users**. Do not send team e-mails or tracked documents to individuals who are not a part of the team. Team members should be cautioned not to reveal their passwords to anyone.
6. **Require all team members to sign off on and agree to the complete, final document**. Doing this will create a system of checks and balances to ensure quality control.

Revision Checklist

☐ Researched my topic carefully to obtain enough information to answer all my readers' questions—conferences with colleagues/managers, online searches, evaluations of websites, feedback from customers.

☐ Before writing, determined how much and what types of information are needed to complete writing task to meet my readers' needs.

☐ Spent necessary time planning—brainstorming, outlining, clustering, or a combination of those techniques. Produced substantial material from which to shape a draft.

☐ Prepared enough drafts to decide on major points in message to readers. Made major changes and deletions where necessary in drafts to strengthen overall document.

☐ Revised drafts carefully to successfully answer reader's questions about content, organization, and tone.

☐ Made time to edit work so that style is clear and concise and sentences are readable and varied. Checked punctuation, sentences, and words to make sure they are spelled correctly and used appropriately for audience.

☐ Eliminated sexist and other biased language that unfairly stereotypes individuals because of race, ethnicity, age, sexual orientation, or disability.

☐ Followed necessary steps of the writing process to take advantage of team effort and feedback.

☐ Attended all group meetings and understood and agreed to responsibilities of the group as well as my own obligations.

☐ Finished research, planning, and/or drafting expected of me.

☐ Succeeded in being a team player by putting the success of my group over the needs of my own ego.

☐ Shared my research, ideas, and suggestions for revision through constructive criticism.

☐ Participated honestly and politely in discussions with colleagues.

☐ Was open to criticism and suggestions for change.

☐ Took advantage of e-mail, instant messaging, and groupware technology to communicate with my collaborative team.

☐ Investigated the research, drafting, revising, and editing benefits available through computer software.

☐ Answered questions and responded to requests promptly from team leader and collaborative team members.

☐ Attached pertinent documents in e-mails to collaborative team.

☐ Avoided technical problems with online collaboration by adhering to company/group policies.

☐ Respected confidentiality and used computer-assisted editing technologies responsibly and ethically.

Exercises

1. Following is a writer's initial brainstormed list for a report on stress in the workplace. Revise the brainstormed list, eliminating repetition and combining related items.

> leads to absenteeism
> high costs for compensation for stress-related illnesses
> proper nutrition
> numerous stress-reduction techniques
> good idea to conduct interviews to find out levels, causes, and extent of stress in the workplace
> low morale caused by stress
> higher insurance claims for employees' physical ailments
> myth to see stress leading to greater productivity
> various tapes used to teach relaxation
> environmental factors—too hot? too cold?
> teamwork intensifies stress
> counseling
> work overload
> setting priorities
> wellness campaign
> savings per employee add up to $6,150 per year
> skills to relax
> learning to get along with co-workers
> need for privacy
> interpersonal communication
> employee's need for clear policies on transfers, promotion
> stress management workshops very successful in California
> physical activity to relieve stress
> affects management
> breathing exercises

2. Prepare a suitable outline from your revised list in Exercise 1 for a report to a decision maker on the problems of stress in the workplace and the necessity of creating a stress-management program.

3. From the revised brainstormed list in Exercise 1, write a one-page memo to a decision maker about how the problems of stress negatively affect workplace production.

4. Assume you have been asked to write a short report (two to three pages) to a decision maker (the manager of a business you work for or have worked for; the director of your campus union, library, or security force; a city official) about one of the following topics. Write the report alone or as part of a collaborative team.
   a. recruitment of more specialists in your field
   b. Internet resources
   c. security lighting
   d. food service
   e. health care plans

f. public transportation
g. sporting events/activities
h. team building/morale
i. greening the workplace/community
j. hiring more part-time student workers

Then do relevant research and planning about one of those topics and the audience for whom it is intended by answering the following questions:

- What is my precise purpose in writing to my audience?
- What do I know about the topic?
- What information will my audience expect me to know?
- Where can I obtain relevant information about my topic to meet my audience's needs?

5. Using one or more of the planning strategies discussed in this chapter (clustering, brainstorming, outlining), generate a group of ideas for the topic you chose in Exercise 4. Work on your planning activities for about 15–20 minutes or until you have about 10–15 items. At this stage, do not worry about how appropriate your ideas are or even if some of them overlap. Just get some thoughts down on paper.

6. The following paragraphs are wordy and full of awkward, hard-to-read sentences. Edit these paragraphs to make them more readable and user-friendly by using clear and concise words and sentences.

a. It has been verified conclusively by this writer that our institution must of necessity install more bicycle holding racks for the convenience of students, faculty, and staff. These parking modules should be fastened securely to walls outside strategic locations on the campus. They could be positioned there by work crews or even by the security forces who vigilantly and constantly patrol the campus grounds. There are many students in particular who would value the installation of these racks. Their bicycles could be stationed there by them, and they would know that safety measures have been taken to ensure that none of their bicycles would be apprehended or confiscated illegally. Besides the precaution factor, these racks would afford users maximized convenience in utilizing their means of transportation when they have academic business to conduct, whether at the learning resource center or in the instructional facilities.

b. On the basis of preliminary investigations, it would seem reasonable to hypothesize that among the situational factors predisposing the Smith family toward showing pronounced psychological identification with the San Francisco Giants is the fact that the Smiths make their domicile in the San Francisco area. In the absence of contrariwise considerations, the Smiths' attitudinal preferences would in this respect interface with earlier behavioral studies. These studies, within acceptable parameters, correlate the fan's domicile with athletic allegiance. Yet it would be counterproductive to establish domicility as the sole determining factor for the Smiths' preference. Certain sociometric studies of the Smiths disclose a factor of atypicality, which enters into an analysis of their determinations. One of these factors is that a younger Smith sibling is a participant in the athletic organization in question.

7. Following are very early drafts of memos that businesspeople have sent to their bosses or co-workers. Revise and edit each draft, referring to the revising and editing sections of the checklists on pages 42–47. Turn in your revision and the final, reader-ready copy. As you revise, keep in mind that you may have to delete and add information, rearrange the order of information, and make the tone suitable for your readers. As you edit, make sure your sentences are clear and concise and your words are professional.

a. TO: All workers
FROM: B.J. Blackwell
DATE: February 3, 2012
RE: Parking

The parking violations around here have gotten very, very bad. And the administration is provoked and wants some action taken. I don't blame them. I have been late for meetings several times in the last month because inconsiderate folks from other divisions have parked their cars in our zone. That just is not fair, and so I must not be the only one who is upset. No wonder the management finds things so bad they have asked me to prepare this memo.

A big part of the problem it seems to me is that employees just cannot read signs. They park in the wrong zones. They also park in visitors' spots. The penalties are going to be stiff. The administration, or so I was led to believe, is thinking of fining any employee who does not obey the parking policies. I know for a fact that I saw someone from the research department pull right into a visitor parking area last week just because it was 8:55 and he did not want to be late for work. That gives our business a bad name. People will not want to do business with us if they cannot even find a parking spot in the area that the company has reserved for them.

Vice President Watson has laid the law down to me about all this and told me to let each and every one of you know that things have to improve. One of the other big problems around here is that some employees have even parked their cars in loading zones, and security had to track them down to move.

As part of the administration's new policy, each employee is going to be issued a company parking policy and will have to come in and sign for it verifying that he received it. I think things really have gotten out of hand and that some drastic action has to be taken. We will all have to shape up around here.

b. TO: All Employees
FROM: George Holmes
DATE: October 20, 2012
RE: Travel

Every company has its policies regarding travel and vouchers. Ours strike me as important and fairly straightforward. Yet for the life of me I cannot fathom why they are being ignored. It is in everyone's best interest. When you travel, you are on company time, company business. Respect that, won't you. Explain your purpose, keep your receipts, document your visits, keep track of meals. Do the math.

If you see more than one client per day, it should not be too hard or too much to ask you to keep a log of each, separate, individual visit. After all, our business does depend on these people, and we will never know your true contributions on company trips unless you inform us (please!) of whom you see, where, why, when, and how much it costs you. That way we can keep our books straight and know that everything is going according to company policy.

8. The following sentences contain sexist and other biased language. Edit them to remove these errors.
   a. Every intern had to record his readings daily for the spokesman.
   b. Although Marcel was an amputee, he still could hunt and peck at the keyboard.
   c. She saw a woman doctor, who told her to take an aspirin every day.
   d. Our agency was founded to help mankind.
   e. John, who is a diabetic, has an excellent attendance record.
   f. Every social worker found her schedule taxing—not enough days in the week to help out man-to-man.
   g. Maria, who is Cuban, always adds spice to company events.
   h. To be a policeman, each applicant had to pass a rigorous physical and prove himself in the manly art of self-defense.
   i. It's a wise man who can rise to the top in this cutthroat, volatile stock market.
   j. Sandy Frain, a middle-aged Irish woman, came by this morning wanting an appointment about the new policies on energy efficiency.
   k. Mrs. Johnson is in change of safety issues.
   l. He made his PowerPoint presentation as emphatically as an Italian opera singer on stage.
   m. Sanji, a longtime member of the IT department, is naturally adept at crunching the figures.
   n. Team B tried to disable our proposal by introducing irrelevant references to the many foreigners living on the north side.
   o. The average consumer spends at least two to three hours a week before her computer looking for coupons and other bargains.
   p. How many of our customers don't speak English well?

9. A new manager will be coming to your office park in the next month, and you and five other employees have been asked to serve on a committee that will submit a report about safety problems at your office park and what should be done to solve them. You and your team must establish priorities and propose guidelines that you want the new manager to put into practice. After two very heated meetings, you realize that what you and two other employees have considered solutions, the other half of your committee regards as the problems. Here is a rundown of the leading conflicts dividing your committee:
   - **Speed bumps**. Half the committee likes the way they slow traffic down in the office park, but the other half says they are a menace because they jar car CD players.
   - **Sound pollution**. Half your team wants Security to enforce a noise policy preventing employees from playing loud music while driving in and out of the office park, but the other half insists that policy violates employee rights.

■ **Van and sport utility vehicle parking**. Half the committee demands that vans and sport utility vehicles park in specially designated places because they block the view of traffic for any vehicle parked next to them; the other members protest saying that people who drive these vehicles will be singled out for less desirable parking places.

Clearly your committee has reached a deadlock and will be unproductive as long as those conflicts go unresolved. Based on the above scenario, do the following:

a. Have each person on the committee e-mail the other five committee members suggesting a specific plan on how to proceed—how the group can resolve their conflicts. Prepare your e-mail message and send it to the other five committee members and to your instructor. What's your plan to get the committee moving toward writing the report to the incoming manager?

b. Assume that you have been asked to convince the other half of the committee to accept your half's views on the three areas of speed bumps, noise control, and parking. Send the three opposition committee members an attachment via email persuading them to your way of thinking. Your message must assure them that you respect their point of view.

c. Assume that the committee members reach a compromise after seeing your plan put forth in (a) above. Collaboratively draft a three-page report to the new manager.

10. The following is an early draft of a memo written by a collaborative team informing employees about a company's new policies on recycling as part of its commitment to conservation and greening the workplace. The company's vice president for human resources has asked the team to prepare a small two-page memo to be sent to all employees about the new policy and how and why it should be implemented. Team up with two or three other students to revise this draft. Expect to supplement, correct, rearrange, and reformat anything in this initial draft. Then submit the final, edited copy to your instructor.

CAMDEN COMPANIES

TO: All Employees
DATE: February 10, 2012
RE: Improving Our Recycling Program

An in-house study has shown that Camden sends approximately 26,000 pounds of paper to the landfill. The landfill charge for this runs about $2,240, which we could save by recycling. Camden Companies is conscious of our responsibility to save and protect the environment. Accordingly, starting March 1 we will begin a more intensive paper recycling program. Our program, like many others nationwide, will use the latest degradable technology to safeguard the air, trees, and water in our community. It has been estimated that of the 250 million tons of solid waste, three-quarters of goes to landfills. These landfills across the country are becoming dangerously overcrowded. Such a practice wastes our natural resources and endangers our air and drinking water. For example, it takes 10 trees to make 1 ton of paper, or roughly the amount of paper Camden uses in four weeks. If we could recycle that amount of paper, we could save those trees. Recycling old paper into new paper involves less energy than making paper from new trees. Moreover, waste sent to landfills can, once broken down, leach, seep into our water supply, and contaminate it. The dangers are great.

By enhancing our recycling, we will not be sending so much to the Springfield Landfill and so help alleviate a dangerous condition there. We will keep it from overflowing. Camden will also be contributing to trans-forming waste products into valuable reusable materials. Recycling paper in our own office shows that we are concerned about the environmental clutter. By having an improved paper recycling program, we will establish our company's reputation as an environmentally conscious industry and enhance our company's image.

Camden is primarily concerned with recycling paper. The 200 old phone books that otherwise would be tossed away can get our recycling program off to a good start.

We encourage you to start thinking about the additional kinds of paper around your office/workspace that needs to be earmarked for recycling. When you start to think about it, you will see how much paper we as a company use.

(continued)

Page 2

Starting the last week of February, paper bins will be placed by each office door inside the outer wall. These bins will be green—not unsightly and blending with our decor. Separate your waste paper (white, colored, and computer) and put it into these bins. You do not need to remove paper clips and staples, but you must remove rubber bands, tape, and sticky labels. They will be emptied each day by the clean-up crew. There will also be large bins at the north end of the hallway for you to deposit larger paper products. The crucial point is that you use these specially marked bins rather than your wastebasket to deposit paper.

Camden Companies will deeply appreciate your cooperation and efforts. Thanks for your cooperation.

3 Writing Routine Business Correspondence

Memos, Faxes, E-Mails, IMs, and Blogs

Memos, faxes, e-mails, IMs, and blogs are the types of writing you will do most frequently on the job. These five forms of business correspondence are quick, easy, and effective ways for a company to communicate internally as well as externally. You can expect to send one or more of these routine forms of correspondence each day to co-workers in your department, to colleagues in other divisions of your company, to decision makers at all levels, and to clients and customers worldwide as well.

Visit www. cengage.com/ english/kolin/ writingatwork concise3e for this chapter's online exercises, ACE quizzes, and Web links.

What Memos, Faxes, E-Mails, IMs, and Blog Posts Have in Common

Although memos, faxes, e-mails, IMs, and blog posts are very different types of correspondence, they share the following characteristics:

1. **Each of these types of correspondence is streamlined for the busy world of work**. Unlike letters, proposals, or reports, which can be long and detailed and contain formal parts and sections, these routine types of correspondence give writers and readers a particularly fast way to communicate. Even though memos do have to be formatted (see pages 71–73), they, like blog posts, IMs, or e-mails, are ready-made to send and receive shorter messages.

2. **They give busy readers information quickly**. While the messages they contain can be about any topic in the world of work, most often they focus on day-to-day activities and operations at your company—sales and product information, policy and schedule changes, progress reports, orders, troubleshooting problems, and so forth.

3. **They are informal**. Compared to letters, proposals, and reports, these kinds of routine correspondence are not as formal. They emphasize a conversational, yet professional, style of writing.

4. **Even though they are routine, they still demand a great deal of thought and time**. Although memos, e-mails, blogs, and IMs are less formal than, for instance, a letter to a client, they all must be written clearly and with correct grammar and punctuation, even when the correspondence is between two employees in the same department of a company. Always plan and review what you are going to write.

Memos

Memorandum, usually shortened to *memo*, is a Latin word for "something to be remembered." The Latin meaning points to the memo's chief function: to record information of immediate importance and interest in the busy world of work, as in Figure 3.1. Memos are brief and informal but can contain official announcements that serve a variety of functions, including:

- making an announcement
- providing instructions
- clarifying a policy, procedure, or issue
- changing a policy or procedure
- alerting employees to a problem or issue
- delegating responsibilities
- making a request
- offering suggestions or recommendations
- providing a record of an important matter
- confirming an outcome

Memos are usually written for an in-house audience, although the memo format can be used for documents sent outside a company, such as short reports or proposals (see Chapter 8) or for cover notes for longer reports (see Chapter 9).

Memos keep track of what jobs are done where, when, and by whom; they also report on any difficulties, delays, or cancellations and what your company or organization needs to do about correcting or eliminating them.

Memo Protocol and Company Politics

Memos reflect a company's image—its politics, policies, and organization. Note how Janet Hempstead's memo in Figure 3.2 reminds workers about a crucial safety policy at Dearborn Equipment Company. A company's logo may even appear on the top, as in Figures 3.2 and 3.3. These memos reflect the company's corporate culture and interests and the ways in which they build employee morale and encourage productivity.

Regardless of where you work, your employer will expect your memos to be timely (don't wait until the day of a meeting to announce it), professional, and tactful. Just because a memo is an informal, in-house communication does not mean you can be gruff, curt, or bossy. Politeness and diplomacy count a lot at work. Learning effective memo writing is vital to your success in any organization.

Most companies have their own memo *protocol*—accepted ways in which in-house communications are formatted, organized, written, and routed. In the corporate world, protocol determines where your memo will go. Know your company's/department's chain of command, and use common sense. Don't send copies of memos to people who don't need them (it wastes paper and unnecessarily wastes the time of those who don't need to see them) or to high-ranking company personnel instead of your immediate supervisor, who may think you are going over his or her head. Respect your boss's, or department's, routing list.

Standard memo format.   Figure 3.1

MEMO Health Systems Inc
4008 Washington
Reno, NY 89501
www.hsincorp.com

TO:	Lucy
FROM:	Roger
DATE:	November 11, 2011
SUBJECT:	Review of Successful Website Seminar

As you know, I attended the "How to Build a Successful Website" seminar on
November 7 and learned the "rules and tools" we will need to redesign our
own site.

Here is a review of the major topics covered by the presenter, Jackie Chen:

1. Keep your website content-based—identify your target audience.
2. Visualize and "map out" your site ahead of time.
3. Design your website to look the way you envision it—make it aesthetically
   pleasing.
4. Be sure your site easy to navigate, especially for global readers.
5. Be sure the site easy to find by search engines such as Google for maximum
   exposure.
6. Create hot links and image maps to move users from page to page.
7. Encourage visitor interaction by soliciting feedback.
8. Complete your site with appropriate sound and animation.
9. Keep your site updated.

Could we meet in the next day or two to discuss recreating our website in light
of these guidelines? I would really appreciate your suggestions about this
project as well.
Thanks.

Header

Memo parts

*Introduction
gives
background
and tells reader
what memo
will do*

Preview

*Body:
Numbered
list helps
readers follow
information
quickly*

*Conclusion:
Asks for
comments*

Memo Format

Memos vary in format and the way they are sent. Some companies use standard,
printed forms (Figure 3.1), while others have their names (letterhead) printed on their
memos (as in Figures 3.2 and 3.3). You can also create a memo by including the necessary parts in an e-mail, as in Figures 3.4 and 3.5, which appear later in this chapter.

As you can see from looking at Figures 3.1 through 3.3, memos look different
from letters. They are also less formal. Because they are often sent to individuals
within your company, memos do not need the formalities necessary in business
letters, such as an inside address, a formal salutation or complimentary close, or
signature line, as discussed in Chapter 4 (see pages 99–102).

Figure 3.2   **Memo on letterhead with a clear introduction, discussion, and conclusion.**

Dearborn Equipment Company

To: Machine Shop Employees
From: Janet Hempstead, Shop Supervisor *JH*
Date: September 27, 2011
Subject: Cleaning Brake Machines

Introduction gives purpose and importance of memo

During the past two weeks I have received several reports that the brake machines are not being cleaned properly after each use. Through this memo I want to emphasize and explain the importance of keeping these machines clean for the safety of all employees.

Discussion states why problem exists and how to solve it

When the brake machines are used, the cutter chops off small particles of metal from brake drums. These particles settle on the machines and create a potentially hazardous situation for anyone working on or near the machines. If the machines are not cleaned routinely before being used again, these metal particles could easily fly into an individual's face when the brake drum is spinning.

Safety message is boldfaced for emphasis

To prevent accidents like this from happening, please make sure you vacuum the brake machines after each use.

You will find two vacuum cleaners for this purpose in the shop—one of them is located in work area 1-A and the other, a reserve model, is in the storage area. Vacuuming brake machines is quick and easy: It should take no more than a few seconds. This is a small amount of time to make the shop safer for all of us.

Conclusion builds goodwill and asks for questions

Thanks for your cooperation. If you have any questions, please call me at Extension 324, e-mail me, or come by my office.

204 South Mill St., South Orange, NJ 02341-3420 (609) 555-9848 JHEMP@dearco.com

A memo that uses headings to highlight organization.   Figure 3.3

RAMCO TECHNOLOGIES

Where Technology Shapes Tomorrow

ramco@gem.com   http://www.Ramcogem.com

Company logo

TO: Rachel Mohler, Vice President
                      Harrison Fontentot, Public Relations
FROM: Mike Gonzalez
SUBJECT: Ways to Increase Ramco's Community Involvement
DATE: March 2, 2012

At our planning session in early February, our division managers stressed the need to generate favorable publicity for our new Ramco facility in Mayfield. Knowing that such publicity will highlight Ramco's visibility in Mayfield, I think the company's image might be enhanced in the following ways.

Introduction supplies background and rationale

CREATE A SCHOLARSHIP FUND

Ramco would receive favorable publicity by creating a scholarship at Mayfield Community College for any student interested in a career in technology. A one-year scholarship would cost $6,800. The scholarship could be awarded by a committee composed of Ramco executives and staff. Such a scholarship would emphasize Ramco's support for technical education at a local college.

Body offers concrete evidence (costs, personnel, location) that plan can work

OFFER SITE TOURS

Guided tours of the Mayfield facility would introduce the community to Ramco's innovative technology. The tours might be organized for academic, community, and civic groups. Individuals would see the care we take in protecting the environment in our production and equipment choices and the speed with which we ship our products. Of special interest to visitors would be Ramco's use of industrial robots alongside its employees. Since these tours would be scheduled in advance, they should not conflict with our production schedules.

Headings reflect organization

PROVIDE GUEST SPEAKERS

Many of our employees would be excellent guest speakers at civic and educational meetings in the Mayfield area. Possible topics include the technological advances Ramco has made in designing and engineering and how these advances have helped consumers and the local economy.

Thanks for giving me your comments as soon as possible. If we are going to put one or more of these suggestions into practice before the facility opens in mid-April, we'll need to act before the end of the month.

Closing requests feedback and authorization

Memo Parts

Basically, the memo consists of two parts: the identifying information at the top and the message itself. The identifying information includes these easily recognized parts: **To, From, Date,** and **Subject** lines.

TO: Aileen Kelly, Chief Computer Analyst
FROM: Stacy Kaufman, Operator, Level II
DATE: January 30, 2012
SUBJECT: Progress report on the fall schedule

You can use a memo template in your word processing program that will list these headings, as follows, to save time.

TO: (Enter recipient's name here.)
FROM: Thin Ong
DATE: October 4, 2011
SUBJECT: (Enter subject here.)

On the **To** line, type the name and job title of the individual(s) who will receive your memo or a copy of it. If your memo is going to more than one reader, make sure you list your readers in the order of their status in your company or agency, as Mike Gonzalez does in Figure 3.3 (according to company policy the vice president's name appears before that of the public relations director). If you are on a first-name basis with the reader, use just his or her first name, as in Figure 3.1. Otherwise, include the reader's first and last names.

On the **From** line, type your name (use your first name only if that is how your reader refers to you, as in Figure 3.1) and your job title (unless it is unnecessary for your reader). Some writers handwrite their initials after their typed name to verify that the message comes from them.

On the **Subject** line, key in the purpose of your memo. The subject line serves as the title of your memo; it summarizes your message. Vague subject lines, such as "New Policy," "Operating Difficulties," or "Shareware," do not identify your message precisely and may suggest that your message is not carefully restricted or developed. Note how Mike Gonzalez's subject line in Figure 3.3 is so much more precise than just saying "Community Involvement."

On the **Date** line, do not simply name the day of the week—Friday. Give the full calendar date—June 3, 2011.

Questions Your Memo Needs to Answer for Readers

Here are some key questions your audience may ask and that your memo needs to answer clearly and concisely:

1. **When?** When did it happen? Is it on, ahead of, or behind schedule? When does it need to be discussed or implemented? *When* is answered in Figures 3.1 ("November 7," "in the next day or two"), 3.2 ("during the past two weeks," "after each use"), and 3.3 ("in early February," "before the end of the month").

2. **Who?** Who is involved? Who will be affected by your message? How many people are involved? *Who* is answered in Figures 3.1 (Jackie Chen), 3.2 (all machine shop employees), and 3.3 (Ramco Technologies as a whole).

3. **Where?** Where did it take place or will it take place? *Where* is answered in Figures 3.1 (the website seminar), 3.2 (the machine shop), and 3.3 (the Mayfield plant).

4. **Why?** Why is it an important topic? *Why* is clearly answered in Figures 3.1 (because the website is being redesigned), 3.2 (because it's a safety issue), and 3.3 (because favorable publicity could result).

5. **Costs?** How much will it cost? Will the costs be lower or higher than a competitor's costs? Not every memo will answer financial questions, but in Figure 3.3, the specific cost of an individual scholarship ($6,800) is an important issue.

6. **Technology?** What technology is involved? Why is the technology needed? Is the technology available, current, adaptable, safe for the environment? Again, not every one of your memos will answer questions about technology, but note that Figures 3.1, 3.2, and 3.3 all especially refer to technological issues—web design, equipment safety, robots.

7. **What's next?** What are the next steps that should be taken as a result of the issues discussed in the memo? What are the implications for the product, service, budget, staff? A good example of this is in Figure 3.3 (the company needs to implement the change before the new plant opens).

Memo Style and Tone

The audience within your company will determine your memo's style and tone (for a review of identifying audience, see Chapter 1, pages 5–10). When writing to a co-worker whom you know well, you can adopt a casual, conversational tone. You want to be seen as friendly and cooperative. In fact, to do otherwise would make you look self-important, stuffy, or hard to work with. Consider the friendly tone appropriate for one colleague writing to another as in Roger's memo to Lucy in Figure 3.1. Note how he ends in a polite but informal way.

When writing a memo to a manager, though, you will want to use a more formal tone than you would when communicating with a co-worker or peer. Your boss will expect you to show a more respectful, even official, posture. See how formal yet conversationally persuasive Mike Gonzalez's memo to his bosses is in Figure 3.3. His tone and style are a reflection of his hard work as well as his respect for his employers. Here are two ways of expressing the same message, the first more suitable when writing to a co-worker and the second more appropriate for a memo to the boss.

> Co-worker: I think we should go ahead with Marisol's plan for reorganization. It seems like a safe option to me, and I don't think we can lose.
>
> Boss: I think that we should adopt the organizational plan developed by Marisol Vega. Her recommendations are carefully researched and persuasively answer the questions our department has about solving the problem.

When an employer writes to workers informing them about policies or procedures, as Janet Hempstead does in Figure 3.2, the tone of the memo is official and straightforward. Yet even so, Hempstead takes into account her readers' feelings (she does not blame) and safety, which are at the forefront of her rhetorical purpose.

Finally, remember that your employer and co-workers deserve the same clear and concise writing and attention to the "you attitude" (see Chapter 4, pages 00–00) that your customers do. Memos require the same care and should follow the same rules of effective writing outlined in Chapter 2, as letters do.

Strategies for Organizing a Memo

Organize your memos so that readers can find information quickly and act on it promptly. For longer, more complex communications, such as the memos in Figures 3.2 and 3.3, your message might be divided into three parts: (1) introduction, (2) body or discussion, and (3) conclusion. Regardless of how short or long your memo is, recall the three *P*'s for success—*p*lan what you are going to say; *p*olish what you wrote before you send it; and *p*roofread everything.

Introduction

The introduction of your memo should do the following:

- Tell readers clearly about the issue, policy, or problem that prompted you to write.
- Explain briefly any background information the reader needs to know.
- Be specific about what you are going to solve a problem.

Do not hesitate to come right out and say, "This memo explains new password security procedures" or "This memo summarizes the action taken in Evansville to reduce air pollution."

Body (Discussion)

In the body, or discussion section, of your memo, help readers in these ways:

- State why a problem, procedure, or decision is important; who will be affected by it; and what the consequences are.
- Indicate why and what changes are necessary.
- Give precise dates, times, locations, and costs.

See how Janet Hempstead's memo in Figure 3.2 carefully describes an existing problem and explains the proper procedure for cleaning the brake machines.

Conclusion

In your conclusion, state specifically how you want the reader to respond to your memo. To get readers to act appropriately, you can do one or more of the following:

- Ask readers to call you if they have any questions, as in Figure 3.2.
- Request a reply—in writing, over the telephone, via e-mail, or in person—by a specific date, as in Figures 3.1 and 3.3.
- Provide a list of recommendations that readers are to accept, revise, or reject, as in Figure 3.3.

Organizational Markers

Throughout your memo, use the following organizational markers, where appropriate:

- **Headings** organize your work and make information easy for readers to follow, as in Figure 3.3.
- **Numbered** or **bulleted lists** help readers see comparisons and contrasts readily and thereby comprehend your ideas more quickly, as in Figure 3.1.
- **Underlining** or **boldfacing** emphasizes key points (see Figures 3.2 and 3.3). Do not overuse this technique; draw attention only to main points and those that contain summaries or draw conclusions.

Organizational markers are not limited to memos; you will find them in e-mail, letters, reports and proposals as well. (See Chapter 6, pages 198–203.)

Sending Memos: E-mail or Hard Copy?

A memo can be sent as a printed hard copy, as an e-mail, or in an e-mail attachment. Find out your company's policy. Increasingly, e-mail is replacing printed memos, but there are times when a hard-copy memo is preferred as essential.

Consider the level of official importance and confidentiality of your memo. If your memo is an official document, such as the policy outlined in Figure 3.2, you may not want to send it via e-mail, because it could easily be deleted or altered. Moreover, if your memo is confidential (e.g., an evaluation of a co-worker or vendor, a message containing sensitive financial or medical information), you may not want to send it via e-mail, because it could easily be forwarded or fall victim to hackers.

Sending Faxes: Some Guidelines

Even though you will use e-mail attachments and scan and send documents via e-mail, faxes are still used in the world of work. A fax is an original document copied and transmitted over telephone or computer lines. Faxes are particularly helpful either when you have only hard copy to send or when you want to send an original signed letter, contract, blueprint, artwork, or other document that you could not send via an e-mail transmission. Faxes demonstrate exactly what original documents look like and allow recipients to obtain a hard copy quickly.

Cover Page

When you send a fax, make sure your cover page includes the following information:

1. **The name of the sender and his or her fax and phone numbers**. The phone number is important because it enables the recipient to report an incomplete transmittal.

2. **The name of the recipient and his or her fax and phone number.** The recipient's correct fax and phone numbers should be included to ensure delivery of the message.

3. **The total number of pages being faxed.** Note that the total number of pages includes the cover page itself.

4. **A brief explanatory note** that lets the recipient know what the fax is, what its purpose is, and how and when to respond to it.

Sending a Document

Follow these four guidelines to fax a clear and complete document:

1. **Make sure the original documents you send are clear.** An unclear faxed document will be difficult for the recipient to read. For example, penciled comments may be too faint to fax clearly.

2. **Avoid writing on the top, bottom, or edges of the documents to be faxed.** Any comments written on the outer edges may be cut off or blurred during transmittal.

3. **Do not send overly long faxes.** Be careful about sending anything longer than three or four pages because you will tie up both your own and the recipient's fax machines.

4. **Respect the recipient's confidentiality.** Because faxes may be picked up by other employees in your office, don't assume your message will be confidential, unless, of course, the recipient has a private fax machine.

E-Mail: Its Importance in the Workplace

E-mail continues to be one of the most common forms of communication in the workplace. It is the lifeblood of every business or organization because it expedites communications within your firm as well as outside of it. Professionals in the world of work may receive hundreds of e-mails a day from supervisors, colleagues, clients, and vendors. Moreover, these e-mails may contain visuals, soundbites, and tables, lists, and statistical files. As we saw, memos, too, can be sent via e-mail.

Business E-Mail Versus Personal E-Mail

E-mail is an informal, relaxed type of business correspondence, far more so than a printed memo, letter, short report, or proposal. (However, e-mail should never be sent in place of a formal letter; see pages 95–96). Think of business e-mail as a polite, informative, and professional conversation—friendly, to the point, and always accessible, as in Figure 3.4. Yet even though business e-mail is basically informal and casual, sending it does not mean you can forget about your responsibilities as an informed, cautious, and courteous employee and co-worker. Figure 3.5 exemplifies an e-mail used as a diplomatic yet informal way of communicating with a collaborative team.

An example of an effectively written business e-mail.   Figure 3.4

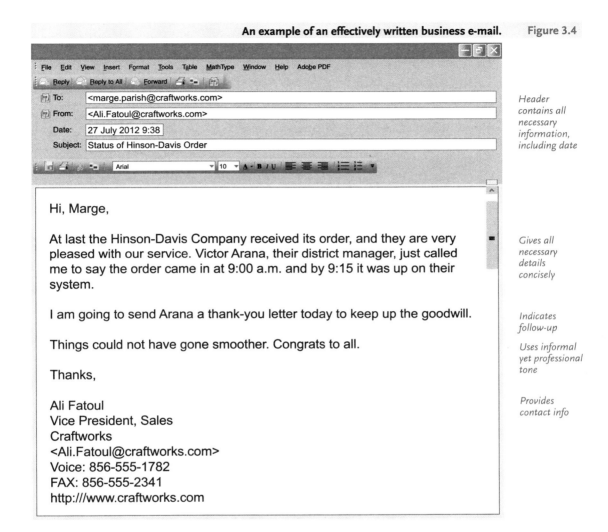

File Edit View Insert Format Tools Table MathType Window Help Adobe PDF

Reply   Reply to All   Forward

To: <marge.parish@craftworks.com>

From: <Ali.Fatoul@craftworks.com>

Date: 27 July 2012 9:38

Subject: Status of Hinson-Davis Order

Header contains all necessary information, including date

Arial   10   A · B I U

Hi, Marge,

At last the Hinson-Davis Company received its order, and they are very pleased with our service. Victor Arana, their district manager, just called me to say the order came in at 9:00 a.m. and by 9:15 it was up on their system.

I am going to send Arana a thank-you letter today to keep up the goodwill.

Things could not have gone smoother. Congrats to all.

Thanks,

Ali Fatoul
Vice President, Sales
Craftworks
<Ali.Fatoul@craftworks.com>
Voice: 856-555-1782
FAX: 856-555-2341
http:///www.craftworks.com

Gives all necessary details concisely

Indicates follow-up

Uses informal yet professional tone

Provides contact info

The e-mail you write on the job will require more from you as a writer than your personal e-mail will. You cannot be sloppy or unorganized. Proofread and, if necessary, revise your e-mail before you send it. You need to follow all the rules of proper spelling, capitalization, punctuation, and word choice, plus the e-mail guidelines in the following section. The **tone** (see pages 75–76) of your business e-mail should be much more professional than the instant messaging you may do with friends or e-conversations you may have in a chat room. Don't assume you can write to your employer or a customer the way you would to an old friend.

Figure 3.5   **E-mail sent to a distribution list of co-workers.**

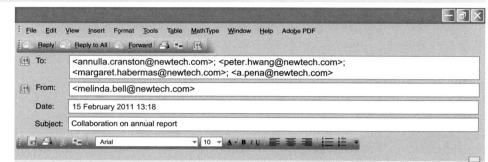

Precise subject line

Starts with context for and confirmation of meeting

Provides clear explanations and instructions

Requests information politely

Ends by building morale

Provides contact information

File   Edit   View   Insert   Format   Tools   Table   MathType   Window   Help   Adobe PDF

Reply   Reply to All   Forward

To: <annulla.cranston@newtech.com>; <peter.hwang@newtech.com>; <margaret.habermas@newtech.com>; <a.pena@newtech.com>

From: <melinda.bell@newtech.com>

Date: 15 February 2011 13:18

Subject: Collaboration on annual report

Arial   10   A · B I U

Hello, Team:

To follow up on our conversation yesterday regarding working on this year's annual report, I'm glad our schedules are flexible. I've checked our calendars, and we are all available next Tuesday the 22nd at 10:30 a.m. Let's meet in Conference Room 410.

Don't forget we have to draft a two- to three-page overview first that explains New Tech's strategic technology objectives for fiscal year 2012. Not an easy assignment, but we can do it, gang.

It would be a big help if Annulla would bring copies of the reports for the last three years. Would Peter please call Ms. Jhandez in Engineering for a copy of the talk she gave last month to the Powell Chamber of Commerce? If memory serves me correctly, she did a first-rate job summarizing NewTech's accomplishments for 2010.

Thanks for all your splendid work, team. See you Tuesday.

Melinda Bell
Director, Marketing
New Tech
<melinda.bell@newtech.com>
FAX: (603) 555-2162
Voice: (603) 555-1505
http://www.newtech.com

Unlike your personal e-mail, your business e-mail must consider the impact it will have on your company and on your career. When you send a business e-mail, you are representing more than just yourself and your preferences, as in a personal e-mail. You are speaking on behalf of your employer. Because your e-mail must reflect your company's best image, make sure it is businesslike, carefully researched, and polite. Sarcasm, slang, an aggressive tone, name calling and inappropriate icons/clip art do not belong in a company e-mail. As we saw, Figures 3.4 and 3.5 illustrate effectively written business e-mails. Notice that e-mail can be cordial without being unprofessional.

E-Mails Are Legal Records

Keep in mind employers own their internal e-mail systems and thus have the right to monitor what you write and to whom. (See pages 21–22 in Chapter 1.) Any e-mail written at work on the company's server can be copied, archived, forwarded, and, most significantly, intercepted. You can be fired for writing an angry or abusive e-mail. Your e-mail can easily be converted into an electronic paper trail. You never know who will receive and then forward your e-mail—to your boss, a customer, an attorney, or a licensing board. Many companies include disclaimers protecting themselves from legal action against them because of an employee's offensive behavior in a company e-mail. In court, an e-mail can carry the same weight as a printed hard copy.

Here are some helpful rules to follow about using your company's e-mail:

- Do not use it for your personal messages. Send e-mails only for appropriate company business, and make sure you are professional and conscientious.
- Never write an e-mail to discuss a confidential subject—a raise, a grievance, or a complaint about a co-worker. Meet with your supervisor in person.
- Make sure of your facts before sending an e-mail to customers. If it gives them wrong or misleading information about prices, warranties, or safety features, your company can be legally liable.

Guidelines for Using E-Mail on the Job

When you prepare and organize your e-mail message, always consider your reader's specific needs as well as those of your company. Following the guidelines below will help you to write effective business e-mails.

1. **Make sure your e-mail is confidential and ethical**.
   - Avoid *flaming*, that is, using strong, angry language that mocks, attacks, or insults your employer, a colleague, a customer, government agency, or a company, as in Figure 3.6 (page 84). Abusive, obscene, or racially/culturally offensive language in an e-mail constitutes grounds for dismissal.
   - Send nothing through e-mail that you would not want to see on your company's website or on the front page of your newspaper.

- Do not forward a co-worker's or employer's e-mail without that person's approval.
- Do not change the wording of a message that you are expected to read and forward.

2. **Make your e-mails easy to read.**
   - **Provide a clear, precise subject line.** Avoid one-word subjects: "Report," "Meeting." Instead, write "Meeting to boost declining April sales." A subject line like "Bill" leaves readers wondering if your e-mail is about a person or an unpaid account.
   - **Try to limit your e-mails to one screen.** Longer messages are better sent in an attachment rather than in the body of an e-mail.
   - **Do not send e-mails written in all capital or all lowercase letters.** All capital letters look as if you are screaming. Conversely, e-mails in all lowercase imply you do not know how to punctuate.
   - **Break your message into paragraphs.** A screen filled with one dense block of text is intimidating. Make each paragraph no more than three to four lines long, always double-space between paragraphs, and use no more than 60 characters per line.
   - **Use plain text.** Because different e-mail programs can garble your message, avoid overusing typefaces like italic script, or bold complex formatting (such as long numbered and bulleted lists) and symbols (monetary, accents, etc.) within the text of an e-mail.
   - **Avoid long strings of e-mails.** Delete strings of previously answered e-mails when you reply so the reader will not waste time scrolling to find your most recent answer.

3. **Observe the rules of netiquette (Internet + etiquette).**
   - **Respond promptly to an e-mail.** Check for new messages several times each day. If you will be offline for an extended period, use the auto-reply feature to tell all readers when you expect to return.
   - **Give your readers reasonable time to respond.** Consider time zone differences between you and your reader. It may be 2:00 a.m. when your e-mail arrives for an international recipient.
   - **Do not keep sending the same e-mail over and over.** This is discourteous and will only antagonize your recipient.
   - **Avoid unfamiliar abbreviations, jargon, and emoticons.** Don't use abbreviations found in your personal e-mails (BTW, LOL) or that are used in text messaging. Include only those abbreviations and jargon that your recipients will understand, e.g., FYI. Also, do not use emoticons (smiley faces, sad faces, etc.) in your professional communications.
   - **Don't use red flag words unnecessarily.** Stay away from words like "Urgent," "Crucial," "Top Priority," along with accompanying red exclamation marks, in your subject line just to get your reader's attention. Your tactic will backfire, potentially upsetting readers or, worse yet, causing them to ignore any genuinely urgent messages you may send in the future.

- **Include a signature block.** A signature block includes your name, title, and contact information at the end of your message (see Figures 3.4 and 3.5). Such information is crucial when you are part of a large organization or when you are communicating with someone outside of your company or agency.

4. **Adopt a professional style.**
   - **Use a salutation (greeting), but always follow your company's policy.**
     - to a colleague—Hi, Hello
     - to a customer—Dear Ms. Pietz; Dear Bio Tech
   - **Get to the point right away.** Because readers receive a lot of e-mail, they may look only at the first few lines you write.
   - **Keep your messages concise.** Cut wordy phrases and send only the information your reader needs. Exclude unnecessary details and chatter.
   - **Don't turn your e-mail into a telegram.** "Send report immediately; need for meeting" is rude as is a reply only with "Yes," "No," or "Sure." Save words like "Nope," "Yeah," and "Huh" for your personal e-mails.
   - **Respect the cultural traditions of international readers.** Avoid using abbreviations, symbols, or measurements that your reader may not know, and do not use first names unless the reader approves. Refer to the "Ten Guidelines for Communicating with International Readers" in Chapter 4 (pages 134–138).
   - **End politely.** Let readers know in your last sentence that you appreciate their help/cooperation and look forward to their reply (see Figure 3.5).
   - **Use a complimentary close, but always follow your company's policy.**
     - to a colleague—Thanks, Later, Take care,
     - to a customer—Sincerely yours, Sincerely, Best regards,

5. **Ensure that your e-mail is safe and secure.**
   - **Use e-mail protection services and software.** Always consult with your company's information technology (IT) department.
   - **Avoid contracting e-mail viruses** by deleting unopened, unsolicited e-mail attachments.
   - **Don't be a victim of identity theft, or "phishing."** Companies you do business with will never ask for personal information, such as your bank account or Social Security number.
   - **Never provide company financial information unless you are sure that it will be relayed over a safe connection.** Always check with your boss before providing such information.
   - **Create an e-mail password that is not easy to guess.** Do not use a password such as "ABCDE" or "123456." Change your password regularly, and do not use the same password for all your accounts.
   - **Back up important files, including e-mails.** If you contract a computer virus or your computer crashes, be sure you have saved your most important and current files.

Figure 3.6 shows an example of a poorly written e-mail that violates many of the preceding guidelines. Figure 3.7 on page 85 contains an effective revision that reflects the professional and courteous way the writer and his company conduct business.

Figure 3.6	A poorly written e-mail, with annotations.

No specific reader

Vague subject line

To: `<newtech@widedoor.com>`
From: `<sammy@dataport.com>`
Date: 11/16/11
Subject: Upgrades

Arial   10

Unprofessional greeting

Discourteous tone

All caps shout

HEY GUYS------------

ARE YOU AWAKE OUT THERE? THIS IS THE THIRD TIME I HAVE SENT THIS MESSAGE. AND I NEED YOU TO GET BACK TO ME STAT. MY BOSS IS ON MY BACK. : P ———————— *Confusing emoticon*

Insufficient information

Unclear abbreviation, flaming

I NEED THE UPGRADES YOUR SALES FOLKS--ROBERT T., JAN W., AND GRAF H.--PROMISED BUT NEVER MADE GOOD ON.

FWIW YOU HAVE MISSED THE BOAT.☹ ———————— *Unprofessional emoticon*

SAMMY

No signature block

Instant Messages (IMs) for Business Use

IMs are streamlined textual conversations online and sent close to real time. Think of a professional IM as somewhere between a phone call and an e-mail, or a chat with a colleague in the hallway. IM conversations are almost as instantaneous as phone conversations, but at the same time they provide written records of communications, just as e-mails do. Keep in mind, though, that IMs are not just for communications with your friends; they play a vital role in workplace correspondence. In fact, researchers estimate that 90 percent of all businesses have or will use

A revised, effective version of the poor e-mail in Figure 3.6.   Figure 3.7

| File | Edit | View | Insert | Format | Tools | Table | MathType | Window | Help | Adobe PDF |

Reply   Reply to All   Forward

To:	<MWood@widedoor.com>
From:	<sammy@dataport.com>
Date:	11/16/11
Subject:	Upgrades for service contract #4552

Arial   10   A · B / U

Uses email address of specific person

Precise subject

Hello, Mary:

I would appreciate your delivering the upgrades for our service contract #4552 by Thursday afternoon, the 17th of November, if at all possible.

We need to proceed to the next phase of our operation, and the upgrades are crucial to that task.

I am attaching a copy of our service agreement with Wide Door for your convenience.

If you run into any problems with the delivery date, please give me a call this afternoon or e-mail me.

Thanks,

Sammy

Samuel Atherton
Operations Assistant
Data Port
4300 Morales Highway
San Padre, CA 95620-0326
Voice mail: 723-555-1298
http://dataport.com

Polite salutation

Gets to the point concisely but diplomatically

Provides explanation and documentation

Ends with clear-cut directions

Professional close

Includes signature block

IMs for routine workplace correspondence. IMs allow you to communicate with co-workers and managers in the same office, at remote sites, or around the globe. Crossing time zones, IMs give you access to anyone around the world who is online and connected to the same service. Figure 3.8 is an example of a workplace IM conversation.

When to Use IMs Versus E-Mails

Like e-mails, IMs promote collaboration, provide a written record, and further global communication. But they are used for very different kinds of messages. E-mails are more detailed than IMs. By answering the following questions, you will be better able to determine when to send an IM or an e-mail.

1. **How quickly does my message need to be sent/received?** If you need information right away, use IM rather than an e-mail because recipients will most likely reply at once if they are online.
2. **How long or complex is my message?** If you need to transmit a message that, say, is more than a line or two (e.g., even extending to a paragraph or more) or contains multiple points, send an e-mail.
3. **If your message requires more time than a few brief back-and-forth communications**, start an e-mail exchange that can extend over several hours or days.

Guidelines on Using IMs in the Workplace

Your IMs may be instantaneous and informal, but that does not mean that you can send them with little thought about their content, tone, and punctuation. Again, keep in mind that your company can monitor, trace, record, and archive your IM conversations just as it can with e-mails. In addition to the guidelines for writing workplace e-mails (pages 81–83), observe these safe rules for your IMs.

1. **Stay connected.** Indicate your messenger status—"Away," "Busy," "Offline. Please e-mail me at tjones@comcast.com." If you are away, tell those on your buddy list (contact list) when you will be back or give them alternate contact information ("Back in 30 minutes").
2. **Keep your messages short.** Get to the point right away. A line or two at most is enough for your IMs. Don't inject unnecessary pleasantries, e.g., "How was your weekend?"
3. **Write about one topic at a time.** Don't include information about two or three different topics in one IM exchange. Keep the conversation flowing in one direction, not two or three.
4. **Avoid unfamiliar abbreviations.** As in an e-mail, commonly understood abbreviations such as "FYI" or "ASAP" are fine—even encouraged—in IMs. But avoid "textspeak" abbreviations such as "CUL8R" for "See you later," especially when writing to an international reader who may not understand them. Moreover, your boss might not appreciate a textspeak message such as "np gtg ttyl" for "No problem. Got to go. Talk to you later."
5. **Make sure your message is professional.** Even though IMs are the most informal business messages you can send, don't disregard professional courtesy and ethics. Never send personal messages, tell jokes, spread office gossip, or attack a co-worker or boss in an IM. Also, choose an appropriate, professional online name, not "Go-Getter Pete."

6. **Organize your contact (buddy) lists into separate groups,** such as business, co-workers, friends/family, etc., so you do not embarrassingly send someone the wrong message.
7. **Don't use IMs to send sensitive/confidential information** about personnel, financial, hiring, or medical/legal issues.
8. **Watch for viruses.** Be careful about sharing files or opening attachments via IMs.
9. **Safeguard the privacy of your contact lists** as well as the contents of any IM attachments.

An IM exchange between co-workers   Figure 3.8

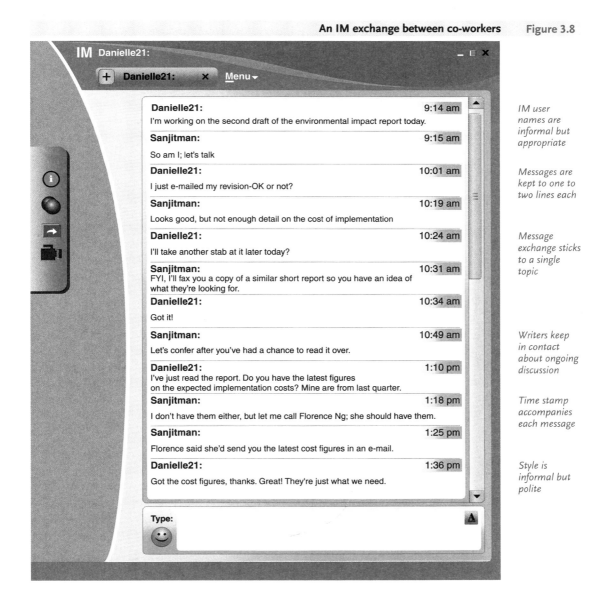

IM user names are informal but appropriate

Messages are kept to one to two lines each

Message exchange sticks to a single topic

Writers keep in contact about ongoing discussion

Time stamp accompanies each message

Style is informal but polite

Blogs

Like e-mails and IMs, blogs (an abbreviation of "Web logs") are important tools for e-correspondence for employees, managers, and customers. A blog is more than just a casual journal or diary in the world of work. Think of a business blog as an evolving website or daily business newspaper for which employees and management write regular columns, or posts. A blog provides a fast, informal way to give readers information and news on a variety of issues of vital concern to your company or organization. Blogs are also highly interactive—readers can write comments in response to blog posts, allowing for a two- or multiple-way conversation. Customers can give feedback and employees can communicate with one another via a blog post. These responses are often posted on a blog.

Internal/External Blogs

Blogs can be either internal or external, depending on the intended audience. Internal blogs are designed exclusively for employee readers. Employees can interact by posting their comments and asking questions, express their views on new policies, announce events and introduce new staff, and communicate up and down the corporate ladder.

External blogs allow companies to express their side of the story—their interpretation of events and their clarification of the issues—quickly and publicly. But besides being a forum for the company's viewpoints, external blogs are a fast way to get the information out about new or updated product and services, expanded locations, changes in technology, product recalls, efforts at greening an environment, community service, or just about anything related to a company's mission and activities. Figure 3.9 shows an example of an external blog about a company's reason for discontinuing one of its popular products but replacing it with a better alternative.

Guidelines for Writing a Business Blog

An external, or company, blog is your employer's official, and many times daily, publication. You may have to get your business blog approved by a blog administrator to make sure it meets your company's expectations. One way to make sure it is approved is by always honoring the confidentiality of company business. Moreover, keep your blog concise and make it readable.

Following the guidelines below will help you write an effective (external) business blog. As you read these guidelines, refer to Figure 3.9 on page 89.

1. **Respect your company's protocols**. Don't post anything discriminatory or confidential about an employer, co-workers, product, or service.
2. **Positively project your company's image**. Keep your company's history, mission, and reputation in mind when you prepare your blog. Be enthusiastic about its products, services, and workforce. Note how Clay Denton-Tyler shows how consumer-oriented Powerhouse, Inc. is. Never make your company look bad by undermining management or criticizing a vendor.
3. **Avoid making your company liable for false or misleading information**. Don't make promises, make guarantees, or provide additional warranties unless approved by upper management.

An external blog.   Figure 3.9

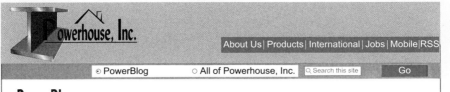

Search engine
and categories
fields simplify
navigation

About Us | Products | International | Jobs | Mobile | RSS

○ PowerBlog   ○ All of Powerhouse, Inc.   Search this site   Go

PowerBlog

| All | Company News | Product News | Distribution | Manufacturing |

<< Previous Post   >> Next Post

*"Previous,"
"next," and
"recent,"
posts allow
for further
navigation*

Today's Post (August 15, 2012):

>> A Power Tool Even Better than the PH450?
by Clay Denton-Tyler, District Manager

Recent Posts:

International
Dyanamics, SE
CFO announces
retirement.

PowerHouse
announces the
discontinuation of
the PH-450.

PowerHouse opens
new retail outlets in
Sydney, Australia,
and Dunedin, New
Zealand.

International
Dyanamics, SE
acquires quality
Swiss Home
Products company.

PowerHouse goes
international.

*Provides a
clear and
concise title
and date*

*Writes to
a general
audience and
avoids jargon*

*Thanks
customers for
feedback*

*Uses a
conversational
but
professional
tone*

*Acknowledges
customers'
disappointment*

Many of our loyal customers disagree with Powerhouse's decision to discontinue manufacturing the PH-450 All-in-One Power Tool. It is always great to hear from our customers and to receive their feedback, even when they believe we've done something wrong. To help our customers to better understand our perspective, let me fill in some of the background about why the PH-450 is not being manufactured any more.

Discontinuing the PH-450 was not an easy decision. After all, this was the product that first brought our company to national attention and widespread customer acceptance. Also, I know from many complimentary customer e-mails and replies to earlier posts, as well as from my personal experience as a proud owner of the PH-450, that our customers have always applauded its price, compact design, durability, and all-weather usability. As one of you put it, "Why kill a popular product that has worked so well for 20 years?"

Figure 3.9	(Continued)

More

Attempts to persuade customer to switch to a new model

All of the responses are true but the good news is that even though the PH-450 is being discontinued, our customers now have a very similar but improved alternative. Recently, our parent company, International Dyanamics, SE, acquired a new multipurpose tool from Swiss Home Products. The SHP-1000 is not only just as compact, durable, and weather-friendly as the PH-450, but it offers several additional features, such as a nail gun attachment and lifetime limited warranty. And due to an excellent distribution deal negotiated between International Dyanamics, SE, and Swiss Home Products, we can sell it at less than 30 percent of the retail cost of the PH-450.

Describes benefits of new model and why it is being marketed

Sympathizes with readers but offers personal endorsement characteristic of bloggers at same time

I know it is hard to say goodbye to a reliable helper but like many products in our increasingly technical age, the PH-450 is being replaced by a more efficient alternative for our customers. I will miss the old PH-450, but I have found that the SHP-1000 is just as effective in my shop at home. Adapting to a new model has never come easier for me. Why not give it a try? Thanks. I would like to hear from you.

Tone is sincere and friendly

Comments link allows for further discussion

Comments: (21)
- Sign in to add a comment
- First time users, please register first, in order to add a comment

4. **Target your audience.** Try to reach current and potential customers by anticipating their questions and needs. Determine if your blog will reach an international audience as well as customers for whom English is their native language.

5. **Date every blog posting so readers can follow a conversation.** Update your blog to make sure your information is current.

6. **Write an attention-grabbing headline.** Attract readers with a title that tells them how and why they can profit from reading your blog and encourage them to respond to your post: e.g., "A Power Tool Even Better than the PH450." Avoid vague headlines such as "Important News," "Any Further Ideas?" and so on.

7. **Share your story**. Draw from your experiences. Offer your own perspectives. Observe how in Figure 3.9 Denton-Tyler provides his own views as an owner of a popular discontinued product.

8. **Adopt a casual, conversational style**. Be upbeat and succinct. Emphasize how and why your comments will help readers. Regard your blog as a fruitful chat between you and your readers. Don't weigh readers down with long, windy paragraphs that make your comments sound boring and complex.

9. **Structure your blog so that your first paragraph comes to the point at once and tells readers what you are blogging about and why**. For example, Denton-Tyler clearly states he has information on "why this product will no longer be manufactured," a question his audience is eager to see him answer.

10. **Make it easy for readers to respond to your posting**. Tell them where and how to reply. As in Denton-Tyler's blog, refer to customer postings that show how interested you and your company are in your audience's opinions

11. **Document your sources**. If you use someone else's statistic, surveys, illustrations, or ideas, first get permission from the individual or the company that owns the copyright.

12. **Include relevant visuals**. The blog in Figure 3.9 clearly shows what the new model looke like.

Dentor-Tyler had to carefully consider his audience and their reasons for reading his comments in Figure 3.9. His readers had been loyal customers who naturally were not happy that a well-received product was being taken off the market, so he had to reassure them that the replacement was better and cheaper without losing any of the features of the discontinued model. To do this, he had to be credible yet cautious in responding to the posts he had received from this group of readers.

But in the interactive world of blogging, he recognized that he was also writing to potential customers in the blogosphere and wanted them to see the company as customer oriented and driven by the needs of this audience. Finally, he realized that his post would be a part of an ongoing public discussion about the new model and his company, and so he wanted to answer as many questions as he could while still keeping the conversation going—all in a positive direction.

Routine Correspondence: A Final Word

In today's global marketplace, you can expect to write the types of correspondence discussed in this chapter as a regular and vital part of your job. They are quick, brief, to-the-point, and informal, the workhorses of the business world. Always remember, though, that they are professional business communications that represent your company and should portray you as a proficient and ethical employee. As this chapter has stressed, to write successful memos, faxes, e-mails, IMs, and company blogs, follow these essential guidelines:

(1) Plan your correspondence—don't just write and click.
(2) Politely address your reader's questions and problems.
(3) Make sure your information is accurate and up-to-date.
(4) Adhere to your company security/privacy rules.
(5) Proofread for spelling, grammar, punctuation, style, and tone.

Revision Checklist

Memos
☐ Used appropriate and consistent format.
☐ Announced purpose of memo early and clearly.
☐ Organized memo according to reader's need for information, putting main ideas up front, providing necessary documentation, and supplying conclusion.
☐ Wrote clearly and concisely.
☐ Included bullets, lists, and underscoring where necessary to reflect logic and organization of memo and make it easier to read for the audience.
☐ Refrained from overloading reader with unnecessary details.

Faxes
☐ Verified reader's fax number and sent cover sheet with phone number to call in the event of transmission trouble.
☐ Excluded anything confidential or sensitive if reader's fax machine is not secure.

E-Mail
☐ Considered difference between business e-mail and personal e-mail.
☐ Formatted e-mail with acceptable margins and spacing.
☐ Observed netiquette; avoided flaming.
☐ Kept paragraphs short but used full—not telegraphic—sentences.
☐ Avoided unfamiliar abbreviations or terms that would confuse readers.
☐ Safeguarded employer's confidentiality and security by excluding sensitive or privileged information.
☐ Began with friendly greeting; ended politely.
☐ Included sufficient information and documentation for reader's purpose.
☐ Considered needs of international audience.

Instant Messaging
☐ Used IMs only for professional, job-related communications.
☐ Kept responses/questions short—not over a line or two.
☐ Avoided "textspeak" in business IMs.
☐ Notified readers when you would be offline and back online.
☐ Did not send anything confidential through an IM exchange.

Blogs
☐ Posted nothing critical of employers, products and services, co-workers, or anything that would be embarrassing, offensive, or confidential to an employer or customer.
☐ Made posts sincere, conversational, and informal yet professional.
☐ Posted only current and relevant information.
☐ Included relevant visuals.
☐ Provided a comments area for readers to give feedback.

Exercises

1. Write a memo to your boss saying that you will be out of town two days next week and three days the following week for one of the following reasons: (a) to inspect some land your firm is thinking of buying, (b) to investigate some claims, (c) to look at some new office space for a branch your firm is thinking of opening in a city 500 miles away, (d) to attend a conference sponsored by a professional society, or (e) to pay calls on customers. In your memo, be specific about dates, places, times, and reasons.

2. Send a memo to your public relations department informing it that you are completing a degree or work for a certificate. Indicate how the information could be useful for your firm's publicity campaign.

3. Write a memo to the director of your school's library asking for one of the following. Be sure you include specific reasons for such a change.
   a. extended weekend hours
   b. more vending machines
   c. more computers
   d. more group study rooms
   e. increased journal subscriptions in your field of study

4. Select some change (in policy, schedule, or personnel assignment) you encountered in a job you held in the last two or three years and write an appropriate memo describing that change. Write the memo from the perspective of your former employer explaining the change to employees.

5. Send a fax to a company or an organization requesting information about the products or services it offers. Include an appropriate cover sheet.

6. Write an e-mail to a business that provides daily or weekly information to interested customers and submit its response along with your e-mail request to your instructor. Choose one of the following:
   a. an airline: an up-to-date schedule along a certain route and information about any bonus-mile or discount programs
   b. a catalog order company: information about any specials for Internet users
   c. a stock brokerage firm: free quotes or research about a particular stock
   d. a resort: special rates for a given week

7. Write an e-mail with one of the following messages, observing the guidelines discussed in this chapter.
   a. You have just made a big sale, and you want to inform your boss.
   b. You have just lost a big sale, and you have to inform your boss.
   c. Tell a co-worker about a union or national sales meeting.
   d. Notify a company to cancel your subscription to one of its publications because you find it to be dated and no longer useful in your profession.
   e. Request help from a listserv about research for a major report you are preparing for your employer.

    f. Advise your district manager to discontinue marketing one of the company's products because of poor customer acceptance.

    g. Send a short article (about 200 words) to your company's online newsletter about some accomplishment your office, department, or section achieved in the last month.

    h. Write to a friend studying finance at a German, Korean, or South American university about the biggest financial news in your town or neighborhood in the last month.

8. Rewrite the following e-mail to your boss to make it more professional.

Hi—

This new territory is a pain. Lots of stops; no sales. Ughhhh. People out here resistant to change. Could get hit by a boulder and still no change. Giant companies ought to be up on charges. Will sub. reports asap as long as you care rec.

The long and short of it is that market is down. No news=bad news.

9. As a collaborative venture, join with three or four classmates to prepare one or more of the e-mail messages for Exercise 7. Send each other drafts of your messages for revision. Submit the final copy of the group's effort.

10. Send your instructor an e-mail message about a project you are now working on for class, outlining your progress and describing any difficulties you are having.

11. You have just missed work or a class meeting. E-mail your employer or your instructor explaining the reason and telling how you intend to make up the work.

12. As a group activity, instant message two or three other members of your collaborative writing team while you are drafting on a project you are working on. Print out your IM exchanges during this time, and submit them to your instructor.

13. As a collaborative project, write external blog posts for three or four days about some aspect of the team's current job or a previous job in which they share with readers news about the company's products/services, technology professional travel, community service, working with international colleagues, and so forth. Be sure that your posts put the company, department, or agency in a good light.

14. Your company has just issued a product recall of a small appliance because of suspected faulty wiring. Write a company blog explaining employee response to the recall, which may result in furloughing 20–25 employees.

4 Writing Letters
Some Basics for Communicating with Audiences Worldwide

Letters are among the most important writing you will do on your job. Businesses worldwide take letter writing very seriously, and employers will expect you to prepare and respond to your correspondence promptly and diplomatically. Your signature on a letter tells readers that you are accountable for everything in it. The higher up the corporate ladder you climb, the more letters you will be expected to write.

Visit www. cengage.com/ english/kolin/ writingatwork concise3e for this chapter's online exercises, ACE quizzes, and Web links.

Because letter writing is so significant to your career, this chapter introduces you to the entire process, provides guidelines and problem-solving strategies, and shows you how to prepare the most frequently written types of business letters. It also shows you how to write for international readers.

Letters in the Age of the Internet

Even in this age of the Internet, letters are still vital in the world of work. A professional-looking letter is one of the most significant symbols in the business world for the following reasons:

1. **Letters represent your company's public image and your competence.** A firm's corporate image is on the line when it sends a letter. Carefully written letters can create goodwill; poorly written letters can anger customers, cost your company business, and project an unfavorable image of you.

2. **Letters are far more formal—in tone and structure—than any other type of business communication.** Memos, e-mails, and IMs are the least formal communications.

3. **Letters constitute an official legal record of an agreement.** They state, modify, or respond to a business commitment. When sent to a customer, a signed letter constitutes a legally binding contract. Be absolutely sure that what you put in a letter about prices, guarantees, warranties, equipment, delivery dates, and/or other

issues is accurate. Your readers can hold you and your company accountable for such written commitments.

4. Unlike e-mails, many businesses require letters to be routed through channels before they are sent out. Because they convey how a company looks and what it offers to customers, letters often must be approved at a variety of corporate levels.

5. Letters are more permanent than e-mails. They provide a documented hard copy. Unlike e-mails that can be erased, letters are often logged in, filed, and bear a written, authorized signature.

6. A letter is the official and expected medium through which important documents and attachments (contracts, specifications, proposals) are sent to readers. Sending such attachments via e-mail or with a memo lacks the formality and respect readers deserve and expect.

7. A letter is still the most formal and approved way to conduct business with many international audiences. These readers see a letter as more polite and honorable than an e-mail for initial contacts and even for subsequent business communications.

8. A hard-copy letter is confidential. It is more likely to be delivered to the proper recipient in its sealed envelope and less likely to be forwarded to unintended readers, as an e-mail might be.

Letter Formats

Letter format refers to the way in which you print a letter—where you indent and where you place certain kinds of information. Several letter formats exist. Two of the most frequently used business letter formats are full-block and modified-block, but you should also be familiar with a third format, the semi-block.

Full-Block Format

In full-block format all information is flush against the left margin, double spaced between paragraphs. Figure 4.1 shows a full-block letter. Many employers prefer this format when your letter is on **letterhead stationery** (specially printed paper giving a company's name and logo, business and Web addresses, fax and telephone numbers, and sometimes the names of its executives).

Modified-Block Format

In modified-block format (see Figure 4.2), the writer's address (if it is not imprinted on a letterhead), the date, the complimentary close, and the signature are positioned at the center point and then keyed toward the right side of the letter. The date aligns with the complimentary close. The inside address, the salutation, and the body of the letter are flush against the left margin.

Full-block letter format with appropriate margins.   Figure 4.1

NIRA Nevada Insurance Research Agency
7500 South Maplewood Drive, Las Vegas, NV 89152-0026
(702) 555-9876 **http://www.NIRA.org**

1"-1.5"

Letterhead with company name

April 6, 2011

2 spaces

Ms. Molly Georgopolous, C.P.A.
Business Manager
Meyers, Inc.
3400 South Madison Road
Reno, NV 89554-3212

All text lined up against the left-hand margin

2 spaces

Dear Ms. Georgopolous:

2 spaces

As I promised in our telephone conversation earlier this afternoon, I am enclosing a study of the Nevada financial responsibility law. I hope that it will help you prepare your report.

Professional, businesslike font choice

2 spaces

Let me emphasize again that probably 95 percent of all individuals who are involved in an accident obtain reimbursement for medical bills and for damages to their automobiles. If individuals have insurance, they can receive reimbursement from their own carrier. If they do not have insurance and the other driver is uninsured and judged to be at fault, the Nevada Bureau of Motor Vehicles revokes that party's driver's license until all costs and damages are paid.

Text of letter balanced on letterhead page

Generous margins on all sides of the letter

2 spaces

Please call me again if I can help you.

2 spaces

Sincerely yours,

2 spaces

Carmen Tredeau

Signature written in black ink

Carmen Tredeau, President

2 spaces

Encl.

Bradley Fuller, CPCU
Chairperson

Carmen Tredeau, CPCU
President

Theodore Kendrick
Vice President
Public Affairs

Iping Li, CPCU
Vice President
Research

Dora Salinas-Diego, CPCU
Vice President
Actuary

Figure 4.2   **Modified-block letter format.**

7239 East Daphne Street
Mobile, AL 36608-1012

September 30, 2011

Date is indented at center point of letter.

Mr. Travis Boykin, Manager
Scandia Gifts
703 Hardy St.
Hattiesburg, MS 39401-4633

Inside address is single-spaced

Dear Mr. Boykin:

I am writing to see if you currently stock the Crescent pattern of model 5678 and how much you charge per model number. I would also like to know if you offer special prices for multiple-box orders.

Paragraphs can be indented or not indented.

Your store has been highly recommended to me by several colleagues, who have praised your service and the excellent quality of your products.

I look forward to working with you.

Sincerely yours,

Arthur T. McCormack

Complimentary close and writer's name are indented and aligned under date line

Arthur T. McCormack

Semi-Block Format

The semi-block format (see Figure 4.7, page 111) looks just like the modified-block format in terms of aligning the date line with the complimentary close, signature, and any enclosures at the center point of the letter. But the paragraphs in the semi-block format are always indented five to seven spaces. Though not as frequently used as the full-block or modified-block formats, the semi-block format is a template an employer may ask you to use.

Continuing Pages

To indicate subsequent pages if your letter runs beyond one page, use one of these two conventions. Note the use of the recipient's name.

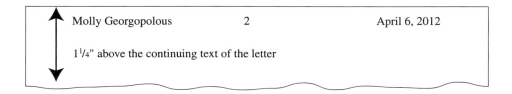

Molly Georgopolous   2   April 6, 2012

1 1/4" above the continuing text of the letter

2
T. Boykin
September 30, 2012

Parts of a Letter

A letter contains many parts, each of which contributes to your overall message. The parts and their placement in your letter form the basic conventions of effective letter writing. Readers look for certain information in key places.

The parts of a letter discussed in the following sections will appear in every letter you write. Figure 4.3 is a sample letter containing all of the parts discussed here. Note where each part is placed in the letter.

Heading

The heading of a letter may be either your company's letterhead or your full return address. (See Figure 4.2 for an example of a full return address when letterhead is not used.)

Date Line

Spell out the name of the month in full—"September" or "March" rather than "Sept." or "Mar." The date line is usually keyboarded this way: November 12, 2012. Different cultures express dates in different ways. Most countries, including those in Europe, list the day first, then the month, and then the year. See Figure 4.23 on page 141 for an example of how to correctly provide a date for international readers.

Inside Address

The inside address, the address of the recipient, is always placed against the left margin, two lines below the date line. It contains the name, title (if any), company, street address, city, state, and ZIP code of the person to whom you are writing. Single-space the inside address, and do not use any punctuation at the end of the lines.

Figure 4.3   A sample letter, full block format, with all parts labeled.

Heading
(letterhead)

M& **Madison and Moore, Inc.**
Professional Architects

7900 South Manheim Road
Crystal Springs, NE 71003-0092
Phone 402-555-2300 **http://www.MMI.com**

Date line

July 12, 2011

Inside address

Ms. Paula Jordan
Systems Consultant
Broadacres Development Corp.
12 East River Street
Detroit, MI 48001-0422

Salutation

Dear Ms. Jordan:

Thank you for your letter of July 6, 2011. I have discussed your request
with the staff in our planning department and have learned that the design
modules we used for our Vestuvia project are no longer available.

Body of letter

In searching through our files, however, I have come across the enclosed
catalog from a California firm that might be helpful to you. This firm,
California Concepts, offers plans very similar to the ones you are interested
in, as you can tell from the design I checked on page 23 of their catalog.

I hope this will help you and I wish you every success in your project.

Complimentary
close

Sincerely yours,

Company name

MADISON AND MOORE, INC.

Signature

William Newhouse

Writer's name
and title

William Newhouse
Office Manager

Enclosure
Copy to

Encl.: Catalog
cc: Planning Department

Dr. Mary Petro
Director of Research
Midwest Laboratories
1700 Oak Drive
Rapid City, SD 56213–3406

Always try to write to a specific person rather than just "Sales Manager" or
"President." To find out the person's name, check previous correspondence, e-mail
lists, or the company's or individual's website, or call the company. Make sure you

use an abbreviated courtesy title (Ms., Mr., Dr., Prof.) before the recipient's name for the inside address (e.g., Capt. María Torres; Mr. A. T. Ricks; Rev. Siam Tau). Use Ms. when writing to a woman unless she has expressly asked to be called Miss or Mrs.

The last line of the inside address contains the city, state, and ZIP code.

Salutation

Begin with *Dear*, and then follow with a courtesy title, the reader's last name (unless you are on a first-name basis), and a colon (Dear Mr. Brown:). **Never use a comma for a formal letter**. Avoid the sexist "Dear Sir," "Gentlemen," or "Dear Madam" and the stilted "Ladies and Gentlemen" or "Dear Sir/Madam." (For a discussion of sexist language and how to eliminate it, see Chapter 2, pages 48–50.)

Sometimes a first name does not reveal whether the reader is masculine or feminine. There are women named Stacy, Robin, and Lee, and men named Leslie, Kim, and Kelly. If you aren't certain, you can use the reader's full name: "Dear Terry Banks." If you know the person's title, you might write "Dear Credit Manager Banks."

Avoid casual salutations such as "Hello," "Good Morning," "Greetings," or "Happy Tuesday"; these are best reserved for e-mails or IM's. And never begin a letter with "To Whom It May Concern," which is old fashioned, impersonal, and trite.

Body of the Letter

The body of the letter contains your message. Some of your letters will be only a few lines long, while others may extend to three or more paragraphs. Keep your sentences concise, and try to hold your paragraphs to under six or seven lines. (Refer to pages 42–47 in Chapter 2.)

Complimentary Close

A close is the equivalent of a formal goodbye. For most business correspondence, use one of these standard closes:

Sincerely,
Respectfully,
Sincerely yours,
Yours sincerely,

Capitalize only the first letter of the first word. The entire close is followed by a comma. If you and your reader know each other well, you can use

Cordially,
Best wishes,
Regards,

But avoid flowery closes, such as

Forever yours,
Devotedly yours,
Faithfully yours,

These belong in a romance novel, not in a business letter.

Signature

Allow four spaces between the complimentary close and your typed name/title so that your signature will not look squeezed in. Always sign your name in black ink. An unsigned letter indicates carelessness or, worse, indifference toward your reader. A stamped signature tells readers you could not give them personal attention.

Some firms prefer using their company name along with the employee's name in the signature section. If so, type the company name in capital letters two line spaces below the complimentary close and then sign your name. Add your title underneath your typed name. Here is an example:

Sincerely yours,

THE FINELLI COMPANY

Helen Stravopoulos

Helen Stravopoulos
Cover Coordinator

Enclosure(s) Line

The enclosure line informs the reader that additional materials (such as brochures, diagrams, forms, contracts, a proposal) accompany your letter.

Enclosure (only one item is enclosed)
Enclosures (2)
Encl.: Spring Quarter Sales Report

Copy Notation

The abbreviation *cc:* (courtesy copy) informs your reader that a copy of your letter has been sent to one or more individuals.

cc: Service Dept.
cc: Hannah Pittman
    Ivor Vas

Letters are copied and sent to third parties for two reasons: (a) to document a paper trail and (b) to indicate to other readers who else needs the information contained in the letter. Unless your boss instructs you otherwise, tell your reader if others will receive a copy of your letter.

The Appearance of Your Letter

The way your letter looks can determine how readers will respond to your message. Here are some tips on how to format and print professional-looking letters:

- **Use a letter-quality printer** and check toner cartridge levels to avoid sending a fuzzy, faint, or messy letter.

- **Stay away from fancy fonts and scripts.** Use the business-like Times New Roman or Arial. (For a discussion of typography, see Chapter 6, pages 203–206.)
- **Consider using letter wizards to help format and design your letters.** Most word processing programs, such as Microsoft Word or Corel Word Pefect, have them. Some organizations, however, may prefer not to use standard letter formatting. Always check with your company before using a letter wizard.
- **Leave generous margins of at least 1 to 1 and 1/4 inches all around your message.** For a shorter letter, as in Figure 4.5, don't expand your margins to 3 or 4 inches or increase the font size, which will only make your letter look unprofessional.
- **Leave double spaces between key parts of a letter**—the letterhead and the dateline, the salutation, copy notations, and enclosure—but leave four spaces between the complimentary close and your typed signature.
- **Single space within each paragraph, but double space between paragraphs.**
- **Avoid crowding too much text onto one page.** Squeezing too many characters on a line by using overly small fonts will make your letter look cramped and hard to read. Also, don't cram a long letter onto one page; instead, allow your letter to flow to a second page, as in Figure 4.8 (pages 113–114).
- **Be careful about lopsided letters.** Don't start a brief letter at the top of the page and then leave the lower three-fourths blank. Begin a shorter message near the center of the page.
- **Used Print Preview to see an image of your letter before you print a hard copy** so you can make any necessary changes or adjustments to correct it. Never print over your company's letterhead or any addresses or company logos printed across the bottom of the letter.
- **Always print your letter on high-quality white bond paper (20-pound, 8-1/2 × 11) and matching standard-size (#10) business envelopes.** Avoid colored paper, which can look unprofessional.

Organizing a Standard Business Letter

Like the memo in Figure 3.3, a standard business letter can be divided into an introduction, a body, and a conclusion, each section responding to or clarifying a specific issue for your recipient. These three sections can each be one paragraph long, as in Figures 4.1, 4.2, and 4.3, or the body of your letter may be two or more paragraphs, as in Figure 4.4.

To help readers grasp your message clearly and concisely, follow this simple plan for organizing your business letters:

- In your first paragraph start with a friendly opening and tell readers why you are writing and why your letter is important to them. Acknowledge any relevant previous meetings, correspondence, or telephone calls early in the paragraph (as in Figures 4.1 and 4.3).
- Put the most significant point of each paragraph first to make it easier for the reader to find. Never bury important ideas in the middle or at the end of your paragraph.
- In the second (or subsequent) paragraph, develop the body of your message with factual support, the key details, and descriptions readers need. For instance, note how Figure 4.4 refers to the specific changes to improve security that the reader had requested.

▪ In your last paragraph, thank readers and be very clear and precise about what you want them to do or what you will do for them. You can let them know what will happen next, what they can expect, or any combination of these messages. Don't leave your readers hanging. End cordially and professionally.

Figure 4.4   **Organization of a business letter.**

Office Property Management Associates
2400 South Lincoln Highway
Livingston, NJ 07040-9990
(201) 555-3740   www.opma.com

August 15, 2011

Mr. W. T. Albritton
Albritton & Sharp Accounting Services
Suite 400
Suburban Office Complex
Livingston, NJ 07038-2389

Dear Mr. Albritton:

Introduction comes to point quickly, cordially by referencing reader's earlier request

Thank you for your recent suggestions on improving security at the Suburban Office Complex. You will be pleased to learn that at our July meeting OPMA has agreed to make the following improvements in services, which will go into effect within 45 days.

Body describes changes with specific details

Starting September 7, you will have an on-site manager, Thomas Vasquez, who will be happy to answer any questions you may have about the Complex and help you with any problems you may encounter. His ten years of experience in managing commercial office parks will benefit you and other businesses at the Complex.

The new outdoor security system you asked for will be installed by September 21. It will offer you and your employees greater protection through seven additional security cameras around the perimeters of the parking lot, and movement sensors will monitor every outside door.

Conclusion builds goodwill by promising reader what will be done and how

I can reassure you that none of these changes will inconvenience the operation of your firm. We are honored to have Albritton & Sharp as residents. I welcome your comments as these changes are implemented as well as any additional suggestions you may have.

Sincerely yours,

Cheryl Hu

Cheryl Hu
Vice President

Making a Good Impression on Your Reader

You have just learned about formatting and organizing your letters. Now we turn to the content of your letters—what you say (your message) and how you say it (your style and tone). Writing letters means communicating to influence your readers, not to alienate or antagonize them. Keep in mind that writers of effective letters are like successful diplomats; they represent both their company and themselves. You want readers to see you as courteous, well informed, and professional.

First, put yourself in the reader's position. What kinds of letters do you like to receive: vague, impersonal, sarcastic, pushy, and condescending; or polite, business-like, and considerate? If you have questions, you want them answered honestly, courteously, and fully.

To send such effective letters, adopt the **"You Attitude"** in other words, signal to readers that they and their needs are of utmost importance. Incorporating the "you attitude" means you should be able to answer "Yes" to these two questions:

1. Will my readers receive a positive image of me?
2. Have I chosen words that convey both my respect for the readers and my concern for their questions and comments?

Figures 4.5 (page 106) and 4.6 (page 107) contain two versions of the same letter. Which one would you rather receive?

Achieving the "You Attitude": Four Guidelines

As you draft and revise your work, pay special attention to the following four guidelines for making a good impression on your reader.

1. **Never forget that your reader is a real person**. Avoid writing cold, impersonal letters that sound as if they were form letters or voice mail instructions. Let the readers know that you are writing to them as individuals. The letter below violates every rule of personal and personable communications.

> It has come to our attention that policy number 342q–765r has been delinquent in payment and is in arrears for the sum of $302.35. To keep the policy in force for the duration of its life, a minimum payment of $50.00 must reach this office by the last day of the month. Failure to submit payment will result in the cancellation of the aforementioned policy.

The example displays no sense of one human being writing to another, of a customer with a name, personal history, or specific needs. Revised, this letter contains the necessary personal (and human) touch.

> We have not yet received your payment for your insurance policy (342q–765r). By sending us your check for $50.00 within the next two weeks, you will keep your policy in force and can continue to enjoy the financial benefits and emotional security it offers you.

Figure 4.5   A letter lacking the "you attitude."

Brown County • Office of the Tax Assessor

County Building, Room 200, Ventura, Missouri 56780-0101

712-555-3000

February 4, 2011

Mr. Ted Ladner
451 West Hawthorne Lane
Morris, MO 64507-3005

Dear Mr. Ladner:

Tone is sarcastic and uncooperative

You have written to the wrong office here at the County Building. There is no way we can attempt to verify the kinds of details you are demanding from Brown County.

Use of "you" alone does not signal a positive image of reader

Simply put, by carefully examining the 2010 tax bill you said you received, you should have realized that it is the Tax Collector's Office, not the Tax Assessor's, that will have to handle the problem you claim exists.

In short, call or write the Tax Collector of Brown County.

Insulting and curt ending and close

Thank you!

Tracey Kowalski

Tracey Kowalski

http://www.browncounty.gov

The benefits to an individual reader are stressed, and the reader is addressed directly as a valued customer.

Don't be afraid of using "you" in letters. Readers will feel more friendly toward you and your message. Of course, no amount of "yous" will help if they appear in a condescending context, such as the letter in Figure 4.5 above.

Brown County • Office of the Tax Assessor

County Building, Room 200, Ventura, Missouri 56780-0101

712-555-3000

February 4, 2011

Mr. Ted Ladner
451 West Hawthorne Lane
Morris, MO 64507-3005

Dear Mr. Ladner:

Thank you for writing about the difficulties you encountered with your 2010 tax bill. I wish I could help you, but it is the Tax Collector's Office that issues your annual property tax bill. Our office does not prepare individual homeowners' bills. *(Thanks reader and gives polite explanation)*

If you will kindly direct your questions to Paulette Sutton at the Brown County Tax Collector's Office, County Building, Room 100, Ventura, Missouri 56780-0100, I am sure that she will be able to assist you. Should you wish to call her, the number is 458-3455, extension 212. Her e-mail address is **psutton@bctc.gov**. *(Helps reader solve problem with specific information)*

Respectfully,

Tracey Kowalski
Tracey Kowalski

(Uses appropriate close)

http://www.browncounty.gov

2. Keep the reader in the forefront of your letter. Make sure the reader's needs control the tone, message, and organization of your letter—the essence of the "you attitude." Stress the "you," not the "I" or the "we." Below is a paragraph from a letter that forgets about the reader.

I-Centered Draft

```
I think that our rug shampooer is the best on the market.
Our firm has invested a lot of time and money to ensure that
it is the most economical and efficient shampooer available
today. We have found that our customers are very satisfied
with the results of our machine. We have sold thousands of
these shampooers, and we are proud of our accomplishment.
We hope that we can sell you one of our fantastic machines.
```

The draft spends its time on the machine, the company, and its sales success. Readers are interested in how *they* can benefit from the machine, not in how much profit the company makes from selling it.

To win the readers' confidence, the writer needs to show how and why they will find the product useful, economical, and worthwhile at home or at work. Here is a reader-centered revision.

You-Centered Revision

```
Our rug shampooer would make cleaning your Comfort Rest Motel
rooms easier for you. It is equipped with a heavy-duty motor
that will handle your 200 rooms with ease. Moreover, that
motor will give frequently used areas, such as the lobby or
hallways, the fresh and clean look you want for your motel.
```

3. Be courteous and tactful. Refrain from turning your letter into a punch through the mail. Don't inflame your letter or e-mail readers, as the example in Figures 3.6 does. Again, when you capture the reader's goodwill, your rewards will be greater. The following words can leave a bad taste in the reader's mouth.

it's defective	makes no sense
I insist	unprofessional (job, attitude, etc.)
we reject	your failure
that's no excuse for	you contend
totally unacceptable	you should have known
unavoidable delay	your outlandish claim

Compare the following discourteous sentences with the courteous revisions.

Discourteous	Courteous Revision
We must discontinue your service unless payment is received by the date shown.	Please send us your payment by November 4 so your service will not be interrupted.
You are sorely mistaken about the contract.	We are sorry to learn about the difficulty you experienced over the terms in your contract.
The new notebook you sold me is third-rate and you charged first-rate prices.	Since the notebook is still under warranty, I hope you can make the repairs easily and quickly.
It goes without saying that your suggestion is not worth considering.	It was thoughtful of you to send me your suggestion, but unfortunately we are unable to implement it right now.

4. Don't sound pompous or bureaucratic. Write to your reader as if you were carrying on a professional conversation with her or him. A business letter should be upbeat, clear, and to the point. It needs to be reader-friendly and believable, not stuffy and overbearing.

Avoid using phrases that remind readers of "legalese"—language that smells of contracts, deeds, and stuffy rooms. In the following list, the words and phrases on the left are pompous expressions that have crept into letters for years; the ones on the right are contemporary equivalents you should use:

Pompous	Contemporary
aforementioned	previously mentioned
as per your request	as you requested
at this present writing	now
I am in receipt of	I have
attached herewith	enclosed
I am cognizant of	I know
be advised that	for your information
endeavor	try
forthwith	at once
henceforth	after this
herewith, heretofore, hereby	(drop these three "h's")
immediate future	soon
in lieu of	instead of
pursuant	concerning
remittance	payment
your letter arrived and I have same	I have your letter
under separate cover	I'm also sending you
this writer	I
we regret to inform you that	we are sorry that

The Five Most Common Types of Business Letters

The following section discusses the common types of business correspondence that you will be expected to write on the job.

1. Inquiry letters
2. Cover letters
3. Special request letters
4. Sales letters
5. Customer relations letters

   - Follow-up letters
   - Complaint letters
   - Adjustment letters
   - Collection letters

These letter types involve a variety of formats, writing strategies, and techniques. Business letters can be classified as **positive, neutral**, or **negative**, depending on their message and the anticipated reactions of your audience. Inquiry and special request letters are examples of neutral, routine letters. Letters can be positive or negative, depending on your message.

- Neutral letters request specific information about a product or service, place an order, or respond to some action or question.
- Sales letters promoting a product carry good news, according to the companies that spend millions of dollars a year preparing them.
- Customer relations letters can be positive (responding favorably to a writer's request or complaint) or negative (e.g., refusing a request, saying no to an adjustment, denying credit, seeking payment, critiquing poor performance, or announcing a product recall).

Inquiry Letters

An inquiry letter asks for information about a product, service, or procedure. Businesses frequently exchange such letters. As a customer, you too have occasion to ask in a letter about a service or a special line of products, the price, the size, the color, delivery arrangements or recent technological changes. The clearer your letter, the quicker and more helpful your answers are likely to be.

Figure 4.7 illustrates a letter of inquiry from Michael Ortega to a real estate office managing a large number of apartment complexes. Note that it follows these five rules for an effective inquiry letter:

- states exactly what information the writer wants
- indicates clearly why the writer requests the information
- keeps questions short and to the point
- specifies when the writer must have the information
- thanks the reader

Had Michael Ortega simply written the following very brief letter to Hillside Properties, he would not have received information he needed about size, location, and price of apartments: "Please send me some information on housing in Roanoke. My family and I plan to move there soon."

Cover Letters

A cover letter accompanies a document (a proposal, a report, a catalogue, a portfolio) that you send to your readers. It identifies the type of document you are sending and prepares your audience to read it. Figure 4.1 is an example of a cover letter used to send a smaller document while Figure 9.1 contains a cover letter sent with a copy of a long report. A cover letter should do the following:

- Provide a written record that you have transmitted a document.
- Tell readers why you are sending them the document.
- Briefly summarize what the document contains—number of sections, visuals, statistics, appendices, etc.

A letter of inquiry written in Semi-Block Format.   Figure 4.7

Michael Ortega
403 South Main Street
Kingsport, TN 37721-0217
mortega@erols.com

April 7, 2011

Mr. Fred Stonehill
Property Manager
Hillside Properties
701 South Arbor St.
Roanoke, VA 24015-1100

Dear Mr. Stonehill:

    Please let me know if you will have any two-bedroom furnished apartments available for rent during the months of June, July, and August. I am willing to pay up to $750 a month plus utilities. My wife, one-year-old son, and I will be moving to Roanoke for the summer so I can take classes at Virginia Western Community College.

    If possible, we would like to have an apartment that is within two or three miles of the college. We do not have any pets.

    I would appreciate hearing from you within the next two weeks. My e-mail address is **mortega@erols.com**, or you can call me at home (606-555-8957) any evening from 6–10 p.m.

    Should you have any suitable vacancies, we would be happy to drive to Roanoke to look at them and give you a deposit to hold an apartment. Thanks for your help.

    Sincerely yours,

Michael Ortega

Michael Ortega

States precise request

Explains need for information

Identifies area of interest

Specifies exact date when a reply is needed

Offers to confer with and then thanks the reader

- Explain why the document is of interest to readers.
- Express a willingness to answer questions about the document.
- Thank readers for their time.

Special Request Letters

Special request letters make a special demand, not a routine inquiry. For example, these letters can ask a company for information that you as a student will use in a paper, an individual for a copy of an article or a speech, or an agency for facts that your company needs to prepare a proposal or sell a product. The person or company being asked for help stands to gain no financial reward for supplying the information; the only reward is the goodwill a response creates.

Make your request clear and easy to answer. Supply readers with an addressed, postage-paid envelope, an e-mail address, and fax and telephone numbers in case they have questions.

Follow these seven guidelines when asking for information in a special request letter.

1. Address your letter to the appropriate person.
2. State who you are and why you are writing—e.g., student doing a paper, employee compiling information for a report, and so on.
3. Indicate clearly your reason for requesting the information. Mention any individuals who may have suggested you write for help and information.
4. State precisely and succinctly the questions you want answered; list and number your questions.
5. Specify exactly when you need the information. Allow sufficient time—at least three weeks. Be reasonable; don't ask for the impossible.
6. Offer to forward a copy of your report, paper, or survey in thanks of the anticipated help.
7. Thank the reader for helping.

Figure 4.8 gives an example of a letter that follows these guidelines.

Sales Letters

A sales letter is written to persuade the reader to buy a product, try a service, support some cause, or participate in some activity. No matter what profession you have chosen, there will always be times you have to sell a product, a service, a community or charitable program, a point of view, or yourself! In fact, an application letter for a job (see page 179–186) as an introduction to a new or prospective customer is a sales letter.

The Four A's of Sales Letters

Successful sales letters follow a time-honored and workable plan—what can be called the "Four A's":

1. It gets the reader's *attention*—with a question or a how-to statement (e.g., "We can show you how to save $100 on your next credit card purchase").
2. It highlights the product's or service's *appeal*—emotionally or financially, or both. Focus on benefits to the reader.

1505 West 19th Street
Syracuse, NY 13206
phone 315-555-1214
jkawatsu@webnet.com

October 5, 2011

Ms. Sharonda Aimes-Worthington
Research Director
Creative Marketing Associates
198 Madison Ave.
New York, NY 10016-0092

Dear Ms. Aimes-Worthington:

I am a junior at Monroe College in Syracuse, and I am writing a report on "Internet Marketing Strategies for the Finger Lakes Region of New York" for my Marketing 340 class. Several of my professors have spoken very highly of Creative Marketing Associates, and in my own research I have learned a great deal from reading your article on Web designs and local economies posted on your website last month.

Given your extensive experience in developing Internet sites to promote regional businesses and tourism, I would be grateful if you would share your responses to the following three questions with me:

1. What have been the most effective design components in websites for a regional marketplace such as the Finger Lakes?

2. How can area chambers of commerce and various municipalities help generate Web traffic to a regional marketplace website for the Finger Lakes area?

3. Which other regional area(s) do you see having the same or very similar marketing goals and challenges as the Finger Lakes?

Explains reason for letter and how the writer learned about the firm

Proves writer has done research

Acknowledges reader's expertise

Lists specific numbered questions on the topic

3. It shows the customer the product's or service's *application*—descriptions, special features, guarantees.
4. It ends with a specific request for *action*—call, visit, participate, register online. Motivate reader to act promptly.

Figure 4.8	(Continued)

Aimes-Worthington   2   October 5, 2011

As an incentive, offers to share report

Your answers to these questions would make my report much more authoritative and useful. I would be happy to send you a copy and will, of course, be honored to cite you and Creative Marketing Associates in my work.

Indicates when needed

Because my report is due by December 2, I would greatly appreciate having your answers within the next month so that I can include them. Would you kindly send your responses, or any questions you may have, to my e-mail address, *jkawatsu@webnet.com.* Many thanks for your help.

Makes contact easy through e-mail

Sincerely yours,

Julie Kawatsu

Julie Kawatsu

These four goals can be achieved in fewer than four or five paragraphs. Look at the sales letter in Figure 4.9 (page 115) in which these parts are labeled. Figure 4.23 (page 141) also contains a model sales letter using the "Four A's."

Do I Mention Costs?

As a general rule, do not bluntly state the cost. Relate prices, charges, or fees to the benefits provided by the services or products you are selling. Let customers see how much they are getting for their money, as Cory Soufas does in paragraph 3 in Figure 4.9. Similarly, a dealer who installs steel shutters did not tell readers the exact price of the product but instead stressed in her sales letter that they will save money by buying it: "Virtually maintenance free, your Reel Shutters also offer substantial savings in energy costs by reducing your heat loss through radiation by as much as 65% . . . and that lowers your utility bills by at least 35%."

Customer Relations Letters

Much business correspondence deals explicitly with establishing and maintaining friendly working relations. Customer relations letters show how you and your company regard the people with whom you do business. The letters should reveal your sensitivity to their needs. The first lesson to learn is that you cannot look at your letter only from your (the writer's) perspective. You have to see the letter from the reader's perspective and anticipate his/her needs and reactions. Customer

A sales letter sent to a business reader. Figure 4.9

Workwell Software
3700 Stewart Avenue   Chicago IL 60637-2210
Phone: (312) 555-3720 **Fax:** (312) 555-7601 **E-mail:** sales@workwell.com
http://www.workwell.com

August 12, 2011

Ali Jen
Circuit Systems
7 Tyler Place
Oklahoma City, OK 73101

Dear Ali Jen:

Do you know how much money your company loses from repetitive strain injury? Each year employers spend millions of dollars on employee insurance claims because of back pains, fatigue, eye strain, and carpal tunnel syndrome.

Gets reader's attention with a question

Workwell can solve your problems with its easy-to-use Exercise Program Software that automatically monitors the time employees spend at their computers and also measures keyboard activity. After each hour (or the specified number of keystrokes), **Workwell** software will take your employees through a series of exercises that will help prevent carpal tunnel syndrome and muscle strains.

Emphasizes the product's appeal using precise language

Workwell's Exercise Program Software will not interfere with your busy schedule. Each of the 27 exercises is demonstrated on screen with audio instructions. The entire program takes less than 3 minutes and is available for Windows XP and Windows Vista. For only $1499.00, you can provide a networked version of this valuable software to all your Circuit Systems' employees.

Shows specific application of the product

Links costs to benefits

To help your employees stay at peak efficiency in a safe work environment, please call us at 1-800-555-WELL or contact us at **http://www.workwell.com** to order your software today.

Ends with a call for prompt action

Thank you,

Cory Soufas

Cory Soufas
Sales Manager

relations letters send readers good news or bad news, acceptances or refusals. Good news tells customers one or more of the following:

- You agree with them about a problem they brought to your attention.
- You are solving their problem exactly the way they want.
- You are approving their loan or request for a refund.
- You are grateful to them for their business.

Thank you letters, congratulations letters, and adjustment letters saying "Yes" with these messages are all examples of good news messages.

Bad news messages, however, inform readers that:

- You do not like their work or the equipment/technology they sold you.
- You do not have the equipment or service they want or you cannot provide it at the price they want to pay.
- You cannot refund their purchase price or perform a service.
- You are raising their rent or not renewing their lease.
- You want them to pay what they owe you now.

Bad news messages often come to readers through complaint letters, adjustment letters that say "No," and collection letters.

Being Direct or Indirect

Not every customer relations letter starts by giving the reader the writer's main point, judgment, conclusion, or reaction. Whether you are sending good news or bad news, determine what to say and where. *Where you place your main idea is determined by the type of letter you are writing.* Good news messages require one tactic; bad news ones, another.

Good News Message

If you are writing a good news letter, use the direct approach. Start your letter with the welcome, pleasant news that the reader wants to hear. Don't postpone the opportunity to put your reader in the right frame of mind. Then, provide any relevant supporting details, explanations, or commentary. Being direct is advantageous when you have good news to convey.

Bad News Message

If you have bad news to report, do *not* open your letter with it. Be indirect. Prepare your reader for the bad news; keep the tension level down. If you throw the bad news at your reader right away, you jeopardize the goodwill you want to create and sustain. Consider how you would react to a letter that begins with these slaps:

- Your order cannot be filled.
- Your application for a loan has been denied.
- It is our unfortunate duty to report …

Having been denied, disappointed, or even offended in the first sentence or paragraph, the reader is not likely to give you his or her attentive cooperation thereafter.

Two Versions of a Bad News Message

Notice how A. J. Griffin's bad news letter in Figure 4.10 (page 118) curtly starts off with the bad news of a rent increase. Receiving such a letter, the owner of Flowers by Dan certainly could not be blamed for looking for a new location for his business.

Compare the curt version of Griffin's letter in Figure 4.10 with his revised message in Figure 4.11 (page 119). In the revised version, Griffin begins tactfully with pleasant, positive words designed to put his reader in a good frame of mind. Then Griffin gives some background information that the owner of Flowers by Dan can relate to. Griffin makes one more attempt to encourage Sobol to recall his good feelings about the mall—last year they did not raise rents—before introducing the bad news of a rent increase.

Even after giving the bad news, Griffin softens the blow by saying that the Mall knows it is bad news. Griffin's tactic here is to defuse some of the anger that Sobol will inevitably feel. Griffin then ends on a positive, upbeat note: a prosperous future for Flowers by Dan.

Follow-Up Letters

A follow-up letter is sent by a company after a sale to thank the customer for buying a product or using a service and to encourage the customer to buy more products and services. A follow-up letter is a combination thank-you note and sales letter. The letter in Figure 4.12 (page 121) shows how an income tax preparation service attempts to obtain repeat business by doing the following:

1. begins with a brief and sincere expression of gratitude
2. discusses the benefits (advantages) the customer already knows about and then transfers the firm's dedication to the customer to a continuing sales area
3. ends with a specific request for future business

Complaint Letters

Each of us, either as customers or businesspeople, at some time has been frustrated by a defective product, inadequate or rude service, or incorrect billing. When we get no satisfaction from calling an 800 number and are routed through a series of menu options, our frustration level goes up. Usually our first response is to write a letter in an intimidating font (LIKE THIS), dripping with juicy insults. But an angry letter, like a piece of flaming e-mail (see Figure 3.6, page 84), rarely gets positive results and can hurt your company's image.

A complaint letter is a delicate one to write. First off, avoid the following:

- name calling
- sarcasm
- insults
- threats
- unflattering clip art
- intimidating type fonts (e.g., using all capital letters)

Figure 4.10   An ineffective bad news letter.

River Road Mall

December 1, 2011

Mr. Daniel Sobol
Flowers by Dan
Lower Level 107
River Road Mall

Dear Mr. Sobol:

Blunt opening disregards audience's needs and feelings

This is to inform you of a rent increase. Starting next month your new rent will be $3,500.00, resulting in a 15 percent increase.

Ends with a demand

Please make sure that your January rent check includes this increase.

Sincerely,

A. J. Griffin
Manager
ajg@rrmall.com

300 First Street
Canton, Ohio 44701
(216) 555-6700
www.RRMall.com

A diplomatic revision of the bad news letter in Figure 4.10.   Figure 4.11

River Road Mall

December 1, 2011

Mr. Daniel Sobol
Flowers by Dan
Lower Level 107
River Road Mall

Dear Mr. Sobol:

It has been a pleasure to have you as a tenant at the Mall for the past two years, and we look forward to serving you in the future.

Opens with positive association

Over these last two years we have experienced a dramatic increase in costs at River Road Mall for security, maintenance, landscaping, pest control, utilities, insurance, and taxes. Last year we absorbed those increases and so did not have to raise your rent. We wish we could do it again but, unfortunately, we must increase your rent by 15 percent, to $3,500.00 a month, effective January 1.

Prepares reader for bad news to follow

States bad news in most concise, up-beta way

Although no one likes a rent increase, we know that you do not want us to compromise on the quality of service that you and your customers expect and deserve here at River Road Mall.

Links bad news to reader benefits

Please let us know how we can assist you in the future. We wish you a very successful and profitable 2012. If you have any questions, please call or visit my office.

Does not apologize but ends with respectful tone

Cordially,

A. J. Griffin

A. J. Griffin
Manager
ajg@rrmall.com

Worm, complimentary close

300 First Street
Canton, Ohio 44701
(216) 555-6700
www.RRMall.com

The key thing to keep in mind is that you can disagree without being disagreeable. Be rational, not hostile. Just to let off steam, you might want to write an angry letter but then tear it up, replacing all the heat with courteous and diplomatic language.

Establishing the Right Tone

A complaint letter is written for more reasons than just blowing off steam. You want some specific action taken. The "you attitude" is especially important here to maintain the reader's goodwill. Complaint letters that are professional and considerate are more likely to receive positive attention than letters bristling with angry words. An effective complaint letter can be written by an individual consumer or by a company. Figure 4.13 (page 122) shows Michael Trigg's complaint about a defective fishing reel; Figure 4.14 (page 123) expresses a restaurant's dissatisfaction with an industrial dishwasher.

Writing an Effective Complaint Letter

To increase your chances of receiving a speedy settlement, follow these seven steps in writing your letter of complaint. They will help you build your case.

  1. **Sending your letter to the right person further ensures your success**. But, as we saw, never address it "To Whom It May Concern." Do your homework—search the company's website or go to Hoover's business directory (http://www.hoovers.com) to get contact information. But don't send your letter to the CEO. Find the appropriate person or office that responds to customer problems.

  2. **Be concise**. Keep your letter to one page. Your reader wants essential details, not a saga of your troubles.

  3. **Begin with a detailed description of the product or service**. Give the appropriate model and serial numbers, size, quantity, color, and cost. Specify check and invoice numbers. Indicate when, where (specific address), and how (through a vendor, the Internet, at a store) you purchased it and also the remaining warranty. If you are complaining about a service, give the name of the company, the date of the service, the personnel providing it, and their exact duties.

  4. **State exactly what is wrong with the product or service**. Be factual. Precise information will enable the reader to understand and act on your complaint.

  - How many times did the product work before it stopped?
  - What parts were malfunctioning?
  - What parts of a job were not done or were done poorly?
  - When did all this happen? How many times?
  - Where and how were you inconvenienced?
  - Was the service late, incomplete, rude?

Stating that "the brake shoes were defective" tells very little about how long they were on your car, how effectively they may have been installed, or what condition they were in when they ceased functioning safely.

  5. **Briefly describe the inconvenience you have experienced**. Show that your problems were directly caused by the defective product or service. To build your

A follow-up letter to encourage repeat business.   Figure 4.12

Taylor Tax Service
Highway 10
North Jennings, TX 78326
phone **(888) 555-9681** e-mail **taylor@aol.com**
http://www.taylor.com

December 2, 2011

Ms. Laurie Pavlovich
345 Jefferson St.
Jennings, TX 78326

Dear Ms. Pavlovich:

Thank you for using our services in February of this year. We were pleased to help you prepare your 2011 federal and state income tax returns. Our goal is to save you every tax dollar to which you are entitled. If you ever have questions about your return, we are open all year long to help you.

Links business goal to customer advantage

We are looking forward to serving you again next year. Several new federal tax laws enacted this year will change the types of deductions you can declare. These changes might appreciably increase your refund. Our consultants know the new laws and are ready to apply them to your return.

Stresses reasons/ benefits for customer to return for service

Another important tax matter influencing your 2011 returns will be any losses you may have suffered because of the hailstorms and tornadoes that hit our area three months ago. Our consultants are specially trained to assist you in filing proper damage claims with your federal and state returns.

Makes it easy and profitable for customer to act soon

To make using our services even more convenient, we file your tax return electronically to speed up any refund. Please call us at (888) 555-9681 or e-mail us at **taylor@aol.com** as soon as you have received all your forms to set up an appointment. We are waiting to serve you seven days a week from 9:00 a.m. to 9:00 p.m.

Ends with commitment to customer convenience

Sincerely yours,

TAYLOR TAX SERVICE

Demetria Taylor

Demetria Taylor

Figure 4.13   **A complaint letter from a consumer.**

17 Westwood Drive

Magnolia, MA 02171

mtrigg@roof.com

October 10, 2011

*Identifies
appropriate
person to
resolve
problem*

Mr. Ralph Montoya
Customer Relations Department
Smith Sports Equipment
P.O. Box 1014
Tulsa, OK 74109-1014

Dear Mr. Montoya:

*Documents all
relevant details
about the
product*

On September 21, 2011, I purchased a Smith reel, model 191, at the Uni-Mart
Store on Marsh Avenue in Magnolia. The reel sold for $94.95 plus tax. The reel is
not working effectively, and I am returning it to you under separate cover by
first-class mail.

*Explains
politely what is
wrong*

I had made no more than five casts with the reel when it began to malfunction.
The button that releases the spool and allows the line to cast will not spring back
into position after casting. In addition, the gears make a grinding noise whenever I
tried to retrieve the line. Because of these problems, I was unable to continue my
participation in the Gloucester Fishing Tournament last week.

*Clearly states
what should be
done*

*Specifies an
acceptable
time frame*

I request that a new reel be sent to me free of charge in place of the defective one
I returned. I would also like to know what was wrong with the defective reel.

Thank you for processing my claim within the next two weeks.

Sincerely yours,

Michael Trigg

A complaint letter from a business.   Figure 4.14

The LoFT

Camerson and Dale, Sunnyside, California 91793-4116   213-555-7500

June 21, 2011

Ms. Priscilla Dubrow
Customer Relations Department
Superflex Products
San Diego, CA 93141-0808

Dear Ms. Dubrow:

On September 15, 2010, we purchased a Superflex industrial dishwasher, model 3203876, at the Hillcrest store at 3400 Broadway Drive in Sunnyside, for $5,000. In the last three weeks, our restaurant has had serious and repeated problems with this machine. Three more months of warranty remain on the unit.

The machine does not complete a full cycle; it stops before the final rinsing and thus leaves the dishes dirty. It appears that the cycle regulators are not working properly because they refuse to shift into the next necessary gear. Attempts to repair the machine by the Hillcrest service team on June 4, 11, and 15 have been unsuccessful.

The Loft has been greatly inconvenienced. Our kitchen team has been forced to sort, clean, and sanitize utensils, dishes, pans and pots by hand, resulting in additional overtime. Moreover, our expenses for proper detergents have increased.

We want your main office to send another repair crew to fix this machine. If your crew is unable to do this, we want a discount worth the amount of the warranty life on this model to be applied to the purchase of a new Superflex dishwasher. This amount would come to $1,000, or 20 percent of the original purchase price.

So that our business is not further disrupted, we would appreciate your resolving this problem promptly within the next four to five business days.

Sincerely yours,

Emily Rashon

Emily Rashon
Co-owner

▶ Browse our menu, which changes daily, at www.theloft.com

Writes to specific reader

Gives all product's facts and warranty information

Describes what's happened and when

Documents problem and reason for adjustment

Provides clear description of how problem should be solved

Concludes politely with justification for prompt action

case, give precise details about the time and money you lost. Don't just say you had "numerous difficulties." Did you have to pay a mechanic to fix your car when it was stalled on the road? Did you have to buy a new printer or Blu-Ray DVD player? Where appropriate, refer to any previous telephone calls, e-mails, or letters. Give the names of the people you have written to or spoken with and the dates.

6. Indicate precisely what you want done. Be realistic. Don't inflate costs or damages. And do not simply write that you "want something done." State precisely that you want one or more of the following:

- your purchase price refunded in full
- a credit made to your account
- a credit toward the purchase of another model
- your exact model repaired or replaced
- a new repair crew assigned to the job
- an apology from the company for discourteous or late service

If you are asking for damages, state your request in dollars and cents and always include copies of bills documenting your expenses related to the problem.

7. Ask for prompt handling of your claim. In your concluding paragraph, ask the reader to answer any question you may have (such as finding out where calls came from that you were billed for but did not make). Also specify a reasonable time by which you want to hear from the reader or need the problem fixed. Note how the writer does this in the last paragraph in Figure 4.14.

Adjustment Letters

Adjustment letters respond to complaint letters by telling customers dissatisfied with a product or service how their claim will be settled. Adjustment letters should reconcile the differences that exist between a customer and a company and restore the customer's confidence in that company.

Adjustment Letters That Tell the Customer "Yes"

It is easy to write a "Yes" letter if you remember a few useful suggestions. As with a good news message, start with the favorable news the customer wants to hear; that will put him or her in a positive frame of mind to read the rest of your letter. Let the customer know that you sincerely agree with him or her—don't sound as if you are begrudgingly honoring the request.

The two examples of adjustment letters saying "Yes" show you how to write this kind of correspondence. The first example, Figure 4.15, says "Yes" to Michael Trigg's letter in Figure 4.13. Reread the Trigg complaint letter to see what problems Ralph Montoya faced when he had to write to Mr. Trigg. The second example of an adjustment letter that says "Yes" is in Figure 4.16. It responds to a customer who has complained about an incorrect billing.

An adjustment letter saying "Yes" to complaint letter in Figure 4.13   Figure 4.15

Smith Sports Equipment
P.O. Box 1014   Tulsa, Oklahoma 74109-1014
(918) 555-0164 ▪ www.smithsport.com

October 19, 2011

Mr. Michael Trigg
17 Westwood Drive
Magnolia, MA 02171

Dear Mr. Trigg:

Thank you for alerting us in your letter of October 10 to your problems with one of our model 191 spincast reels. I am sorry for the inconvenience the reel caused you. A new Smith reel is on its way to you.

We have examined your reel and found the difficulty. It seems that a retaining pin on the button spring was improperly installed by one of our new soldering machines on the assembly line. We have thoroughly inspected, repaired, and cleaned this machine to eliminate the problem from happening again.

Since we began making quality reels in 1955, we have taken pride in helping loyal customers like you who rely on a Smith reel. We hope that your new Smith reel brings you years of pleasure and many good catches, especially next year at the Gloucester Fishing Tournament.

Thank you for your business. Please let me know if I can assist you again.

Respectfully,

SMITH SPORTS EQUIPMENT

Ralph Montoya

Ralph Montoya, Manager
Customer Relations Department

Responds within time frame specified in complaint letter

Apologizes and announces good news

Explains what happened and why problem will not recur

Expresses respect for customer

Closes with friendly offer to help again

Guidelines for Writing a "Yes" Letter

The following four steps will help you write a "Yes" adjustment letter.

1. Admit immediately that the customer's complaint is justified and apologize. Briefly state that you are sorry and thank the customer for writing to inform you.

2. State precisely what you are going to do to correct the problem. Let the customer know that you will

- extend warranty coverage
- credit the account with funds, more air miles, or the like
- offer a discount on the next purchase
- cancel a bill or give credit toward another purchase
- repair damaged equipment
- enclose a free pass, coupon, or waiver
- upgrade a product or service

Do not postpone the good news the customer wants to hear. That way the rest of your letter will be much more appreciated and convincing. In Figure 4.15 Michael Trigg is told that he will receive a new reel; in Figure 4.16 Kathryn Brumfield learns she will not be charged for parts or service.

3. Tell customers exactly what happened. They deserve an explanation for the inconvenience they suffered. Note that the explanations in Figures 4.15 and 4.16 (pages 125 and 127) give only the essential details; they do not bother the reader with side issues or petty remarks about who was to blame. But assure customers that the mishap is not typical of your company's operations.

4. End on a friendly—and positive—note. Don't remind customers about their trouble. Leave customers with a positive feeling about your company. You want them to purchase your product or service again.

Adjustment Letters That Tell the Customer "No"

Writing to tell customers "No" is obviously more difficult than agreeing with them. You are faced with the sensitive task of conveying bad news, while at the same time convincing the reader that your position is fair, logical, and consistent. Do not bluntly start off with a "No." Do not accuse or argue. Avoid remarks such as the following that blame, scold, or remind customers of a wrongdoing:

- You obviously did not read the instruction manual.
- Our records show that your purchased the equipment after the policy went into effect.
- The company policy plainly states that such refunds are not allowed.
- You were negligent in running the machine.
- You claim that our scanner was poorly constructed.
- Your complaint is unjustified.

Brunelli Motors

Route 3A, Giddings, Kansas 62034-8100 (913) 555-1521

August 6, 2012

Ms. Kathryn Brumfield
34 East Main
Giddings, KS 62034-1123

Dear Ms. Brumfield:

We appreciate your notifying us, in your letter of July 30, about the problem you experienced regarding warranty coverage on your new Phantom Hawk GT. The bills sent to you were incorrect, and I have cancelled them. Please accept my apologies. You should not have been charged for a shroud or for repairs to the damaged fan and hose, since all those parts, and labor on them, are fully covered by your warranty.

The problem was the result of an error in the way the charges were listed. Our firm has begun using new billing software to give customers better service, and the technician apparently entered the wrong code for your account. We have since programmed our system to flag any bills for vehicles still under warranty. We hope that this new procedure will help us serve you and our other customers more efficiently.

We value you as a customer of Brunelli Motors. When you are ready for another Phantom, I hope that you will once again visit our dealership. Happy motoring!

Sincerely yours,

Susan Chee-Saafir
Susan Chee-Saafir
Service Manager

Experience virtual reality: Drive a new Phantom at
http://www.brunelli.com

Responds promptly

Thanks customer and complies with request

Explains why problem occurred and how it has been resolved

Ends courteously and leaves reader with good feeling about the dealership

Guidelines for Saying "No" Diplomatically

The following five suggestions will help you say "No" diplomatically. Practical applications of these suggestions can be found in Figures 4.17 (page 129) and 4.18 (page 130). Contrast the refusal of Michael Trigg's complaint in Figure 4.17 with the favorable response to it in Figure 4.15.

1. **Thank customers for writing.** Open with a polite, respectful comment, called a buffer, to soften your reader's response before he or she sees your "No." Don't put them on the defensive by beginning with "We regret to inform you." The letter writers in Figures 4.17 and 4.18 use buffers to thank the customers for bringing the matter to their attention and sympathize with them about their inconvenience. As with other bad news letters, never begin with a refusal. Telling them "No" in the first sentence or two will negatively color their reactions to the rest of your letter. Use the indirect approach discussed earlier in this chapter (page 116), and avoid these reader-hostile openings:

- I was surprised to learn that you found our product unsatisfactory.
- We have been in business for years and nothing like this has ever happened.
- There is no way we could give you what you demand.

2. **State the problem carefully to reassure the customer that you understand the complaint.** You thereby prove that you are not trying to misrepresent or distort what the customer has told you.

3. **Explain what happened with the product or service before you give the customer a decision.** Provide a factual explanation to show the customer that he or she is being treated fairly. Rather than focusing on the customer's misunderstanding the instructions or a failure to observe details of a service contract, state the proper ways of handling a piece of equipment.

> Poor: By reading the instructions on the side of the paint can, you would have avoided the streaking condition that you claim resulted.
>
> Revised: Hi-Gloss Paint requires two applications, four hours apart, for a clear and smooth finish.

The revision reminds the customer of the right way to apply the paint without pointing an accusing finger. Note how the explanations in Figures 4.17 and 4.18 emphasize the appropriate way of using the product equipment.

4. **Give your decision without hedging.** Do not say, "Perhaps some type of restitution could be made later" or "Further proof would have been helpful." Indecision will infuriate customers who believe that they have already presented a sound, convincing case. Never apologize for your decision.

5. **Leave the door open for better and continued business.** Whenever possible, help customers solve their problem by offering to send them a new product or part, or installing or repairing a product free of charge or at a discount. Note how the second-to-last paragraphs in Figures 4.17 and 4.18 do that diplomatically.

An adjustment letter saying "No" to complaint letter in Figure 4.13.   Figure 4.17

Smith Sports Equipment

P.O. Box 1014   Tulsa, Oklahoma 74109-1014
(918) 555-0164 ▪ www.smithsport.com

October 19, 2011

Mr. Michael Trigg
17 Westwood Drive
Magnolia, MA 02171

Dear Mr. Trigg:

Thank you for writing to us on October 10 about the trouble you experienced with our model 191 spincast reel. We are sorry to hear about the difficulties you had with the release button and gears.

We have examined your reel and found the difficulty. It seems that a retaining pin in the button spring was pushed into the side of the reel casing, thereby making the gears inoperable. The retaining pin is a vital yet delicate part of your reel. In order to function properly, it has to be pushed gently. Since our replacement guarantee does not cover the use of the pin in this way, we cannot refund your purchase price.

However, we want you to have many more hours of fishing pleasure and so we would be happy to repair your reel for $19.98 and return it to you within 5–7 days. Please let us know your decision.

I look forward to hearing from you.

Respectfully,

SMITH SPORTS EQUIPMENT

Ralph Montoya

Ralph Montoya, Manager
Customer Relations Department

Buffer—thanks and sympathizes with reader

Explains problem whiteout directly blaming the reader; gives firm decision

Turns a "No" into a "Yes" for customer

Ends politely without any reference to the problem

Figure 4.18   Another adjustment letter saying "No."

Health AIR

4300 Marshall Drive
Salt Lake City, Utah 84113-1521
(801) 555-6028
www.healthair.com

August 28, 2011

Inside address and salutation recognize reader's title

Ms. Denise Southby, Director
Bradley General Hospital
Bradley, IL 60610-4615

Dear Director Southby:

Professional you-centered opening

Thank you for your letter of August 20 explaining the problems you encountered with our Puritan MAII ventilator. We were sorry to learn that you were unable to get the high-volume PAO_2 alarm circuit to work.

Justifies firm decision by explaining causes of problem and conditions of sale

Our ventilator is a high-volume, low-frequency unit that can deliver up to 40 ml. of water pressure. The ventilator runs with a center of gravity attachment on the right side of the diode. The trouble you had with the high oxygen alarm system is due to an overload on your piped-in oxygen. Our laboratory inspection of the ventilator you returned indicated that the high-pressure system had blown a vital adaptor in the MAII. An overload in an oxygen system is not covered by the warranty on the ventilator, and so we cannot replace it free of charge.

Provides practical alternative with financial incentive to keep customer's business

We would, however, be pleased to send you another model of the adapter, which would be more compatible with your system, as soon as we receive your order. The price of the adapter is $600, but for being a valued customer, our service representative for your area will install it at no charge.

Ends with goodwill

Please let me know your decision. I look forward to hearing from you.

Sincerely yours,

R. P. Gifford

R. P. Gifford
Customer Service Department

Collection Letters

Collection letters require the same tact and fairness as do complaint and adjustment letters. Each nonpayment case needs to be evaluated separately. A nasty collection letter sent to a customer who is a good credit risk after only one month's nonpayment can send the customer elsewhere. On the other hand, three easygoing letters to a customer who is a poor credit risk may encourage that individual to postpone payment, perhaps indefinitely.

Types of Collection Letters

Many businesses send several letters to customers before turning matters over to a collection agency. Each letter in the series employs a different technique, ranging from giving compliments and offering flexible credit terms to issuing demands for immediate payment and threats of legal consequences. One hospital uses the collection letters illustrated in Figures 4.19 (page 132) and 4.20 (page 133) to encourage patients to pay their bills. Figure 4.19 is a letter sent early in the collection process when a client is only a month or two late. The collection letter in Figure 4.20, however, is sent much later to a client who has ignored earlier notices.

The tone of Figure 4.19 is cordial and sincere—now is not the time to say "Pay up or else." Instead it stresses how valuable the patient is and underscores how pleased the hospital is to have provided the care he needed. The second-to-last paragraph makes a request for payment, offering: (1) a flexible payment schedule and (2) an escape from the inconvenience (or embarrassment) of receiving past due notices. The bottom of the letter conveniently lists payment options available to the patient. The last paragraph leaves the door open for communication with the hospital.

The late collection letter in Figure 4.20, on the other hand, points out that the time for concessions is over and reminds the patient of all the efforts that the hospital has expended to collect its bills. Appealing to the reader's need to maintain his good credit record, the last paragraph then announces what unfortunate consequences will result if he still does not pay.

International Business Correspondence

After e-mails, letters are the most frequent type of communication you are likely to have with international readers. Formal letter writing is a highly prized skill in the global marketplace. But, as we saw in Chapter 1, you cannot assume that every culture writes letters the way we do in the United States. The conventions of letter writing—formats, inside addresses, salutations, dates, complimentary closes, signature lines—are as diverse as international audiences are. Sometimes your international client may reside in the United States. Then you have to exercise the same diplomacy when communicating with an international audience within the United States as you would when communicating with audiences living in other countries. For example, Patrice St. Jacques, in Figure 4.21, writes an effective sales letter by zeroing in on his reader's, Etienne Abernathy's, ethnic pride and heritage. Although

Figure 4.19   **A first, or early, collection letter.**

SABINE COUNTY HOSPITAL
7200 Medical Blvd.
Sabine, TX 77231-0011
512-555-6734
www.sabine.org

May 15, 2012

Re: Inpatient Services
Date of Hospitalization: March 11–12, 2012
Balance Due: $4,725.48

Mr. Cal Smith
24 Mulberry Street
Valley, TX 77212-3160

Dear Mr. Smith:

Links hospital mission to patient's payment

We are honored that we were able to serve your health care needs during your recent stay at Sabine County Hospital. It is our continuing goal to provide the best possible care for residents of Sabine County and its vicinity. To do so we must keep our finances up-to-date.

Diplomatic reminder to pay now

Our records indicate that your account is now overdue, and that we have not received a payment from you for two months. If you have recently sent one in, kindly disregard this letter and accept our thanks.

Offers options to maintain goodwill

If for any reason you are unable to pay the full amount at this time, we will be happy to set up a convenient payment schedule. Just fill in the appropriate blanks below, and return this letter to us. That will enable us to avoid billing you on a "Past Due" basis. Thank you for your cooperation.

Please call me if you have any questions.

Sincerely,

Morris T. Jukes

Closes with offer of assistance if customer cannot pay now

Morris T. Jukes
Accounts Department

( ) I will pay $ _____   ( ) monthly   ( ) quarterly on my account.

( ) Enclosed is a check for full payment in the amount of $ _____.

Signature

_____

A final collection letter. Figure 4.20

SABINE COUNTY HOSPITAL
7200 Medical Blvd.
Sabine, TX 77231-0011
512-555-6734
www.sabine.org

September 21, 2012

Mr. Cal Smith
24 Mulberry Street
Valley, TX 77212-3160

Re: Inpatient Services
Date of Hospitalization: March 11–12, 2012
Balance Due: $4,725.48

Dear Mr. Smith:

During the past few months we have written to you several times about your balance of $4,725.48 for services you received on March 11 and 12. Your account is more than 190 days overdue, and we cannot allow any further extensions in receiving a payment from you.

Direct opening about history and current status of the account

As you will recall, we have tried to help you meet your obligations by offering several options for paying your bill. You could have arranged for installment payments that would be due each month or even each quarter, whichever would be more convenient. Because you have not replied, we must ask for full payment now.

Reminds patient of goodwill and insists on payment

If we do not hear from you within the next ten days, we will have no alternative but to turn your account over to our collection agency, which will seriously hurt your credit rating. Neither of us would find this a welcome alternative.

States final option in a respectful yet firm tone

Sincerely,

Morris T. Jukes

Morris T. Jukes
Accounts Department

Chapter 4 Additional Activities, located at college.cengage.com/pic/kolinconciseze, include a global-focused activity titled "Writing Letters for International Readers."

Abernathy's company is located in the United States, St. Jacques persuasively sees her reader from a much broader cultural perspective.

It would be impossible to give you information about how to write letters to each international audience. There are at least 5,000 major languages representing diverse ethnic and cultural communities around the globe. But here are some of the most important culturally sensitive questions you need to ask about writing to readers whose cultures are different from yours:

- What is your status in relationship to the reader (client, vendor, salesperson, or international colleague)?
- How should you format and address your letter?
- What is an appropriate salutation?
- What is the most appropriate tone to use?
- How should you begin and conclude your letter?
- What types and amount of information will you have to give?

To answer these and similar questions about proper letter protocol for your international readers, you need to learn about their culture on country-specific websites, on a travel site such as *cyborlink.com*, from multinational colleagues, or through foreign language instructors.

Ten Guidelines for Communicating with International Readers

When you communicate with international readers, you need to be aware of cultural differences between you and your reader. The conventions of writing—the words, sentences, even the type of information you offer—can and do change from one culture to another.

The following ten guidelines will help you communicate more successfully with an international audience, significantly reducing the chances of their misunderstanding you.

1. **Use common, easily understood vocabulary**. Write basic, international English (see pages 4–5). Choose words that are widely understood as opposed to those that are not used or comprehended by many speakers. Whenever you have a choice, use the simpler word; for example, use *stop*, not *refrain; prevent*, not *forestall; discharge*, not *exude; happy*, not *exultant*.

2. **Keep your sentences simple and easy to understand**. Short, direct sentences will cause a reader whose native language is not English the least amount of trouble. A good rule of thumb is that the shorter and less complicated your sentences, the easier they will be for a reader to process. Long (more than 15 words) and complex (multiclause) sentences can be so difficult for readers to unravel that they may skip over them or guess at your message. Do not, however, be insultingly childish as if you were writing for someone in kindergarten. Always try to avoid the passive voice; it is one of the most difficult sentence patterns for a non-native speaker to comprehend. Stick to the common subject-verb-object pattern as often as possible (see Chapter 2, page 43).

A sales letter that appeals to an international audience in the U.S.   Figure 4.21

ISLAND JACQUES
4700 Cyprus Avenue
Philadelphia, PA 19172

6 July 2011

Mr. Etienne Abernathy, President
Seagrove Enterprises
1800 S. Port Haven Road
Philadelphia, PA 19103-1800

Dear Mr. Abernathy:

Congratulations on winning the Hanover Award for Community Service. We in the Port Haven area are proud that a business with Caribbean roots has received such a distinguished honor.

To celebrate your and Seagrove's success, as well as all your business entertaining needs (annual banquet, monthly meetings, etc.), I invite you to Island Jacques. We are a family-owned business that for 30 years has offered Philadelphia residents the finest Caribbean atmosphere and food west of the Islands. Our black pepper shrimp, reggae or mango chicken, and steak St. Lucie—plus our irresistible beef and pork jerk—are the talk from here to Kingston. You and your guests can also savor our original Caribbean art and enjoy our steel drum music.

Island Jacques can also offer Seagrove a variety of dining options. With separate rooms, we are small enough for an intimate party of 4 yet large enough to accommodate a group of 250. We can do early lunches or late dinners, depending on your schedule. And we even cater, if that's your style. Our chefs—Diana Maurier and Emile Danticat—will prepare a special calypso menu just for you. Also a benefit, our prices are generously competitive for the Philadelphia area.

www.islandjacques.netdoor.com
856-555-3295

Distinctive, functional letterhead

Compliments reader on award and respects his heritage

Appeals to reader's ethnic background in art, music, and food

Features flexible dining options, including a specialized ethnic menu

Continued

Figure 4.21   (Continued)

Gives reader
incentive, tied
to his heritage,
to act soon

Etienne Abernathy   2   6 July 2011

Please call me soon so you can see Island Jacques's unique hospitality.
For your convenience, I am enclosing a copy of this week's menu
delights. Check out our website, too, for a taste of Caribbean sound. We
would love to feature Seagrove as Island Jacques's "Guest of the Week"!

Stay Cool, Mon

Patrice St Jacques

Patrice St. Jacques
Manager

3. Avoid ambiguity. Words that have double meanings force non-native readers
to wonder which one you mean. For example, "We fired the engine" would baffle
your readers if they were not aware of the multiple meanings of *fire*. Unfamiliar
with the context in which *fire* means "start up," a non-native speaker of English
might think you're referring to "setting on fire or inflaming," which is not what
you intend. Such misinterpretation is likely because most bilingual dictionaries
would probably list only those two meanings.

4. Be careful about technical vocabulary. While a reader who is a non-native
speaker may be more familiar with technical terms than with other English words,
make sure the technical word or phrase you include is widely known and not a
word or meaning used only at your facility or office. Double check by consulting
the most up-to-date manuals and guides in your field, but steer clear of technical
terms in fields other than the one with which your reader is familiar. Be especially
careful about using business words and phrases an international reader may not
know, such as "lean manufacturing," "reverse mortgage," "toxic asset," and so
forth.

5. Avoid idiomatic expressions. The following colorful idiomatic expressions
may confuse or even startle a non-native reader:

I'm all ears	think outside the box
throw cold water on it	sleep on it
hit the nail on the head	give a heads up to
new blood	land in hot water
easy come, easy go	touch and go
get a handle on it	pushed the envelope
right under your nose	it was a rough go

The meanings of these and similar phrases are not literal but figurative—a reflection of the U.S. culture, not necessarily your reader's. A non-native speaker of English will approach such phrases as combinations of the separate meanings of the individual words, not as the collective meaning of these expressions.

A non-native speaker of English—a potential customer in Asia or Africa, for example—might be shocked if you wrote about a sale concluded at a branch office this way: "Last week we made a killing in our office." Substitute a clear, unambiguous translation easily understood in international English: "We made a big sale last week." For "Sleep on it," you might say, "Please take a few days to make your decision."

6. Delete sports and gambling metaphors. These metaphors, which are often rooted in American popular culture, do not translate word for word for non-native speakers and so again can interfere with communication with your readers. Here are a few examples to avoid:

out in left field	a ballpark figure
struck out	won by a nose
dropped the ball	out of bounds
down for the count	made a pass
long shot	beat the odds

Use a basic English dictionary and your common sense to find nonfigurative translations for the preceding and similar expressions.

7. Don't use unfamiliar abbreviations, acronyms, or contractions. While these shortened forms of words and phrases are a part of U.S. business culture, they might easily be misunderstood by a non-native speaker. Avoid abbreviations such as pharm., gov., org., pkwy, rec., hdg., hr., mfg., or w/o. The following acronyms can also cause your international reader trouble: ASAP, PDQ, p's and q's, IRA, SUV, RV, DOB, DOT, SSN. Moreover, contractions such as the following might lead readers to mistake them for the English word they look like: I've (ivy), he'll (hell), he'd (held), I'll (ill), we'll (well), can't (cant), won't (wont, want).

8. Watch units of measure. Do not fall into the cultural trap of assuming that your reader measures distances in miles and feet (instead of kilometers and meters as most of the world does), records temperature on a Fahrenheit scale (instead of Celsius), buys gallons of gasoline (instead of liters), and spends dollars (rather than euros, pesos, rupees, or yen). Always respect the monetary unit your reader uses. Also, many countries follow a 24-hour clock without a.m. or p.m. In Paris, for instance, 13:30 is 1:30 p.m. in New York, while 9 p.m. in the United States translates into 21:00 in Sweden. Adapt your measurements to the readers' culture.

9. Avoid culture-bound descriptions of place and space. For example, when you tell a reader in Hong Kong about the "Sunbelt" or a potential client in Africa about the "Big Easy," will he or she know what you mean? When you write from California to a non-native English speaker in India about the eastern seaboard, meaning the East Coast of the United States, the directional reference may not mean the same thing to your audience as it does to you. Be respectful of your readers' cultural (and physical) environment as well. Thanksgiving is celebrated in the United States in November, but in Canada it falls on the second Monday in October; elsewhere around the world it may not be a holiday at all. Calling February a winter month does not make sense to someone in New Zealand for whom it is a summer one.

10. **Use appropriate salutations, complimentary closes, and signature lines.** Find out how individuals in the recipient's culture are formally addressed in a salutation (e.g., Señor, Madame, Frau, Monsieur). Unless you are expressly asked to use a first name, always use your reader's surname and include proper titles and other honorifics (e.g., Doctor, Sir, Father). For a complimentary close, choose an appropriately formal one, such as *Respectfully*, which is acceptable in almost any culture.

Respecting Readers' Nationality and Ethnic/Racial Heritage

Do not risk offending any of your readers, whether they are native speakers of English or not, with language that demeans or stereotypes their nationality or ethnic and racial background. Here are some precautions to take.

1. **Respect your reader's nationality.** Always spell your reader's name and country properly, which may mean adding diacritical marks (e.g., accent marks) not used in English—e.g. Muñoz. If your reader has a hyphenated last name (e.g., Arana-Sanchez), it would be rude to address him or her by only part of the name (e.g., only Arana or only Sanchez). In addition, be careful not to use the former name of your reader's country or city, for instance, the Soviet Union (now Russia), Malaya (now Malaysia), Czechoslovakia (now the Czech Republic), Rhodesia (now Zimbabwe) or Bombay (now Mumbai). Not only is it rude, but it also demonstrates a lack of interest about your reader's nationality.

2. **Observe your reader's cultural traditions.** Cultures differ widely in the way they send and receive information and how they prefer to be addressed, greeted, and informed in a letter. What is acceptable in one culture may be offensive in another. A sales letter to an Asian business executive, for example, needs to employ a very different strategy from one intended for an American reader. The best strategy for an American audience would be hard-hitting and to the point, stressing your product's strengths versus the competitor's weaknesses. But the East Asian way of writing such a letter would be more subtle, indirect, and complimentary.

> American: Our Imaging 500 delivers much more extensive internal imaging than any of our competitors' models.
>
> East Asian: One of the ways we may be able to serve you is by informing you about our new Imaging 500 MRI (Magnetic Resonance Imaging) equipment.

The hard sell in the American example would be a sign of arrogance, suggesting inequality for a reader in China, Japan, Malaysia, or Korea who is more comfortable with a compliment or a wish for prosperity.

3. **Honor your reader's place in the world economy.** Phrases such as "third-world country," "emerging nation," and "undeveloped/underprivileged area" are derogatory. Using such phrases signals that you regard your reader's country as inferior. Use the name of your reader's country instead. Saying that someone lives in the Far East implies that the United States, Canada, or Europe is the center of culture, the hub of the business community. Never use the word "Oriental," which is insulting. Simply say "East Asia."

4. **Avoid insulting stereotypes.** Expressions such as "oil-rich Arabs," "time-relaxed Latinos," and "aggressive foreigners" unfairly characterize particular groups. Similarly, prune from your communications any stereotypical phrase that insults one group or singles it out for praise at the expense of another—"Mexican standoff," "Russian roulette," "Chinaman's chance," "Irish wake," "Dutch treat," "Indian giver." Also note that the word *Indian* refers to someone from India; use *Native American* to refer to the indigenous people of North America, who want to be known by their tribal affiliations (e.g., the Lakota).

5. **Be visually sensitive to the cultural significance of colors.** Do not offend your audience by using colors in a context that would be offensive. Green and orange have a strong political context in Ireland. In China, white does not symbolize purity and weddings but mourning and funerals. Similarly, in India if a married woman wears all white, she is inviting widowhood. While red symbolizes good luck in China, it has just the opposite meaning in Korea.

6. **Be careful, too, about the symbols you use for international readers.** Triangles are associated with anything negative in Hong Kong, Korea, and Taiwan. Political symbols, may have controversial implications as well (e.g., the hammer and sickle, a crescent). Avoid using the flag of a country as part of your logo or letterhead for global audiences. Many countries see this as a sign of disrespect, especially Saudi Arabia, whose flag features the name of Allah.

Writing to Readers from a Different Culture: Some Examples

Figures 4.22 and 4.23 (pages 140–141) show two versions of a letter to an international business executive. The letter in Figure 4.22 violates the guidelines discussed above when addressed to such an audience because it:

- uses the incorrect format for the dateline
- misspells the name of the reader's city
- leaves out important postal information
- contains U.S. idioms (e.g., *drop you a line*)
- includes troubling abbreviations
- disregards how the reader's culture records time and temperature

Even more disrespectful, the writer's overall tone is condescending (*south of the border*) and inappropriately casual. At the end of the second paragraph, for instance, the writer tells his reader that his U.S. firm is superior to the Argentine company.

Figure 4.22   An inappropriately written letter for an international reader.

Pro-Tech, Ltd. 452 West Main St.   Concord, MA 01742   978-634-2756
www.protech.com

Misleading date line

5-9-11

Incorrect, misspelled address

Mr. Antonio Guzman
Canderas
Mercedes Ave.
Bunos Aires, ARG.

Salutation is too informal

Dear Tony,

Impolite opening disrespecting reader's status

Culturally condescending

I wanted to drop you a line before the merger hits and in doing so touch base and give you the lowdown on how our department works here in the good old U.S. of A.

None of us had a clue that Pro-Tech was going to go south of the border but your recent meeting about the Smartboard T-C spoke volumes to the tech people who praised your operations to the hilt. So it looks like you and I both will be getting a new corp. name. Olé. I love moving from Pro-Tech, Ltd. to Pro-Tech International. We are so glad we can help you guys out.

Filled with American idioms and abbreviations

At any rate, I'm sending you an e-mail with all the ins and outs of our department struc., layout, employees, and prod. eff. quotas. From this info, I'm hoping you'll be able to see ways for us to streamline, cooperate, and soar in the market. I understand that all of this is in the works that you and I need to have a face-to-face and so I'd appreciate your reciprocating with all the relevant data stat.

Disregards time differences and reader's 24-hour clock

Consequently, I guess I'll be flying down your way next month. Before I take off, I would like to give you a ring. How does after lunch next Thursday (say, 1:00–1:30) sound to you? I hope this is doable.

Ignores, differences between Fahrenheit/ Celsius scales

We've had a spell of great weather here (can you believe it's in the low 80s today!). So, I guess I'll just sign off and wait 'til I hear from you further.

I send you felicitations and want things to go smoothly before the merger is a done deal.

Close sounds insincere

Adios,

Frank Sims

Frank Sims

Pro-Tech, Ltd. 452 West Main St.   Concord, MA 01742   978-634-2756
www.protech.com

9 May 2011

Clear date line

Señor Antonio Mosca-Guzman
Director, Quality Assurance
Tecnología Canderas, S.A.
Av. Martin 1285, 4° P.C.
C1174AAB BUENOS AIRES
ARGENTINA

Uses appropriate title for reader

Complete and correct address

Dear Señor Mosca-Guzman:

Courteous salutation

As our two companies prepare to merge, I welcome this opportunity to write to you. I am the manager for the Quality Assurance division at Pro-Tech, a title I believe you have at Tecnología Canderas. I am looking forward to working with you both now and after our companies merge in two months.

Clear and diplomatic opening

Allow me to say that we are very honored that your company is joining ours. Tecnología Canderas has been widely praised for the research and production of your Smartboard T-C systems. I know we have much to learn from you, and we hope you will allow us to share our systems analyses with you. That way everyone in our new company, Pro-Tech International, will benefit from the merger.

Respectful view of reader's company

Later this week, I will send you a report about our division. I will describe how our division is structured and the quality assurance inspections we make. I will also give you a brief biography of our staff so that you can learn about their qualifications and responsibilities.

Explanation of business procedures in plain English

The director of our new company, Dr. Suzanne Nknuma, asked me to meet with you before the merger occurs to discuss how we might help each other. I would very much like to travel to Buenos Aires in the next month to visit you and take a tour of your company.

Would you please let me know by e-mail when it may be convenient for us to talk on the telephone so we might discuss the agenda for our meeting? I am always in my office from 11:00 to 17:00 Buenos Aires time.

Recognition of reader's time zone

I look forward to working with and meeting you.

Respectfully,

Polite close

Frank Sims

Frank Sims
Quality Assurance Manager

Gives business title

Note how the letter in Figure 4.23, on the other hand, respects the reader's cultural conventions because it:

- spells and punctuates the reader's name and address correctly
- incorporates a clear date line
- uses an appropriate salutation and complimentary close
- is written in plain, international English
- recognizes that the reader uses a twenty-four-hour clock
- identifies the writer and makes the reader feel welcome and honored
- courteously informs the reader how the merger will affect his relationship with the writer
- strives to develop a spirit of cooperation and mutuality
- acknowledges the reader's position of authority in his company

Above all, though, the writer honors his audience's culture and role in the business world and seeks to win the reader's confidence and respect, two invaluable assets in the global marketplace. Writing letters with the specific needs of your audience in mind, as in Figure 4.23, is a crucial skill to have in the global world of work.

Sending Letter-Quality Messages: Final Advice to Seal Your Success

Visit www.cengage.com/english/kolin/writingatwork concise3e for an online exercise, "Correcting Online Letters."

Judging your letters in light of the following guidelines will help you to draft, tailor, and evaluate the types of business correspondence you will be asked to write:

- **Identify your reader**: one individual; a group; a company or agency; a new individual customer or a longtime one; a native or non-native speaker
- **Determine your purpose for writing**: explanation; complaint; apology; sell product/service; build goodwill; express thanks; collect a debt
- **Determine reader's reason for writing**: complaint; request information; make an adjustment; seek credit
- **Organize information**: direct or indirect; begin with good news and save negative message for middle of the letter
- **Include essential information**: schedules; dates; prices and expenses; personnel; explanation of services, warranties, products; background and need for credit
- **Use appropriate style and tone**: professional and courteous; concise and focused; sensitive to reader's needs, including his or her culture and traditions

Planning your letter carefully—its purpose, its organization, and its content—will help you to make sure your message begins, continues, and ends professionally and successfully for readers in the U.S. and around the globe.

Revision Checklist

Audience Analysis and Research

☐ Made sure reader's name and job title are correct.

☐ Found out something about my audience—interests and background, well informed or unfamiliar with topic, former clients or new ones.

☐ Determined whether audience will be friendly, hostile, or neutral about my message.

☐ Did sufficient research—in print, through online sources, in discussions with colleagues—to give audience what they need.

Format/Appearance

☐ Followed one letter format (full block, modified block) consistently.

☐ Left margins wide enough to make my letter look attractive and well proportioned.

☐ Included all the necessary parts of a letter for my purpose.

☐ Made sure that my letter looks neat and professional.

☐ Printed my letter on company letterhead or quality bond paper.

☐ Proofread my letter carefully and made sure each correction was made before final copy was printed.

☐ Eliminated any grammatical and spelling errors.

☐ Signed my letter legibly in blue or black ink.

Content/Organization

☐ Began my correspondence with reader-effective strategies. If reporting good news, told the reader right away. If reporting bad news, was diplomatically indirect and considerate of my reader's reactions.

☐ Followed the four A's of effective sales letters. Identified and convinced my target audience.

☐ Wrote complaint letters in a courteous tone. Informed the reader what is wrong, why it is wrong, and how the problem should be corrected.

☐ Wrote adjustment letters that say "Yes" sincerely and to the point. Made those that say "No" fair but did not begin with a "No." Acknowledged reader's point

Style: Words Tone, Sentences, Paragraphs

☐ Emphasized the "you attitude" by seeing things from reader's perspective.

☐ Chose words that are clear, precise, and friendly.

☐ Cut anything sounding flowery, stuffy, or bureaucratic.

☐ Ensured that my sentences are readable, clear, and not too long.

☐ Wrote paragraphs that are easy to read and that flow together.

Writing to International Readers

☐ Did appropriate research about reader's culture, in print, through online sources, and with colleagues who are native speakers (as well as teachers).

☐ Adopted a respectful, not condescending, tone.

☐ Avoided anything offensive to my reader,
☐ Used plain and clear language that my reader would understand.
☐ Tested my sentences for length and active voice.
☐ Chose colors and symbols culturally appropriate for the context of my message.
☐ Selected the right format, salutation, and complimentary close for my reader.
☐ Observed the reader's units of measurement for time, temperature, etc.

Exercises

Additional Activities related to writing letters are located at www. cengage.com/ english/kolin/ writingatwor kconcise3e

1. Write appropriate inside addresses and salutations to (a) a woman who has not specified her marital status, (b) an officer in the armed forces, (c) a professor at your school, (d) an assistant manager at your local bank, (e) a member of the clergy, (f) a government worker.

2. Rewrite the following sentences to make them more personal.
   a. It becomes incumbent on this office to cancel order #2394.
   b. Management has suggested the curtailment of parking privileges.
   c. ALL USERS OF HYDROPLEX: Desist from ordering replacement valves during the period of Dec. 23–31.
   d. The request for a new catalog has been honored; it will be shipped to same address soon.
   e. Perseverance and attention to detail have made this writer important to company in-house work.
   f. The Director of Nurses hereby notifies staff that a general meeting will be held Monday afternoon at 3:00 p.m. sharp. Attendance is mandatory.
   g. Reports will be filed by appropriate personnel no later than the schedule allows.

3. The following sentences from letters are discourteous, boastful, vague, or lacking the "you attitude." Rewrite them to correct those mistakes.
   a. Something is obviously wrong in your head office. They have once more sent me the wrong model number. Can they ever get things straight?
   b. My instructor wants me to do a term paper on safety regulations at a small factory. Since you are the manager of a small factory, send me all the information I need at once. My grade depends heavily on all this.
   c. It is apparent that you are in business to rip off the public.
   d. I have waited for my confirmation for two weeks now. Do you expect me to wait forever or can I get some action?
   e. It goes without saying that we cannot honor your request.
   f. Your application has been received and will be kept on file for six months. If we are interested in you, we will notify you. If you do not hear from us, please do not write us again. The soaring costs of correspondence and the large number of applicants make the burden of answering pointless letters extremely heavy.

g. My past performance as a medical technologist has left nothing to be desired.

h. Credit means a lot to some people. But obviously you do not care about yours. If you did, you would have sent us the $249.95 you rightfully owe us three months ago. What's wrong with you?

4. Write a business letter to one of the following individuals.
   a. your mayor, asking for an appointment and explaining why you need one
   b. your college president, stressing the need for more parking spaces or for additional computers in a library
   c. the local water department, asking for information about fluoride supplements and why you need such information
   d. an editor of an online or print magazine, asking permission to reprint an article in a company newspaper and indicating why.
   e. a disc jockey at a local radio station, requesting more songs by a certain group
   f. a computer vendor, asking about costs and availability of a specific software package; explain your company's special needs

5. Write a letter of inquiry to a utility company, a safety or health care agency, or a business in your town and ask for a brochure describing its services to the community. Be specific about your reasons for requesting the information.

6. Choose one of the following, and write a sales letter addressed to an appropriate audience on why they should
   a. work for the same company you do
   b. live in your neighborhood
   c. be happy taking a vacation where you did last year
   d. dine at a particular restaurant
   e. use a particular software program
   f. have their cars repaired at a specific garage
   g. give their real estate business to a particular agency
   h. use your company's new website when ordering any of replacement parts

7. Rewrite the following sales letter to make it more effective. Add any details you think are relevant.

```
Dear Pizza Lovers:

Allow me to introduce myself. My name is Rudy Moore and I am
the new manager of Tasty Pizza Parlor in town. The Parlor
is located at the intersection of North Miller Parkway and
95th Street. We are open from 10 a.m. to 11 p.m., except on
the weekends, when we are open later.

I think you will be as happy as I am to learn that Tasty's
will now offer free delivery to an extended service area. As
a result, you can get your Tasty Pizza hot when you want it.

Please see your weekly newspapers for our ad. We also are
offering customers a coupon. It is a real deal for you.
```

I know you will enjoy Tasty's pizza and I hope to see you.
I am always interested in hearing from you about our serv-
ice and our fine product. We want to take your order soon.
Please come in.

8. Send a follow-up letter to one of the following individuals:
   a. a customer who informs you that she will no longer do business with your company because your prices are too high
   b. a family of four who stayed at your motel for a week last summer
   c. a wedding party or professional organization that used your catering services last month
   d. a customer who exchanged a coat for the purchase price
   e. a customer who purchased a used car from you and who has not been happy with service
   f. a company that bought software from you nine months ago, alerting them about updates

9. Write a bad news letter to an appropriate reader about one of the following:
   a. Your company has to discontinue Saturday deliveries because of rising labor and fuel costs.
   b. You are the manager of an insurance company writing to tell one of your customers that, because of reckless driving, his or her rates will increase.
   c. You have to refuse to send a bonus gift to a customer who sent in an order after the expiration date for qualifying for the gift.
   d. You have discontinued a model that a business customer wants to reorder.
   e. You have to notify residents of a community that a bus route or hours of operation are being discontinued.
   f. You represent the water department and have to tell residents of a community that they cannot water their lawns for the next month because of a serious water shortage in your town.
   g. You cannot send customers a catalog—which your company formerly sent free of charge—unless they first send $10 for the cost of that catalog.
   h. You cannot repair a particular piece of equipment because the customer still owes your company for three previous service visits.

10. Write a good news letter about the opposite of one of the situations listed in Exercise 9.

11. Write a complaint letter about one of the following:
    a. an error in your utility, telephone, credit card, or Internet provider bill
    b. discourteous service you received on an airplane, train, or bus
    c. a frozen food product of poor quality
    d. a shipment that arrived late and damaged
    e. an insurance payment to you that is $357.00 less than it should be
    f. a public television station's policy of discontinuing a particular series
    g. junk mail or spam that you are receiving
    h. equipment that arrives with missing parts
    i. misleading representation by a salesperson
    j. incorrect or misleading information given on a website

12. This exercise might be done as a collaborative project. You are a section manager at e-Tech. Your company has a service contract with Professional Office Cleaners (POC). However, each morning when you arrive at work you are disappointed with what they've done. POC has overlooked some essential tasks and done a poor job on others. Your staff is also disappointed and has e-mailed or spoken to you about problems with POC. Write the following:

a. a memo or e-mail to your boss, the vice president, about POC's shoddy work

b. a complaint letter to POC that the vice president has asked you to write and to sign his name to it

c. a letter to the vice president from the manager of POC who is responsible for your e-Tech section, apologizing for the problem and offering a solution

d. a letter from POC to the vice president taking issue with the complaint made against his cleaning company, offering proof that the work was done according to contract specifications

e. an e-mail you send to your staff about what's happened with POC

13. Rewrite the following complaint letter to make it more precise, less emotional, and effectively persuasive.

```
Dear Sir:

We recently purchased a machine from your Albany store and
paid a great deal of money for it. This machine, according
to your website, is supposedly the best model in your line
and has caused us nothing but trouble each time we use it.
Really, can't you do any better with your technology?

We expect you to stand by your products. The warranties you
give with them should make you accountable for shoddy work-
manship. Let us know at once what you intend to do about
our problem. If you cannot or are unwilling to correct the
situation, we will take our business elsewhere, and then
you will be sorry.

Sincerely yours,
```

14. The following story appeared recently in a local newspaper. Based on information in this story, which you may want to supplement, write the following complaint/adjustment letters. You can do this alone or as part of a collaborative writing team.

a. a complaint letter to the city from resident Jo Souers

b. a complaint letter to Finicky Pet Food from city officials warning about dangers of odors to the residential area

c. a letter from plant manager Dean Niemann to the residents of Bienville Place subdivision

d. a letter from city officials to the residents of Bienville Place subdivision

Residents concerned about relocation of pet food plant

OCEAN SPRINGS (AP) — Finicky Pet Food is moving its processing plant from Pascagoula to Ocean Springs, a decision that has some residents concerned of possible odor and other problems.

The plant is moving to an industrial area bordering a subdivision of expensive homes.

"The wind doesn't discriminate," said Jo Souers, who lives in the Bienville Place subdivision. "I don't want this in our neighborhood."

City officials said the plant is moving to an area zoned to accommodate it.

"We don't have a lot of control over it," said city planner Donovan Scruggs. "It is a permitted use for this property."

Scruggs said the property was zoned industrial before the subdivision was built. A body shop, cabinet shop, and boat business are located nearby.

The plant will be built in the small industrial area on U.S. 90, directly across the highway from the Super Wal-Mart.

It is moving into a vacant building, the interior of which has been renovated for its new purpose, city officials said.

The plant will process frozen fish and fish parts for bait and pet food. It will employ 10 workers, with that number doubling during fishing season.

Plant manager Dean Niemann said in a statement that the company no longer needed its Pascagoula location near deep water, which was rented from the county.

Scruggs said the city has investigated the possibility that the plant will emit odors.

"We've told Dean (Niemann) from day one, 'You're locating next to a residential area. If you start stinking, action will be taken,'" Scruggs said.

He said the city has a nuisance ordinance that should handle anything that might arise.

Reprinted with permission of The Associated Press.

15. Rewrite the following ineffective adjustment letter saying "No."

```
Dear Customer:

Our company is unwilling to give you a new toaster or to
refund your purchase price. After examining the toaster you
sent to us, we found that the fault was not ours, as you
insist, but yours.

Let me explain. Our toaster is made to take a lot of punish-
ment. But being dropped on the floor or poked inside with a
knife, as you probably did, exceeds all decent treatment.
You must be careful if you expect your appliances to last.
Your negligence in this case is so bad that the toaster
could not be repaired.
```

In the future, consider using your appliances according to the guidelines set down in warranty books. That's why they are written.

Since you are now in the market for a new toaster, let me suggest that you purchase our new heavy-duty model, number 67342, called the Counter-Whiz. I am taking the liberty of sending you some information about this model. I do hope you at least go to see one at your local appliance center.

Sincerely,

16. You are the manager of a computer software company, and one of your sales-people has just sold a large order to a new customer whose business you have tried to obtain for years. Unfortunately, the salesperson made a mistake writing out the invoice, undercharging the customer by $229. At that price, your company would not break even, and so you must write a letter explaining the problem so that the customer will not assume all future business dealings with your firm will be offered at such "below market" rates. Decide whether you should ask for the $229 or just "write it off" in the interest of keeping a valuable new customer.
   a. Write a letter to the new customer, asking for the $229 and explaining the problem while still projecting an image of your company as accurate, profes-sional, and very competitive.
   b. Write a letter to the new customer, not asking for the $229 but explaining the mistake and emphasizing that your company is both competitive and professional.
   c. Write a letter to your boss explaining why you wrote letter a.
   d. Write a letter to your boss explaining why you wrote letter b.
   e. Write a letter to the salesperson who made the mistake, asking him or her to take appropriate action with regard to the new customer.

17. Write a sales letter similar to the one in Figure 4.21 from a manager of one of the following ethnic restaurants or one of your choice. Make sure you include relevant and culturally sensitive details for the particular audience:
   a. Mexican   j. German
   b. Indian   k. Irish
   c. Cuban   l. Chinese
   d. Soul food   m. Pakistani
   e. Czech   n. Italian
   f. Turkish   o. Argentine
   g. Vietnamese   p. French
   h. Thai   q. Afghan
   i. Greek   r. Japanese

18. You just found out that a business that applied for additional credit has missed its last mortgage payment. You have to refuse credit to this local firm, which has been in business successfully for eight years. Write a refusal letter without jeopardizing future business dealings.

19. Rewrite the following letters, making them appropriate for a reader whose native language is not English. As you revise the letters, pay attention to the words, measurements, and sentence constructions you employ. Be sure to consider the reader's cultural traditions by omitting any cultural insensitivity.

a. Dear Pal,

Our stateside boss hit the ceiling earlier today when she learned that our sales quota for this quarter fell precipitously short. Ouch! Were I in her spot, I would have exploded too. Numerous missives to her underlings warned them to get off the dime and on the stick, but they were oblivious to such. These are the breaks in our business, right? We can't all bat 1000.

Let's hope that next quarter's sales take a turn for the best by 12-1-08. If they are as disastrous, we all may be in hot water. Until then, we will have to watch our p's and q's around here. We're freezing here—20° today.

Cheers,

b. Dear Margaret Wong,

It's not every day that you have the chance to get in on the ground floor of a deal so good you can actually taste it. But Off-Wall Street Mutual can make the difference in your financial future. Give me a moment to convince you.

By becoming a member of our international investing group for just under $250, you can just about ensure your success. We know all the ins and outs of long-term investing and can save you a bundle. Our analysts are the hot shots of the business and always look long and hard for the most propitious business deals. The stocks we select with your interests in mind are as safe as a bank

and not nearly so costly for you. Unlike any of your
undertrained local agents, we can save you money by investing your money. We are penny pinchers with our clients' initial investments, but we are King Midas when it
comes to transforming those investments into pure gold.

I am enclosing a brochure for you to study, and I really
hope you will examine it carefully. You would be foolish
to let a deal like Off-Wall Street Mutual pass you by.
Go for it. Call me by 3:00 today.

Hurriedly,

20. Interview a student at your school or a co-worker who was born and raised in a
non-English-speaking country about the proper etiquette in writing a business letter to someone from his or her country. Collaborate with that student to write a
letter (for example, a sales letter or a letter asking for information) to a business
executive from that country.

21. In a letter to your instructor, describe the kinds of adaptations you had to make
for the international reader you wrote to in Exercise 15.

5

How to Get a Job:
Searches, Dossiers, Portfolios, Résumés, Letters, and Interviews

Visit www.
cengage.com/
english/kolin/
writingatwork
concise3e for
this chapter's
online exercises,
ACE quizzes,
and Web links.

Obtaining a job involves a lot of hard work. Before your name is added to a company's payroll, you will have to do more than simply walk into the human resources office and fill out an application form. Finding the *right* job takes time in this difficult and highly competitive job market. And finding the right person to fill that job also takes time for the employer.

Steps the Employer Takes to Hire

From the employer's viewpoint, the stages in the search for a valuable employee include the following:

1. deciding what duties and responsibilities go with the job and determining the qualifications the future employee should possess
2. advertising the job on the company website, in newspapers, in professional publications and conferences
3. scanning and evaluating résumés and letters of application
4. having candidates complete application forms and submit to background checks
5. requesting further proof of candidates' skills (letters of recommendation, transcripts, portfolios)
6. interviewing selected candidates
7. doing further follow-ups and rank ordering candidates to be interviewed
8. offering the job to the best-qualified individual

Sometimes the steps are interchangeable, especially steps 4 and 5, but, generally speaking, employers go through a long and detailed process to select employees. Step 3, for example, is among the most important for employers (and the most crucial for job candidates). At that stage employers often classify job seekers into one of three groups: those they definitely want to interview, those they may want to interview, and those in whom they have no interest.

Steps to Follow to Get Hired

As a job seeker you will have to know how and when to give the employer the kinds of information the preceding eight steps require. You will also have to follow a certain schedule in your search for a job. You can expect to go through the following eight procedures:

1. preparing for your career
2. analyzing your strengths and restricting your job search
3. looking in the right places for a job
4. assembling a dossier and a portfolio
5. preparing a résumé
6. writing a letter of application and filling out a job application
7. going to an interview
8. accepting or declining a job offer

Your timetable should match that of your prospective employer. This chapter shows you how to begin your job search, prepare an appropriate résumé, write a persuasive letter of application, design an effective portfolio, and prepare for an interview.

Analyzing Your Strengths and Restricting Your Job Search

Before you apply for jobs, analyze your job skills, career goals, and interests. Here are some points to consider.

1. Make an inventory of your most significant accomplishments in your major and/or on the job. What are your greatest strengths—writing and speaking, working with people in small groups, organizing and problem solving, managing money, speaking a second language, developing software, designing websites?
2. Decide which specialty within your chosen career appeals to you the most. If you are in a nursing program, do you want to work in a large teaching hospital, for a home health or hospice agency, or in a physician's office? What kinds of patients do you prefer to care for—pediatric, geriatric, psychiatric?
3. What are the most rewarding prospects of a job in your profession? What most interests you about a position—travel, technology, international contacts, on-the-job training, helping people, being creative?
4. What are some of the greatest challenges you face in your career today—or will face in five years?
5. Which specific companies or organizations have the best track record in hiring and promoting individuals in your field? What qualifications will such firms insist on from prospective employees?

Once you answer these questions, you can avoid applying for positions for which you are either overqualified or underqualified. If a position requires 10 years of related work experience and you are just starting out, you will only waste the employer's time and your own by applying. However, if a job requires a certificate or license and you are in the process of obtaining one, go ahead and apply.

Social Networking Sites and Your Job Search

Note that not only do employers monitor their employees on the job (see Chapter 1, page 21), they also consult such popular networking sites as Facebook, Twitter, and MySpace to screen candidates. Be careful about the content and photos you post for public viewing. Project an image of yourself as a professional, someone whom a company wants to hire; avoid posting photos that you might want to share with friends (what you did on your vacation) but not with your future boss. In the "Info" or "Interests" sections on these networking sites, comment about your accomplishments on the job, at school, in the community, or for a charitable organization. Describe how you solved a problem at work, pinpoint skills you learned as a part of a team, or focus on your career goals. For example, if you have worked for Habitat for Humanity or a local food bank, or participated in a professional organization, by all means mention them. To ensure that you present the best image of yourself, make sure you select "private" settings for your personal accounts so that no one can access information or photos that might be detrimental to your job search.

Enhancing Your Professional Image

Whether you are looking for your first job in your career field, returning to the job market, or changing careers, there are several steps you can take to help improve your chances of getting hired. Here are some suggestions of things that will help you develop your professional career plans:

- Attend job fairs and interviewing workshops on campus as well as those sponsored by municipal, state, and federal agencies.
- Join student and professional organizations and societies in your area(s) of interest. Membership rates for students are often reduced.
- Apply for relevant internships and training programs to gain real-world experience and make networking contacts.
- If available, take a temporary job in your profession to gain some experience.
- Ask your instructors to critique your work in light of your career plans.
- Confer with your academic adviser(s) regularly, not just once a semester.
- Find a mentor—someone in the field you might want to join.
- Consider developing a competency in a second (or third) language.
- Do volunteer work to gain or enhance experience working in a group setting, preparing documents, etc.
- Write a blog to increase your networking potential and demonstrate your interest in and knowledge about your profession.

Looking in the Right Places for a Job

One way to search for a job is simply to send out a batch of letters and résumés to companies you want to work for. But how do you know what jobs, if any, those companies have available, what qualifications they are looking for, and what deadlines they might want you to meet? You can avoid these uncertainties by knowing where to look for a job and knowing what responsibilities a specific job entails. Consult the following resources for a wealth of job-related information.

   1. Networking. Networking pays. Most jobs come through consulting with other people. John D. Erdlen and Donald H. Sweet, experts on job searching, cite the following as a primary rule of job hunting: "Don't do anything yourself you can get someone with influence to do for you." Let your professors, co-workers, friends, classmates, neighbors, relatives, and even your clergy know you are looking for a job. They may hear of something and can notify you or, better yet, recommend you for the position. See how the job seekers in Figures 5.10 (page 183) and 5.11 (page 184) have successfully networked with people they know. You can also network with people you don't know personally through professional networking websites such as:

   http://www.linkedin.com

   http://www.ryze.com

   http://www.tribe.net

   http://www.jobster.com

   http://www.meetup.com

Additionally, attend professional and organization meetings related to your field, and community and civic functions to meet the right contact people whom you can ask for advice and also for possible follow-up help and recommendations.

   2. Your campus placement office. Counselors keep an online file of current available positions and can also tell you when recruiters will be on campus to conduct interviews. Placement offices also have recruiting databases, allowing students access to a broad range of contacts and interview information. Counselors can help you locate summer and part-time work as well, both on and off campus, positions that might lead to full-time jobs. Most important, they will give you sound advice on your job search, including strategies for finding the right job, salary ranges, and interview tips. Many placement offices also sponsor career fairs to bring job seekers and employers together in specific professional fields. Finally, your placement office will help you set up and archive your dossier, or credentials (see pages 158–160).

   3. Internet sites posting jobs. A majority of posted jobs can be found on the Internet. You can learn about jobs at a specific company or organization by visiting its website to see what vacancies it has and what the qualifications are for them. Also consult the *Riley Guide: Employment Opportunities and Job Resources* on the Internet (http://www.rileyguide.com). This invaluable resource surveys and classifies job openings on the Web by field, location, and category (private or public), and provides you with links for direct access. In addition, you might want to explore many of the job posting sites such as those in Table 5.1 below, which list positions and sometimes give you advice about applying for them.

TABLE 5.1   Sites Posting Jobs on the Web

Website	URL	Description
CollegeRecruiter.com	http://www.collegerecruiter.com	Aimed at job seekers with little work experience
Monster	http://www.monster.com	Lists thousands of openings, from those at small companies to global ones; offers advice on the job search process
Monster Trak	http://www.monstertrak.com	Posts jobs for new college grads
Monster + Hotjobs	http://www.hotjobs.yahoo.com	This site combines the former Yahoo! HotJobs site and Monster's site to create an enormous job search website.
CareerBuilder	http://www.careerbuilder.com	One of the largest postings linked to more than 200 newspapers around the country
IMDiversity	http://www.imdiversity.com	Job openings from companies whose mission is to promote diversity in their workforces
Net Temps	http://www.nettemps.com	Links to short-term positions
WetFeet	http://www.wetfeet.com	Lists jobs at companies looking for interns in business, technology, health care, etc.

4. Newspapers. Look at local newspapers as well as the Sunday editions of large city papers with a wide circulation, such as the *New York Times Job Market* (http://jobmarket.nytimes.com). The *National Business Employment Weekly* (www.careerjournal.com), published by the *Wall Street Journal*, also lists jobs in different areas, including technical and managerial positions.

5. Federal and state employment offices. The U.S. government is one of the biggest employers in the country. During 2009 and 2010, for instance, the most active career site on the Web was operated by the federal government, with 1.7 million new hires. Counselors at federal and state employment centers also help job seekers find career opportunities. Figure 5.1 shows the homepage for Student Jobs, which helps job seekers find employment opportunities with the U.S. government. Consult the following websites for listings of government jobs:

- USA Jobs — http://www.usajobs.gov
- Federal Jobs — http://www.federaljobs.net
- Student Jobs — http://www.studentjobs.gov

Student Jobs website. Figure 5.1

6. Professional and trade journals plus associations in your major. Identify the most respected periodicals (print and online) in your field and search their ads. Each issue of *Food Technology*, for example, features a section called "Professional Placement," a listing of jobs all over the country. Similarly, *CIO Magazine—IT Professional Research Center* (*http://itjobs.cio.com/a/all-jobs/list*) can help you find jobs in the computer industry, engineering, and technology. Consulting the *Encyclopedia of Associations* (*http://library.dialog.com/bluesheets/html/bl0114.html*) is the quickest way to find out about professional organizations and the journals they publish in your field.

7. The human resources department of a company or agency you would like to work for. Often you will be able to fill out an application even if there is not a current opening. But do not call employers asking about openings; a visit shows a more serious interest.

8. A résumé database service. A number of online services will put your résumé in a database and make it available to prospective employers, who scan the database regularly to find suitable job candidates. Check to see if a professional society to which you belong (or might join) offers a similar service.

9. Professional employment agencies. Some agencies list jobs you can apply for free of charge (because the employer pays the fee) while others charge a stiff fee, usually a percentage of your first year's salary. Be sure to ask who pays the fee for this service. Because employment agencies often find out about jobs through channels already available to you, speak to someone at your campus placement center first.

Dossiers and Letters of Recommendation

Dossiers play a major role in the job search. A *dossier*, French for "bundle of documents," provides a file of information about you and your work—recommendations and so on—that others have supplied.

Basically, your dossier contains the following documents:

- letters of recommendation
- letters that awarded you a scholarship, gave you an academic honor, or acknowledged your community service
- letters that praised your work on the job
- your academic transcript(s)

You may ask your placement office to send your dossier to an employer, or employers may request it themselves if you have listed the placement office address on your résumé.

Obtaining Letters of Recommendation

Should You See Your Letters?

You have a legal right to see your recommendation letters, but some employers believe that if candidates see what is written about them, their references may be overly complimentary and more inclined to withhold information. Also, some of your references may refuse to write a letter of recommendation that they know you will see. On the other hand, you may feel more comfortable knowing what your recommendation letters contain. Before you make any decisions about seeing your recommendation letters, get the advice of your instructors and placement counselors.

Whom Should You Ask?

Whether your recommendation letters are confidential or not, they can sell you or sink your chances, so select your references carefully. Ask the following individuals to be your references and to write enthusiastically about your work qualifications and skills:

- previous employers (even for summer jobs or internships) who commended you and your work
- two or three of your professors who know and like your work, have graded your papers, or have supervised you in fieldwork or laboratory activities
- supervisors who evaluated and praised your work in the military
- community leaders or officials with whom you have worked successfully on civic projects

Recommendations from such individuals will be regarded as more objective—and more relevant—than letters from friends, neighbors, or members of the clergy.

Whoever you do ask, make sure he or she is a strong supporter of yours, someone who has sincerely and consistently complimented your work and encouraged you in your career. Stay clear of individuals who are lukewarm about your work or who might be reluctant to recommend you. Figure 5.2 shows a letter requesting a letter of recommendation.

Tadeus Majeski • 5432 South Kenneth Avenue • Chicago, IL 60651

312-555-7733   **tmajeski@gatenet.com**

March 30, 2012

Mr. Sonny Butler, Manager
Empire Supermarket
4000 West 79th Street
Chicago, IL 606052-4300

Dear Mr. Butler:

I was employed at your store from September 2010 through August 2011. During my employment, I worked part time as a stock clerk and relief cashier, and during the summer months I was a full-time employee in the produce department, helping to fill in while Bill Dirksen and Vivian Ho were away on their vacations.

In enjoyed my work at Empire, and I learned a great deal about the latest inventory tracking systems, ordering stock, calculating and helping to prevent merchandise shrinkage, and assisting customers.

This May, I will receive my A.A. degree from Moraine Valley Community College in retail merchandising and I have already begun preparing for my job search for a position in retail sales. Would you be willing to write a letter of recommendation for me describing what you regard as my greatest strengths as one of you employees? Having your endorsement would be a great help to me.

To assist you, I can send you a letter of recommendation form from the Placement Office at Moraine Valley. Your letter would then become part of my permanent placement file.

I look forward to hearing from you. I thought you might like to see the enclosed résumé, which shows what I have been doing since I left Empire. Thank you for the opportunity to work at your supermarket.

Sincerely yours,

Tadeus Majeski

Tadeus Majeski

Encl.: Résumé

Reviews employment history

Emphasizes skills learned on the job

Diplomatically requests strong letter of recommendation

Explain how letter will be used

Encloses copy of résumé

Always Ask for Permission

Always ask for permission before you list an individual as a reference, as in Figure 5.2. You could jeopardize your chances for a job if a prospective employer called one of your references and that person did not even know you were looking for a job or, worse yet, reveals that you did not have the courtesy to ask to use his or her name.

Should You Ask Your Current Boss?

Asking your current boss, on the other hand, can be tricky. If your present employer is already aware that you are looking for work elsewhere (for instance, if your job is temporary or if your contract is about to run out) or you are working at a part-time job, by all means ask for a letter of recommendation. However, if you are employed full time and are looking for professional advancement or for a better salary elsewhere, you may not want your current employer to know that you are searching for another job. You may want to speak to a job counselor.

Preparing a Résumé

The **résumé,** sometimes called a *curriculum vitae,* may be the most important document you prepare for your job search. It merits doing some careful homework. A résumé is not your life history or your emotional autobiography, nor is it a transcript of your college work. It is a factual and concise summary of your qualifications, showing a prospective employer that you have the education and experience to do the job you are applying for. Regard your résumé as a persuasive ad for your professional qualifications. It is a billboard advertising you.

What you include—your key details, the wording, the ordering of information, and formatting—are all vital to your campaign to sell yourself and land the interview. Employers want to see the most crucial and current details about your qualifications quickly. Accordingly, keep your print résumé short (preferably one page, never longer than two) and hard hitting. The same thing goes for your online résumé. See page 175 for an online résumé. Everything on your résumé needs to convince an employer you have the exact skills and background he or she is looking for.

What Employers Like to See in a Résumé

Prospective employers will judge you and your work by your résumé; it is their first view of you and your qualifications. They will expect to see the following eight characteristics in an applicant's résumé.

- **Honest.** Be truthful about your qualifications—your education, experience, and skills. Distorting, exaggerating, or falsifying information about yourself in your résumé is unethical and could cost you the job. If you were a clerical assistant to an attorney, don't describe yourself as a paralegal. Employers demand trustworthiness.
- **Attractive.** The document should be pleasing to the eye with appropriate spacing, typeface, and use of boldface; it shows you have a sense of proportion

and document design and that you are visually smart. The print should be clear and dark, not faded, and on high-quality paper. Do not use gimmicks like clip art or excessive capitalization. See pages 198–207 on document design.

- **Carefully organized.** The orderly arrangement of information is easy to follow, logical, and consistent; it shows you have the ability to process information and to summarize. Use plenty of white space and bullets to highlight margin points. Employers prize analytical thinking.
- **Concise.** Make sure your résumé is to the point. Generally, keep your résumé to one page, as in Figure 5.3 (page 164). However, depending on your education or job experience, you may want to include a second page. Résumés are written in short sentences that omit "I" and that use action-packed verbs, such as those listed in Table 5.2.
- **Accurate.** Make sure your grammar, spelling, dates, names, titles, and programs are correct; typos, inconsistencies, and math errors say you didn't check your facts and figures.
- **Current.** All information is up-to-date and documented, with no gaps or sketchy areas about previous jobs or education. Missing or incorrect dates or leaving key information out are flags to employers to reject your résumé.

TABLE 5.2   Action Verbs to Use in Your Résumé

accommodated	customized	instituted	reviewed
accomplished	dealt in	instructed	saved
achieved	delivered	interned	scheduled
adapted	designed	interpreted	searched
adjusted	determined	launched	secured
administered	developed	maintained	selected
analyzed	directed	managed	served
arranged	drafted	mapped	settled
assembled	earned	monitored	sold
assisted	economized	motivated	solved
attended	edited	navigated	streamlined
awarded	elected	negotiated	supervised
bridged	established	operated	surveyed
budgeted	estimated	organized	taught
built	evaluated	oversaw	teamed up
calculated	expended	performed	tested
chaired	expedited	planned	tracked
coached	figured	prepared	trained
collaborated	fulfilled	programmed	translated
collected	generated	provided	tutored
communicated	guided	purchased	updated
compiled	handled	ranked	upgraded
completed	headed	reappraised	verified
composed	implemented	received	volunteered
computed	improved	reconciled	weighed
conducted	increased	reduced	wired
converted	informed	re-evaluated	won
coordinated	initiated	reported	worked
created	installed	researched	wrote

- **Relevant.** The information on your resume is appropriate for the job level. It must show that you have the necessary education and experience, and confirm that you can be an effective team player.
- **Quantification.** Include specific data regarding how much revenue you generated for an employer (or how much money you saved, or how many times you did a complex job).

Your goal is to prepare a résumé that shows the employer you possess the sought-after job skills. One that is unattractive, difficult to follow, poorly written, filled with typos and spelling mistakes or that is sketchy, boastful, or not relevant for the prospective employer's needs will not make the first cut.

It might be to your advantage to prepare several versions of your résumé and then adapt each one you send out to the specific job skills a prospective employer is looking for. It pays to customize your résumé. Following the process detailed in the next section will help you prepare any résumé.

The Process of Writing Your Résumé

To write an effective résumé, ask yourself the following important questions:

1. What classes did you excel in?
2. What papers, reports, surveys, or presentations earned you your highest grades?
3. What computer skills have you mastered—languages, software knowledge, designing or writing for Internet resources? Success in e-commerce? Skill at designing a blog and/or website? Understanding collaborative editing online?
4. What other technical skills have you acquired?
5. What jobs have you had? For how long and where? What were your primary duties? Did you supervise other employees?
6. How did you open or expand a business market? Increase a customer base?
7. What did you do to earn a raise or a promotion in a previous or current job?
8. Do you work well with people? What skills do you possess as a member of a team working toward a common job goal (e.g., finishing a report)?
9. Can you organize complicated tasks or identify and solve problems quickly?
10. Have you had experiences/responsibilities managing money—collecting fees or receipts, preparing payrolls, conducting nightly audits, and so on?
11. Have you won any awards or scholarships or received a commendation, and/or other recognition at work?

Pay special attention to your four or five most significant, job-worthy strengths, and work especially hard on listing them concisely and persuasively.

Although not everything you have done relates directly to a particular job, indicate how your achievements are relevant to the employer's overall needs. For example, supervising staff in a convenience store points to your ability to perform the same duties in another business context.

Balancing Education and Experience

If you have years of experience, don't flood your prospective employer with too many details. You cannot possibly include every detail of your job(s) for the last 10 or 20 years.

- Emphasize only those skills and positions most likely to earn you the job.
- Eliminate your earliest jobs that do not relate to your present employment search.
- Combine and condense skills acquired over many years and through many jobs.
- Include relevant military schools or service.

Figure 5.6 shows the résumé of an individual who has a great deal of experience to offer prospective employer.

Many job candidates who have spent most of their lives in school are faced with the other extreme: not having much job experience to list. The worst thing to do is to write "None" for experience. Any part-time, summer, or other seasonal jobs, as well as work in a library or laboratory, apprenticeship or internship, show an employer that you are responsible and knowledgeable about the obligations of being an employee. Figure 5.3 contains a résumé from Anthony Jones, a student with little job experience; Figure 5.4 shows the résumé of María Lopéz, a student with a few years of experience.

What to Exclude from a Résumé

Knowing what to exclude from a résumé is as important as knowing what to include. Because federal employment laws prohibit discrimination on the basis of age, sex, race, national origin, religion, marital status, or disability, do not include such information on your résumé. Here are some other details best left off your résumé:

- salary demands, expectations, or ranges
- preferences for work schedules, days off, or overtime
- comments about fringe benefits
- travel restrictions
- reasons for leaving your previous job
- your photograph (unless you are applying for a modeling or acting job)
- your Social Security number
- comments about your family, spouse, or children
- height, weight, or hair/eye color
- race, sexual orientation, religious and political affiliations
- hobbies, interests (unless relevant to the job you are seeking)

Save comments about salary and schedules for your interview. (See pages 186–190.) The résumé should be written to earn you an interview.

Parts of a Résumé

As with memos, letters, and reports, résumés consist of specific parts listed in boldface. These parts—contact information, career objective, credentials (education and experience), related skills and achievements, and references/portfolios—need to be included in any résumé.

Figure 5.3	**Résumé from a student with little job experience.**

Headlines major achievements and gives contact details

WEBSITE DEVELOPER
DESIGNER
GRAPHIC ARTIST

Anthony H. Jones
73 Allenwood Boulevard
Santa Rosa, California 95401-1074
707-555-6390
ajdesigner@plat.com
www.plat.com/users/ajones/resume.html

CAREER OBJECTIVE

Offers precise, convincing objective

Full-time position as a layout artist with a commercial publishing house using my training in state-of-the-art design technology.

EDUCATION

Starts with most important qualification— education

Santa Rosa Junior College, 2011–2012, A.S. degree to be awarded in 2012

    Dean's List in 2008; GPA 3.25

    Major: Commercial Graphics Illustration, with specialty in design layout

    Related courses included:

- Digital Photography
- Graphics Programs: Illustrator, Photoshop
- Desktop Publishing: QuarkXPress, Adobe InDesign

Stresses job-related activities of apprenticeship

Internship, 2011–2012, McAdam Publishers

    Major projects included:

- Assisting layout editors with page composition and photo archiving.
- Writing detailed reports on digital photography, designs, and artwork used in *Living in Sonoma County* (www.sonomacounty.com) and *Real Estate in Sonoma County* (www.resc.net) magazines.

EXPERIENCE

Includes part-time work experience

Salesperson (part-time), 2009–2011, Buchman's Department Store

    Duties included assisting customers in sporting goods and appliance departments, designing custom window displays each month for the main entrance, and assisting with the inventory database management.

COMPUTER SKILLS

Demonstrates skills in Web design

Know QuarkXPress, InDesign, Illustator, Photoshop, FinalCut Pro.

RELATED ACTIVITIES

Volunteer; designed website and three-fold brochure for the Santa Rosa Humane Society's 2012 fund drive; helped raise $5,600.

REFERENCES/WEBFOLIO

Credentials and portfolio emphasize accomplishments

References, college transcripts, and a webfolio of website designs, photographs, and graphics available upon request.

Contact Information

At the top of the page, provide your name (do not use a nickname), address including your ZIP code, telephone number (use a cell phone number since you always have it with you), and e-mail address. This contact information can either be centered on the page or flush left or right. Avoid unprofessional e-mail addresses such as *toughguy@netfield.com* or *barbiegirl@techscape.com*. Also include a URL for your website and a fax number if you have these for an employer to contact you.

Career Objective

Although some employment counselors advice against using a career objective, or career profile, an objective can help a prospective employer know at once about the specific job you are looking for and in what ways you are qualified. But make sure your career objective is directly related to the position for which you are applying. Phrase it to sell yourself. Put the emphasis on what you can do for your employer, not what your employer can do for you.

To write an effective career objective statement, ask yourself four basic questions:

1. What kind of job am I searching for?
2. What kind of job am I qualified for?
3. What capabilities do I possess that match what a potential employer wants?

Avoid trite, vague, or self-centered goals such as "Looking for professional advancement," "Want to join" a progressive company," or "Seeking high-paying job that brings personal satisfaction." Compare the vague objectives on the left with the more precise ones on the right.

Unfocused	Focused
Job in sales to use my aggressive skills in expanding markets.	Regional sales representative using my proven skills in e-commerce and communication to develop and expand a customer base.
Full-time position as staff nurse.	Full-time position as staff nurse on cardiac step-down unit to offer excellent primary care nursing and patient/family teaching.
Position in cable industry.	Working as part of a service team to provide efficient cable repair service

A career objective can help you when you are applying for a specific job opening. You might want to omit it, though if you are sending the same résumés to many different potential employers.

Credentials

The order of the next two categories—**Education** and **Experience**—can vary. Generally, if you have lots of work experience, list it first. However, if you are a recent graduate short on job experience, list education first, as Anthony Jones did (see Figure 5.3). María Lopéz (Figure 5.4) also decided to place her education before her job experience because the job she was applying for required the formal training she most recently received at Miami-Dade Community College.

Education

Begin with your most recent education first, then list everything significant since high school. Give the name(s) of the school(s); dates attended; and degree, diploma, or certificate earned. Don't overlook relevant military experience or major training programs (EMT, court reporter), institutes, internships, or workshops you have completed.

Remember, however, that a résumé is not a transcript. Simply listing a series of courses will not set you apart from hundreds of other applicants taking similar courses across the country. Avoid vague titles such as Science 203 or Nursing IV. Instead, concentrate on describing the kinds of skills you learned.

> 30 hours in planning and development courses specializing in transportation, land use, and community facilities; 12 hours in field methods of gathering, interpreting, and describing survey data in reports.

> Completed 28 hours in major courses in business marketing, management, and materials in addition to 12 hours in information science, including HTML/Web publishing.

List your grade point average (GPA) only if it is 3.0 or above; otherwise, indicate your GPA in just your major or during your last term, again if it is above 3.0.

Experience

Your job history is the key category for many employers. It shows them that you have held jobs before and that you are responsible. Here are some guidelines about listing your experience.

1. Begin with your most recent position and work backward—in reverse chronological order. List the company or agency name, location (city and state), your job title, and dates of employment. Do not mention why you left a job.

2. For each job or activity, provide short descriptions (one or two lines) of your duties and achievements. If you were a work-study student, don't say that you helped an instructor. Emphasize your responsibilities; for example, you helped to set up a chemistry laboratory, ordering supplies and keeping an inventory of them. Rather than saying you were a secretary, indicate that you wrote business letters and used various software programs, maintained records, designed a company website, prepared schedules for part-time help in an office of twenty-five people, or assisted the manager in preparing accounts.

3. In describing your position(s), emphasize any responsibilities that involved handling money (for example, assisting customers, filing insurance claims, or preparing payrolls); managing other employees; working with customer accounts, services, and programs; or writing letters and reports. Prospective employers are interested in your leadership abilities, teamwork, financial shrewdness (especially if you

Résumé from a student with some job experience. Figure 5.4

María Lopéz

1725 Brooke Street   Miami, Florida 32701-2121   (305) 555-3429   mlopez@eagle.com

Provides easy-to-find contact information

CAREER OBJECTIVE

Position assisting dentist in providing dental care, counseling, and preventive dental treatments, especially in pedodontics.

EDUCATION

A.S. in Dental Hygiene, Miami-Dade Community College
August 2010–May 2012.
GPA: 3.38 (Ranked in the top ten percent)

Major courses:

• Oral pathology   • Dental materials and specialties
• Periodontics   • Community dental health.

Minored: Psychology (twelve hours in child and adolescent psychology)
• Experienced with procedures and instruments used with oral prophylaxis techniques.
• Subject of major project was proper nutrition and dental health for preschoolers.
• Received excellent evaluations in business writing course.
• Will take American Dental Assisting National Board Exams for licensure on June 2.

Lists course and clinical work required for licensure and job

Calls attention to counseling and bilingual skills that benefit employer and patients

Emphasizes professional licensure qualifications

EXPERIENCE

St. Francis Hospital
(Miami Beach, Florida) April 2008–July 2010
Unit secretary on pediatric unit. Maintained medical supply levels using inventory tracking software, assisted with the computerization of all medical records into hospital-wide database, transcribed medical orders and surgical notes, greeted and assisted visitors.

Murphy Construction Company
(Miami, Florida) June 2007–April 2008
Office assistant-receptionist. Did keyboarding, filing, and assisted with billing and creating project schedules in a small office (5 employees).

City of Hialeh, Florida
Summers 2005–2006
Lifeguard. Established safety procedures and tested pool chlorine levels.

Highlights previous job responsibilities in health care setting

COMPUTER SKILLS

DentiMax, Microsoft Office, Excel, FileMaker Pro.

LANGUAGE SKILLS

Fluent in Spanish

Figure 5.4	(Continued)

Lopéz 2

Specifies how employer can obtain further information

REFERENCES

The following individuals have written letters of recommendation for my dossier available from the Placement Center, Miami-Dade Community College, Medical Center Campus, Miami, FL 33127-2225.

Wisely lists professionals well known to prospective/ area employers; received permission to list names on résumé

Sister Mary Pela, R.N. Pediatric Unit St. Francis Hospital 10003 Collins Avenue Miami Beach, FL 33141 (305) 555-5113	Professor Mitchell Pelbourne Department of Dental Hygiene Miami-Dade Community College Medical Center Campus Miami, FL 33127 (305) 555-3872
Tia Gutiérrez, D.D.S. 9800 Exchange Avenue Miami, FL 33167 (305) 555-1039	Mr. Jack Murphy 1203 Francis Street Miami, FL 33157 (305) 555-6767

saved your company money), tact in dealing with the public, and communications skills. They will also be favorably impressed by commendations ("Employee of the Month") and promotions you may have earned.

4. Include any relevant volunteer work you have done, as Anthony Jones did for an animal shelter in Fig. 5.3. Note, too, how Dora Cooper Bolger's volunteer work translates into marketing skills an employer wants to see in a prospective employee's résumé in Figure 5.6 (see page 171).

5. If you have been a full-time parent for 10 years or a caregiver for a family member or friend, indicate the management skills you developed while running a household and any community or civic service you did, as Dora Cooper Bolger lists on her résumé. She skillfully relates her family and community accomplishments to the specific job she seeks.

Related Skills and Achievements
Not every résumé will have this section, but the following are all employer-friendly things to include:

- second or third languages you speak or write
- extensive travel
- certificates or licenses you hold
- memberships in professional associations (e.g., American Society of Safety Engineers, Black Student Association, National Hispanic Business Association, Texas Women Executives, Child Development Organization)

- volunteer work for community service groups (e.g., Habitat for Humanity, Salvation Army); list any offices you held—recorder, secretary, fund drive chairperson

Computer Skills

Knowledge of computer hardware, software, word processing programs, and Web design and search engines is extremely valuable in the job market. Note how Anthony Jones, María Lopéz, and Dora Bolger inform prospective employers about their relevant technical competencies in Figures 5.3, 5.4 and 5.6.

Honors/Awards

List any civic (mayor's award, community service, cultural harmony award) and/or academic honors you have won (dean's list, department awards, scholarships, grants, honorable mentions). Memberships in honor societies in your major and technical/business associations also demonstrate that you are professionally accomplished and active.

References

As a rule, do not list references with personal contact information. Simply say they are available on request. The only exception would be when you are applying for a specific job as María Lopéz did in Figure 5.4, where your references are well known in the community or belong to the same profession in which you are seeking employment—you profit from your association with a recognizable name or title. Here is where networking can help you. Ask your instructors, previous employer, or individuals who have supervised your work. Always give them a copy of your current résumé, as the job seeker did in Figure 5.2.

Career Portfolios/Webfolios

When preparing your résumé, you may want to indicate if a career portfolio (or webfolio if it is online) of your work is available for review, as we saw Anthony Jones do in Fig. 5.3. Your portfolio/webfolio needs to represent your best work. A career portfolio/webfolio is not a personal/family scrapbook but a record of your most professional work.

What to Include in a Career Portfolio/Webfolio

What you include in your career portfolio and how you organize it says as much about your qualifications for the job as the documents themselves. It is never too early to start assembling relevant documents for your portfolio.

Following is a sampling of the kinds of documents you might include in your career portfolio/webfolio, as illustrated in Figure 5.5:

- a mission statement (two or three paragraphs) that outlines your career goals and work skills
- an additional copy or e-version of your résumé

Figure 5.5   **A sample webfolio with annotated Artifacts/Documents page open.**

| Mission Statement | Résumé | Unofficial College Transcript | Artifacts/Documents | References |

Sample Business Documents

To demonstrate my clear and concise writing style, my ability to design a variety of complex workplace documents, and my understanding of international communication, I have provided links to various business documents I created. None of these contains proprietary or confidential information.

VIEW SAMPLE SALES PROPOSAL ON SAVING ENERGY
VIEW SAMPLE RESEARCH REPORT (TABLE OF CONTENTS, ABSTRACT, FIRST 3 PAGES)
VIEW SAMPLE SALES LETTER TO INTERNATIONAL CLIENT
VIEW SAMPLE SURVEY QUESTIONNAIRE AND ANALYSIS

Company Newsletter Article

At the end of 2011, I was named "Employee of the Year" at my firm's Cincinnati office. An article about the award appeared in the company newsletter, outlining the selection criteria and describing the job accomplishments that earned me this award.

VIEW NEWSLETTER ARTICLE

PowerPoint Presentations

I understand that the position advertised will entail frequent oral presentations at company meetings. My colleagues and supervisors have praised my abilities as a speaker who can simplify complex ideas for audiences and zero in on the main ideas quickly. To give you a sense of my ability to organize information and provide visual support for my ideas, I attach a sample PowerPoint presentation that accompanied one of my talks.

VIEW POWERPOINT PRESENTATION

Website

To illustrate both my design skills and my keen understanding of the Web environment, I have provided several screen captures from a company website I created. The construction of the site was handled by my company's Web group, supervised by me.

VIEW WEBSITE SCREEN CAPTURES

- copies or scans of diplomas, certificates, licenses, internships; unofficial transcript
- copies of awards (academic and job-related), promotion letters, or commendations (e.g., for protecting the environment)
- impressive examples of written work you did for college courses such as reports or proposals (include any positive comments provided by instructors)
- newspaper or newsletter stories about your academic, community, and/or on-the-job successes

Dora Cooper Bolger's résumé organized by skill areas.   Figure 5.6

DORA COOPER BOLGER
1215 Lakeview Avenue
Westhampton, MI 46532
Cell: 616-555-4773 • dcbplanner@aol.com

OBJECTIVE Seek full-time position as public affairs officer to promote the goals of a health care, educational, or charitable organization

Restricted objective

SKILLS

Organizational Communication
- Delivered 20 presentations to civic groups on educational issues
- Recorded minutes and helped formulate agenda as president of large, local PTA (800 members) for past $6\frac{1}{2}$ years
- Possess excellent computer skills in PeopleSoft, Microsoft Word, and BusinessPax
- Updated and maintained computerized mailing lists for Teens in Trouble and Foster Parents' Association

Aptly features skill areas before education

Financial
- Spearheaded 3 major fund-raising drives (total of $200,000 collected)
- Prepared and implemented large family budget (3 children, 8 foster children)
- Budget planner, Foster Parents' Association
- Served as financial secretary, Faith Methodist Church, for 4 years

Links achievements from volunteer and home-based activities most important to employer

Administrative
- Organized volunteers for American Kidney Fund (last 5 years)
- Established and oversaw neighborhood carpool (17 drivers; more than 60 children) for 7 years
- Coordinated after-school tutoring program for Teens in Trouble; President since 1998
- Vice-president, Foster Parents' Association, 2010

Chooses strong, active verbs to convey image of a results-oriented professional

HONORS   "Volunteer of the Year"(2011), Michigan Child Placement Agency

EDUCATION   Metropolitan Community College, A.A., 2008
Mid-Michigan College, B.S., expected 2012
Major: Public Administration; Minor: Psychology
GPA: 3.45 | Dean's List: 2010–2012

Places education after experience; includes major and related minor plus strong GPA

WORK EXPERIENCE *Secretary*, 1995–2007 (full- and part-time):
Merrymount Plastics; Foley and Wasson;
Westhampton Health Dept.; G & K Electric

Excludes details about least recent jobs

REFERENCES   Furnished on request

- pertinent examples of media presentations or other graphic work you have done, PowerPoint presentations you have created, a CD of a website you have designed, layout work you completed, career-related photographs you have taken
- a list of your references with contact information

What Not to Include in a Career Portfolio/Webfolio

Be highly selective about what you include. Never include anything that would contradict or call into question information in your résumé or letter of application. Exclude the following types of documents from your career portfolio/webfolio.

- documents or scans of documents that show your memberships in clubs, fraternities/sororities, sports teams, and so on, unless directly relevant to the job (e.g., applying for a job at the national office for Sigma Sigma Kappa or with the Professional Golf Association)
- links to or printouts from personal Web pages, including *Facebook.com* or *MySpace.com* pages (these may contain inappropriate personal information and/or portray you as unprofessional—see page 154)
- pictures of your family, friends, pets, and the like
- scans of newspaper or newsletter stories about you that are not directly related to your job search, such as your wining a cruise or other prize or playing on a bowling team

When you provide a prospective employer with your career portfolio, you can either mail it as hard copy or send it electronically as a webfolio. If you submit a hard-copy portfolio, always make high-quality copies of each document. Also, never include original documents. If you submit a webfolio, make sure that all of your scans are clean and clear. In addition, you can hyperlink parts of your career portfolio to your résumé, making it easy for a job recruiter or prospective employer to find evidence of your qualifications.

Whether you submit your career portfolio in hard copy or electronically, provide short annotations explaining what each document is and why it is important, as shown in Figure 5.5.

Organizing Your Résumé

There are two primary ways to organize your résumé: chronologically or by function or skill area.

Chronologically

The résumés in Figures 5.3 and 5.4 are organized chronologically. Information about the job applicants is listed year by year under two main categories—education and experience. This is the traditional way to organize a résumé. It is straightforward and easy to read, and employers find it acceptable. The chronological sequence works especially well when you can show a clear continuity toward progress in your career through your job(s) and in school work or when you want to apply for a similar job with another company.

A chronological résumé is appropriate for students who want to emphasize recent educational achievements or when there are no major gaps in your career.

By Function or Skill Area

Depending on your experiences and accomplishments, you might organize your résumé according to function or skill area. According to this plan, you would *not* list your information chronologically in the categories "Experience" and "Education." Instead, you would sort your achievements and abilities—whether from course work, jobs, extracurricular activities, or technical skills—into two to four key skill areas, such as:

- Sales
- Public Relations
- Training
- Management
- Research
- Technical Capabilities
- Counseling
- Group Leadership
- Communications
- Network Operations
- Customer Service
- People Skills
- Teamwork
- Troubleshooting
- Opening New Markets
- Multicultural Experiences
- Computer Skills
- Problem-Solving Skills

Under each area you would list three to five points illustrating your achievements in that area. Skills or functional résumés are often called **bullet résumés** because they itemize the candidate's main strengths in bulleted lists. Some employers prefer the bullet résumé because they can skim the candidate's list of qualifications in a few seconds.

Note Dora Cooper Bolger's profitable use of a functional résumé format in Figure 5.6. She delayed school for several years because of family commitments, yet she uses the experiences she acquired during those years to her advantage in her résumé organized by "Skills." No gap of 10 years interrupts her valuable marketable skills.

Who Should Use a Functional Résumé?

The following individuals would probably benefit from organizing their résumés by function instead of chronologically:

- nontraditional students who have diverse job experiences
- individuals who are changing their profession because of downsizing or seeking new professional opportunities
- individuals who have changed jobs frequently over the last 5 to 10 years
- ex-military personnel re-entering the civilian marketplace upon conclusion of their military service

You might want to prepare two different versions of your résumé—one functional and one chronological—to see which sells your talents better. Don't hesitate to seek the advice of your instructor or placement counselor to see which one will work best for you.

The Online Résumé

In addition to preparing a hard copy of your résumé, expect to create one (or more) to be posted online. An online résumé will give you the widest possible exposure to attract prospective employers.

The online résumé can be delivered in several ways. You can post it on database services such as those in Table 5.1, or you can send it to databases sponsored by professional societies. Expect to send your online résumé directly to an employer's website as well where it can be scanned, indexed, and even stored. Or you can simply send your résumé to a prospective employer via e-mail. You can also post your résumé on your own website. But job counselors caution that posting your résumé on your website may entail providing more information than the prospective employer needs, or giving out too much personal information.

Prospective employers may receive hundreds of résumés in response to a single job posting. Because of such volume, employers will often scan résumés in 30 to 40 seconds to make the first cut. You need to prepare your online résumé so that it captures the attention of the maximum number of employers. To accomplish this, design your online résumé so that it can be read clearly and quickly and so that it contains keywords that match your prospective employer's needs.

While an online résumé contains basically the same information as the print résumé, as discussed on pages 163–169, the design of an online résumé is very different. Don't think that you can just paste your hard-copy résumé into an e-mail or send it to a database or employer's Web portal. You will first have to format it in electronically scannable form. Otherwise, your résumé may be blocked or garbled. The following sections will show you how Anthony Jones's hard-copy résumé in Figure 5.3 and Dora Cooper Bolger's hard-copy résumé in Figure 5.6 have been changed into the online résumés in Figures 5.7 and 5.8, respectively.

Formatting an Online Résumé

Here are some guidelines to help you make sure your résumé is scannable:

1. Create your résumé as a MS Word file (the most common file format in the world of work) rather than as a PDF, WordPerfect, or zip file.
2. Save your résumé properly and consistently—as an ASCII plain text file. Follow the directions given by a database service or the prospective employer.
3. Avoid underlining, boldface, italics, boxes, shadowing, or color, which can garble the text of your résumé and make it unscannable. Use all caps instead of

Anthony H. Jones
ajdesigner@plat.com
Phone: (707) 555-6390

KEYWORDS

Web designer, computer graphics, Illustrator, Photoshop, QuarkXPress, InDesign, fundraiser, budgets, sales, virus protection, team player

OBJECTIVE

Position as layout artist with small commercial publisher using my training in state-of-the-art technology.

EDUCATION

Santa Rosa Junior College, A.S. degree to be awarded in June 2012. Commercial Graphics Illustration major. GPA 3.45

COMPUTER SKILLS

Excellent knowledge of computer graphics and design software: Quark XPress, Adobe InDesign, Illustrator, Photoshop, and FinalCut Pro.

EXPERIENCE

* Intern in layout and design department. Preparing page composition, importing visuals, manipulating images, McAdam Publishers, 8 Parkway Heights, Santa Rosa, CA 94211.

* Salesperson; display merchandise coordinator, Buchman's Department Store, Greenview Mall, Santa Rosa.

* Volunteer: designed website, brochures, and other artwork for successful fund drive, Santa Rosa Humane Society.

* Web designer, graphic artist, proofreader, student magazine.

REFERENCES
Available on request

WEBFOLIO
www.plat.com/users/ajones/resume.html

All lines aligned flush with the left margin

Keywords (for search engine readiness) go at the top of the online résumé

Objective appears in the body of the online résumé

All caps rather than bold, italic, fancy fonts used to highlight categories

Uses terminology appropriate for position

Choose nouns rather than action verbs to increase employer matches

Asterisks rather than bullets mark beginning of lines

Webfolio is hyperlinked (see #10 on page 176)

bold or italics for emphasis, and an asterisk (*) or a plus sign (+) in place of a bullet at the beginning of a line, as in Figures 5.7 and 5.8.

4. Do not use hard-to-read or fancy fonts like script. Instead, choose a font like Tahoma, Arial, or Courier that is clear and does not mask letters.

5. Make sure your résumé is easy to scroll. Keep in mind that reading a résumé on a screen is different from seeing it on a printed page. Use plenty of white space between sections so that headings are clear and distinct from one another. Use at least 10 or 12 point. Put no more than 65 characters on a line to make your résumé easy to read.

6. Design your résumé to be concise. Prospective employers will not have the time or patience to scroll through multiple screens to find information. The equivalent of one printed page is still best. Don't use a "page 2" or "continued."

7. Select the automatic wrapping feature to avoid losing characters at the end of a line, but make sure you do not exceed the limit of 65 characters per line.

8. Align everything flush with the left margin. Do not indent anything or use tabs.

9. Never send your résumé as an e-mail attachment. A prospective employer will not open it because of the risk of contracting a virus. Instead, copy and paste it into the body of your e-mail message.

10. Include hyperlinks to your website or one you helped design reviews of your work or professional blogs (see pages 88–91) you wrote.

11. Format and save your résumé before you send it to any database service.

12. Preview your résumé for quality control.

Making Your Online Résumé Search-Engine Ready

The most important section of your online résumé contains the keywords you use. Prospective employers scan résumés to find the keywords they most want to see in the job seeker's description of his or her experience, education, and interpersonal skills. The more matches, or hits, they find between appropriate keywords on your résumé and those on their list, the better your chances are of being interviewed. List keywords at the top of your résumé, as in Figures 5.7 and 5.8, and also insert them throughout your résumé in appropriate places. Keywords should highlight your technical expertise, training and education, knowledge of a field, leadership ability, team work, writing/speaking skills, sales experience, etc.

Here are a few tips to help you select appropriate keywords:

1. Provide a keyword section at the top of your résumé to give employers an immediate snap shot of your skills.

2. Include the descriptive keywords found in the employer's ad and website to increase your chances of landing an interview.

3. Do not be afraid of using shoptalk (or jargon) of your profession. An employer will expect you to be aware of current terminology.

Dora Cooper Bolger
P.O. Box 3216
Westhampton, MI 46532
dcbplanner@aol.com

Avoids giving personal information

KEYWORDS
Activity planner, budget coordinator, fund-raising, community service campaign, child advocacy, grant writer, public affairs, public speaking, interpersonal communication, management internal and external groups, team builder, strategic planner

Uses keywords from job description

OBJECTIVE
Position as public affairs officer for health care, educational, or charitable organization

RELEVANT EXPERIENCE
Presenter at 20 civic group functions
Fund raiser for 3 major campaigns over $200,000 collected
President and coordinator, Teens in Trouble, a tutoring and mentoring program
Budget planner, Foster Parents Association; large urban church

Uses nouns to list accomplishments

VOLUNTEER WORK/AWARDS
Volunteer of the Year, Michigan Child Placement Agency, 2011
Volunteer Coordinator/Leader, American Kidney Fund
President and Secretary, local PTA, 2003–2010
Vice-president, Foster Parents Association, 2010

Summarizes experience carefully

EDUCATION
Metropolitan Community College, A.A., 2008
Mid-Michigan College, B.S., 2012 Major: Public Administration
Minor: Psychology
GPA: 3.45. Dean's List: 2010–2012

Avoids any symbols, boldfacing, or italics that could garble text

COMPUTER SKILLS
Microsoft Word; Peoplesoft; PowerPoint; BusinessPax

REFERENCES
Available on request

4. Use keywords to connect sections or categories of your résumé. Keyword headers will help you emphasize your job strengths and make it easy for employers to scroll back to an appropriate section of your résumé.
5. Replace the action verbs found in conventional résumés (on the left in the following list) with keyword nouns on the right. Here are some examples:

Conventional Résumé	Online Résumé
Wrote technical report	*Technical writer*
Performed laboratory tests	*Laboratory technician*
Solved consumer complaints	*Consumer advocate*
Responsible for managing accounts	*Accounts manager*
Won three awards	*Award winner*
Edited company newsletter	*Newsletter editor*
Solved software problem	*Software specialist*

Cyber-Safing Your Résumé

Whether you use a database service or post your résumé on a website, protect your identity and your current and future jobs. Be careful about revealing personal information.

- Post your résumé only on legitimate sites. Avoid those that say they will flood the market. You don't know where your résumé will end up.
- Do not put personal information in your résumé—home address, Social Security number, birthday, health status, or photograph.
- Use an anonymous e-mail address rather than your personal one. Consider a generic e-mail address that includes a word or phrase that identifies your area of expertise, for example, *JosieProgrammer@aol.com*. Or include an e-mail link with a built-in "mailto" on your résumé.
- Never put the names of your references or their addresses online. Simply say, "References available on request."
- Block certain readers from searching your résumé, such as your current employer or firms that you know send out spam.
- Never use your present employer's company name or business e-mail address.
- The online résumé is *not* the place to allow readers access to a portfolio of your work. Send your portfolio only to employers who request it.

Testing, Proofing, and Sending Your Online Résumé

1. Test your formatting. Send your résumé to a friend to be sure your file is readable and formatted correctly.
2. Proofread carefully just as you would a hard-copy résumé.
3. Don't just put *résumé* as the subject of your file to an employer. List the title, number, or code of the position for which you are applying.

4. Simply posting your résumé online is not enough. Also send a scannable hard copy and a letter of application (see letters discussed below) to prospective employers. But do not fold or staple your résumé. Put it, along with your letter, in a large envelope (8 1/2" x 11") to avoid folding/bending.
5. Always keep a log of where you have sent your résumé online.

Letters of Application

Visit www.cengage.com/english/kolin/writingatwork concise3e for an online exercise, "Locating and Applying for an International Job."

Along with your résumé, you must send your prospective employer a letter of application, one of the most important pieces of correspondence you may ever write. Its goal is to get you an interview and ultimately the job. Letters you write in applying for jobs should be *personable*, *professional*, and *persuasive*—the three P's. Knowing how the letter of application and résumé work together and how they differ can give you a better idea of how to compose your letter.

How Application Letters and Résumés Differ

The résumé is a persuasive record of dates, important achievements, skills, names, places, addresses, and jobs. As noted earlier, you may prepare several different résumés depending on your experience and the job market.

Your letter of application, however, is much more personal. It introduces you to a prospective employer. Because you must write a new, original letter to each prospective employer, you may write (or adapt) many different letters. Each letter of application should be tailored to a specific job. It should respond precisely to the kinds of qualifications the employer seeks.

The letter of application is a sales letter that emphasizes and applies the most relevant details (of education, experience, and talents) in your résumé. In short, the résumé contains the raw material that the letter of application transforms into a finished and highly marketable product—you.

Résumé Facts to Exclude from Letters of Application

The letter of application should not simply repeat the details listed in your résumé. In fact, the following details that you would include in your résumé should *not* be restated in the letter:

- personal data, including license or certificate numbers
- specific names of courses in your major
- names and addresses of all your references

Writing the Letter of Application

The letter of application can make the difference between your getting an interview and being eliminated early from the competition. It should convince a prospective employer that you will use the experience and education listed on your résumé in the job he or she is hoping to fill. You want your letter to be placed in the "definitely interview" category. Limit your letter to one page. As you prepare your letter, use the following general guidelines.

1. Follow the standard conventions of letter writing (see Chapter 4). Print your letter on good-quality, white 8½" × 11" paper. Proofread meticulously; a spelling error, typo, or grammatical mistake will make you look careless. Don't rely only on your spell-check.

2. Supply all contact information as part of your letterhead (see page 99). Include home address, phone numbers, e-mail address, and your website, if you have one.

3. Make sure your letter looks attractive. Use wide margins and don't crowd your page. Keep your paragraphs short and readable—no more than four or five sentences long (see page 101 in Chapter 4).

4. Send your letter to a specific person. Never address an application letter "To Whom It May Concern," "Dear Sir or Madam," or "Director of Human Resources." Get an individual's name by double-checking the company's website or calling the company's main office and be sure to verify the spelling of the person's name and his or her title.

5. Emphasize the "you attitude" (see Chapter 4, pages 105–109). See yourself as an employer sees you. Focus on how your qualifications meet the employer's needs, not the other way around. Employers are not impressed by vain boasts ("I am the most efficient and effective safety engineer"). Convince prospective employers that you will be a valuable addition to their organization—a team player, a problem solver, an energetic representative, a skilled professional.

6. Don't be tempted to send out your first draft. Write and rewrite your letter of application until you are convinced it presents you in the best possible light. Getting the job may depend on it. A first or even second draft rarely sells your abilities as well as a third, fourth, or even fifth revision does.

The sections that follow give you some suggestions on how to prepare the various parts of an application letter successfully.

Your Opening Paragraph

The first paragraph of your letter of application is your introduction. It must get your reader's attention by answering four questions:

1. Why are you writing?
2. Where or how did you learn of the vacancy, the company, or the job?
3. What is the specific job title for which you are applying?
4. What is your most important qualification for the job?

Begin your letter by stating directly that you are writing to apply for a job. Don't say that you "want to apply for the job"; such an opening raises the question, "Why don't you, then?"

Avoid an unconventional or arrogant opening: "Are you looking for a dynamic, young, and talented accountant?" Do not begin with a question; be more positive and professional.

If you learned about the job through a website or journal, make sure you italicize or underscore its title.

> I am applying for the food service manager position you advertised in the May 10 edition of the *Los Angeles Times* online.

Because many companies announce positions on the Internet, check there first to see if their position is listed online, as Anthony Jones did in Figure 5.9.

If you learned of the job from a professor, a friend, or an employee at the firm, state so. Take advantage of a personal (networking) contact who is confident that you are qualified for and interested in the position, as María Lopéz (Figure 5.10, page 183) and Dora Cooper Bolger (Figure 5.11, page 185) did. But first confirm that your contact gives you permission to use his or her name.

The Body of Your Letter

The body of your letter, comprising one or two paragraphs, provides the evidence from your résumé to prove you are qualified for the job. You might want to spend one paragraph on your education and one on your experience or combine your accomplishments into one paragraph.

Follow these guidelines for the body of your letter:

1. **Keep your paragraphs short and readable—four or five sentences.** Avoid long, complex sentences. Use the active voice to emphasize yourself as a doer. Review the action verbs in Table 5.2 and use keywords found in the employer's ad.
2. **Don't begin each sentence with "I."** Vary your sentence structure. Write reader-centered sentences, even those beginning with "I."
3. **Concentrate on seeing yourself as your employer sees you.** Prove that you can help an employer's sales and service, promote an organization's mission and goals, and be a reliable team player.
4. **Highlight your qualifications by citing specific accomplishments.** Tell your reader exactly how your schoolwork and job experience qualify you to perform and advance in the job advertised. Show how you can make a positive contribution to the employer's company. Don't simply say you are a great salesperson. Demonstrate your accomplishments by stressing that you increased the sales volume in your department by 15 percent within the last six months, you won an award or received a promotion for customer service, or you reduced costs by 10 percent.
5. **Mention you are enclosing your résumé.** Put an "encl." notation at the bottom of your letter.

Education

Recent graduates with little work experience will, of course, spend more time discussing their education. But, regardless, emphasize marketing why and how your most significant educational accomplishments—course work, degrees, certificates, licenses, training—are relevant for the particular job. Employers want to know which specific skills from your education translate into benefits for their company and clients.

Figure 5.9   **Letter of application from Anthony Jones, a recent graduate with little job experience.**

Clear, professional-looking letterhead

WEBSITE DEVELOPER
DESIGNER
GRAPHIC ARTIST

Anthony H. Jones
73 Allenwood Boulevard
Santa Rosa, California 95401-1074
707-555-6390
ajdesigner@plat.com
www.plat.com/users/ajones/resume.html

May 24, 2012

Writes to a specific person

Ms. Jocelyn Nogasaki
Human Resources Manager
Megalith Publishing Company
1001 Heathcliff Row
San Francisco, CA 94123-7707

Identifies position and source of ad

Dear Ms. Nogasaki:

I am applying for the layout editor position advertised on your website, which I accessed on 14 May. Early next month, I will receive an A.S. degree in commercial graphics illustration from Santa Rosa Junior College.

Validates necessary education and applies it directly to employer's business

With a special interest in the publishing industry, I have successfully completed more than 40 credit hours in courses directly related to layout design where I gained experience using QuarkXPress, Adobe InDesign, Illustrator, and Photoshop. You might like to know that many Megalith publications were used as models of design and layout in my graphics communications and digital photography courses.

Convincingly cites related job experience

My studies have also given me practical experience at McAdam Publishers as part of my Santa Rosa internship program. While working at McAdam, I was responsible for assisting the design department in page composition and archiving photos. Other related experience I have had includes creating a website for the Santa Rosa Humane Society and designing and executing custom window displays at Buchman's Department Store. As the enclosed résumé indicates, my references and webfolio are available upon request.

Refers to résumé/webfolio

Asks for an interview and gives contact information; thanks employer

I would appreciate the opportunity to discuss my qualifications in graphic design with you. My phone number, e-mail address, and website are listed above. After June 12, I will be available for an interview at any time that is convenient for you. Thank you for your consideration.

Sincerely yours,

Anthony H. Jones

Anthony H. Jones
Encl. : Résumé

Letter of application from María Lopéz, a recent graduate with some job experience.   Figure 5.10

1725 Brooke Street
Miami, Florida 32701-2121
(305) 555-3429 • mlopez@eagle.com

May 15, 2012

Dr. Marvin Henrady
Suite 34
Medical/Dental Plaza
839 Causeway Drive
Miami, FL 32706-2468

Dear Dr. Henrady:

Mr. Mitchell Pelbourne, my clinical instructor at Miami-Dade Community
College, informs me you are looking for a dental hygienist to work in your
northside office. My education and experience qualify me for that position. This
month I will graduate with an A.S. degree in the dental hygienist program, and I
will take the American Dental Assisting National Board Exams in early June.

I have successfully completed all course work and clinical programs in oral
hygiene, anatomy, and prophylaxis techniques. During my clinical training,
I received intensive practical instruction from several local dentists, including
Dr. Tia Gutiérrez. Since your northside office specializes in pedodontal care, you
might find the subject of my major project—proper nutrition and dental care for
pre-schoolers—especially relevant.

I also have related job experience in working with children in a health care
setting. For over two years, I was a unit clerk on the pediatric floor at St. Francis
Hospital, and my experience in greeting patients, transcribing medical orders
and surgical notes, and assisting the nursing staff would be valuable to you in
your office. An additional job strength I would bring to your office is my
bilingual (Spanish/English) communication skills. You will find more detailed
information about my accomplishments in the enclosed résumé.

I welcome the opportunity to talk with you about the position and my interest
in pedodontics. I am available for an interview any time after 2:00 pm until June
11, but after that date I could come to your office any time at your convenience.
Thank you for considering my application.

Sincerely yours,

María Lopéz

María Lopéz

Encl. Résumé

*Begins with
personal
contact*

*Verifies she will
have necessary
licensure*

*Links training
to job
responsibilities;
demonstrates
knowledge of
employer's
office*

*Relates
previous
experience to
employer's
needs; refers to
résumé*

*Ends with
a polite
request for an
interview and
thanks reader*

Figure 5.11   **Letter of application from Dora Cooper Bolger, job candidate with years of community and civic experience.**

Dora Cooper Bolger
1215 Lakeview Avenue   cell: 616-555-4773
Westhampton, MI 46532 dcplanner@aol.com

February 9, 2012

Dr. Lindsay Bafaloukos, Director
Tanselle Mental Health Agency
4400 West Gallagher Drive
Tanselle, MI 46932-3106

Dear Dr. Bafaloukos:

Begins with contact made at professional meeting, highlighting her qualifications

At a recent meeting of the County Services Council, a member of your staff, Homer Steen, told me that you will be hiring a public affairs coordinator. Because of my extensive experience in and commitment to community affairs, I would appreciate your considering me for this opening. I expect to receive my B.S. in Public Administration from Mid-Michigan College later this year.

Relates proven past successes to employer's needs; gives concrete examples of her skills

For the past ten years, I have organized community groups with outreach programs similar to Tanselle's. I have held administrative positions in the PTA and the Foster Parents' Association and served as president of Teens in Trouble, a volunteer group providing assistance to dysfunctional teens. My responsibilities with Teens have included coordinating counseling activities with various school programs, scheduling tutorials, and representing the organization before local and state government agencies. I have been commended for my organizational and communication skills. My twenty presentations on foster home care and Teens in Trouble also demonstrate that I am an effective speaker, a skill I could put to work for Tanselle immediately.

Encourages reader to see her as best-prepared candidate; includes résumé

Because of my work at Mid-Michigan and in Teens and Foster Parents, I have the practical experience in communication and psychology to promote Tanselle's goals. The enclosed résumé provides details of my experience, education, and awards.

Requests interview and gives contact information

Thanks reader

I would enjoy discussing my accomplishments with Teens and the other organizations I have represented with you, and how I might help Tanselle to promote its programs. I am available for an interview at your convenience. Thank you for considering my interest in and qualifications for your public affairs coordinator position.

Sincerely yours,

Dora Cooper Bolger

Dora Cooper Bolger

Encl. Résumé

Simply saying you will graduate with a degree in criminal justice does not explain how you, unlike all the other graduates of such programs, are best suited for a particular job. But when you indicate that you have completed 36 credit hours in software security and have another 12 credit hours in global business, you show exactly how and where you are qualified. Note how Anthony Jones in Figure 5.9 and María Lopéz in Figure 5.10 establish their educational qualifications with specific details about their training.

Experience

After you discuss your educational qualifications, turn to your job experience. But if your experience is your most valuable and extensive qualification for the job, put it before education and stress any previous experience that is similar to what a new position calls for. If you are switching careers or returning to a career after years away from the work force, start the body of your letter with your experience or your community and civic service, as Dora Cooper Bolger does in Figure 5.11. Her volunteer work convincingly demonstrates she has the organizational and communication skills her prospective employer seeks. Never minimize such contributions.

Relate Your Experience/Education to the Job

Link your education and experience as benefits to the particular job you apply for. Persuasively show a prospective employer how your previous accomplishments, especially teamwork and responsibility, have prepared you for future success on the job. Relate your course work in computer science to being an efficient programmer. Indicate how your summer jobs for a local park district reinforced your exemplary skills in customer service. Connect your background to the prospective employer's company. Any homework you can do about the company's history, goals, or structure (see page 186) will pay off.

- By citing Megalith publications as a model in his courses, Anthony Jones in Figure 5.9 stresses he is ready to start successfully from the first day on the job.

- Note how María Lopéz in Figure 5.10 links her major school project and her work on a hospital pediatric floor to Dr. Henrady's specialty.

- Dora Cooper Bolger in Figure 5.11 likewise shows her prospective employer that she is familiar with and can contribute to Tanselle's programs in community mental health though her volunteer work and public speaking experience.

Closing

Keep your closing paragraph short—about two or three sentences—but be sure it fulfills the following four important functions:

1. emphasizes once again briefly your major qualifications
2. asks for an interview or a phone call
3. indicates when you are available for an interview
4. thanks the reader

End gracefully and professionally. Be straightforward. Don't leave the reader with a single weak, vague sentence: "I would like to have an interview at your convenience." That does nothing to sell you. Say that you would appreciate talking with the employer further to discuss your qualifications. Then mention your chief talent. You might also express your willingness to relocate if the job requires it.

After indicating your interest in the job, give the times you are available for an interview and specifically tell the reader where you can be reached. If you are going to a professional meeting that the employer might also attend, or if you are visiting the employer's city soon, say so.

The following samples show how *not* to close your letter and why not.

Pushy:	I would like to set up an interview with you. Please phone me to arrange a convenient time. (That's the employer's prerogative, not yours.)
Too Informal:	I do not live far from your office. Let's meet for coffee sometime next week. (Say instead that since you live nearby, you will be available for an interview.)
Introduces New Subject:	I would like to discuss other qualifications you have in mind for the job. (How do you know what the interviewer might have in mind?)

Note that the closing paragraphs in Figures 5.9 through 5.11 avoid these errors.

Going to an Interview

There are various ways for a prospective employer to conduct an interview. It might be a one-on-one meeting—you and the interviewer—or you may visit with a group of individuals who are trying to decide if you would fit in. You could even have an interview over the telephone or through a videoconference. An interview can last 30 minutes or take all day.

Preparing for the Interview

Before you go to the interview be prepared by doing the following:

1. Do your homework about the company. Click on "About Us" and other links on the corporate or agency website to find out who founded the company, who the current CEO is, if it is a local firm or a subsidiary, its chief products or services, how many years it has been in business, how many employees it has, where its main office and plants are, and who its major clients and competitors are. Read company brochures, catalogs and blogs to get a sense of the corporate culture. Look for recent stories about the company on their website and in leading business publications, such as:

- *Wall Street Journal:* http://online.wsj.com
- *New York Times:* www.nytimes.com
- *USA Today:* www.usatoday.com
- *Business Week:* www.businessweek.com
- *Fast Company:* www.fastcompany.com
- *Fortune:* www.fortune.com
- *Forbes:* www.forbes.com

2. Review the job description carefully. Bring a copy of the job description with you to review just prior to the interview so that you are clear about what the job entails.

3. Prepare a one- or two-minute summary of your chief qualifications. You will most likely be asked to summarize your education, experience, and professional goals during the job interview.

4. Take your portfolio, including three or four extra copies of your résumé, with you. Also bring a notepad and a pen to jot down essential details.

5. Practice your interview skills with a friend or job counselor. Be sure that this person asks tough questions about your education and experience so that you will get practice answering these questions realistically and convincingly.

6. Brush up on business etiquette. Silence your cell phone before your interview. Remember the name(s) of the interviewer(s) and others you may meet. Always be polite and respectful, saying "Thank you," "You're welcome," and so on. Pay special attention to acceptable ways of communicating with international audiences if your interview is with a multinational company or with a non-native speaker of English.

7. Bring your photo ID and Social Security card. Bring any licenses or certificates you may be asked to present to a human resources office. If you are not a U.S. citizen, bring your work visa.

Questions to Expect at Your Interview

The following questions are typical of those you can expect from interviewers, with advice on how to answer them.

- **Why do you want to work for us?** (Recall any job goals you have and apply them specifically to the job under discussion.)
- **What qualifications do you have for the job?** (Point to educational achievements and relevant work experience, especially computer skills.)
- **What could you possibly offer us that other candidates do not have? Why should we hire you?** (Say enthusiasm, being a team player, having problem-solving abilities; meet deadlines. Stress that you are conscientious diplomatic, and goal oriented.)
- **Why did you attend this school?** (Be honest—location, costs, programs.)
- **Why did you major in "X"?** (Do not simply say financial benefits; concentrate on professional goals and interests.)
- **Why did you get a grade of C in a course?** (Don't say that you could have done better if you'd tried. Explain what the trouble was, and mention that you corrected it in a course in which you earned a B or an A.)
- **What extracurricular activities did you participate in while in high school or college?** (Indicate any responsibilities you had—managing money, preparing minutes, coordinating events. If you were unable to participate in such activities, tell the interviewer that a part-time job or community or church

activities prevented you from participation. Such answers sound better than saying that you did not like sports or clubs in school.)

- **Did you learn as much as you wanted from your course work?** (This is a loaded question. Indicate that you learned a great deal but now look forward to the opportunity to gain more practical skill, to put into practice the principles and procedures you have learned.)
- **What is your greatest strength?** (Say being a team player, planning and organizing tasks efficiently, being sensitive to others' needs, concern about the environment, cooperation, willingness to learn, ability to grasp difficult concepts easily, proficiency at managing time or money, taking criticism easily, and profiting from criticism.)
- **What is your greatest shortcoming?** (Be honest here and mention it, but then turn to ways in which you are improving. Don't say something deadly like, "I can never seem to finish what I start" or "I hate being criticized." You should neither dwell on your weaknesses nor keep silent about them. Saying "None" to this kind of question is as inadvisable as rattling off a list of faults.)
- **How do you handle conflict with a co-worker, boss, or customer?** (Stress your ability to be courteous and honest and to work toward a productive resolution. State that you avoid language, tone of voice, or gestures that interfere with healthy dialogue. Describe a specific situation where you resolved a problem with maturity and grace.)
- **Why did you leave your last job?** (Say, "I returned to school full time" or "I moved from Jackson to Springfield" or say that you changed professions. *Never attack your previous employer.* That only makes you look bad.)
- **Why would you leave your current job?** (Again, never attack an individual or an organization. Say your current job has prepared you for the position you are now applying for. Emphasize your desire to work for a new company because of its goals, work environment, and opportunities. Say you want a challenge.)
- **What are your career goals over the next three to five years ?** (Be able to state your career objectives in terms of what you would like to accomplish for the company, your profession, or the community. Be confident, not cocky.)

What Do I Say About Salary?

Again, do your homework. Find out what the salary range is for your professional level in your area. Consult the U.S. Bureau of Labor Statistics' *Occupational Outlook Handbook* at *http://www.bls.gov/oco* as well as *http://www.salary.com.* You can also ask your instructors or individuals you know who work for the company or call a professional organization to which you may belong for information. If the issue of salary comes up, ask if the company has established a salary range for the position and where you stand in relationship to that range. However, because many companies set fixed salaries for entry-level positions, it may be unwise to try to negotiate.

If you are asked what salary you expect for the job, do not give an exact figure. You may undercut yourself if the employer has a higher figure in mind. By doing your homework on salary ranges, you will have a better feel for the market when the employer does mention salary.

Factor other things into your job equation—health insurance, day care, housing, uniform/clothing allowances, product or service discounts, opportunities for travel and/or foreign language instruction, and tuition reimbursement.

Questions You May Ask the Interviewer(s)

You will have a chance to ask the interviewer(s) questions. Watch for appropriate cues and be prepared to say more than "No, I don't have any questions," which suggests either indifference or unpreparedness on your part. Here are some legitimate questions you can ask interviewers:

1. Will there be any safety, security, or proficiency requirements I will need to meet?
2. When is the starting date?
3. Is there a probationary period? If so, how long?
4. How will my work be evaluated (monthly, quarterly, semiannually) and by whom (immediate superior, committee)?
5. What types of on-the-job training are required and/or offered?
6. Are there any mentoring programs in place?
7. Is there any support for continuing my education to improve my job performance?

Also, ask questions about the company's products and services, including a dedication to greening the environment, to express your interest in the position.

What Interviewer(s) Can't Ask You

There are some questions an interviewer may not legally ask you. Questions about your age, marital status, ethnic background, religion, race, or physical disabilities violate equal opportunity employment laws. Even so, some employers may disguise their interest in those subjects by asking you indirect questions about them. A question such as "Will your husband care if you have to work overtime?" or "How many children do you have?" could probe into your personal life. Confronted with such questions, it is best to answer them positively ("My home life will not interfere with my job," "My family understands that overtime may be required") rather than bristling defensively, "It's none of your business if I have a husband or a wife."

Ten Interview Dos and Don'ts

Keep in mind these other interview dos and don'ts.

1. Be on time. In fact, show up about 15 minutes early in case the interviewer or human resources office wants you to complete some forms.
2. Turn off your cell phone! The last thing you want is for it to ring during your interview. Never text during an interview.

3. Dress appropriately for the occasion and be well groomed. **Men:** Wear a dark solid-color or a pinstripe suit and tie. **Women:** Wear a suit (pants or skirt) or equally business-like attire. Avoid using strong perfume or cologne.

4. Be careful about tattoos. Job counselors warn that visible tattoos can hurt a job seeker's chance for success.

5. Greet the interviewer with a friendly and firm, but not vicelike, handshake. Don't be a wimp, either, with a limp, fishy handshake. Thank your interviewer(s) for inviting you.

6. Don't sit down before the interviewer(s) do. Wait for them to invite you to sit and to indicate where.

7. Speak slowly and distinctly; do not nervously hurry to finish your sentences, and never interrupt or finish an interviewer's sentences. Avoid one- or two-word answers, which sound unfriendly or unprepared. Do not use slang (e.g., "Right on," "Way to go," "You go, girl") or overly casual language ("Like . . ." "You know?"). Don't monopolize the discussion by talking too much and always about yourself. And don't act arrogantly thinking the job is yours already. Show a keen interest in the company and it products/services/organization. See Figure 5.12 where a candidate reiterates her interest in a prospective employer.

8. Do not chew gum; click a ballpoint pen; fidget; twirl your hair; or tap your foot against the floor, a chair, or a desk.

9. Maintain eye contact with the interviewer; do not sheepishly stare at the floor or the desk. If you are interviewed by a group of individuals, make eye contact with each one of them. Body language is equally important. For instance, don't fold your arms—a signal that you are closed to the interviewer's suggestions and comments. Sit up straight; do not slouch.

10. When the interview is over, thank the interviewer(s) for considering you for the job and say you look forward to hearing from him, her, or them.

The Follow-up Letter

Within a week after the interview, it is wise to send a follow-up letter, not an e-mail, thanking the interviewer for his or her time and interest in you. In your letter, you can reemphasize your qualifications for the job by showing how they apply to conditions described by the interviewer; you might also ask for further information to show your interest in the job and the employer. The sample follow-up letter in Figure 5.12 accomplishes all of these things.

Searching for the Right Job Pays

As we saw, finding the right job takes a lot of hard work (researching, organizing, and writing). But all your efforts will pay off with your first and subsequent checks. Make all your letters, résumés, portfolios/webfolios, and applications models of successful writing at work.

2739 EAST STREET
LATROBE, PA 17042-0312

610-555-6373
mlb@springboard.com

September 20, 2012

Mr. Jack Fukurai, Director
Human Resources
Global Tech
1334 Ridge Road N.E.
Pittsburgh, PA 17122-3107

Dear Mr. Fukurai:

I enjoyed talking with you last Wednesday and learning more about the security officer position available at Global Tech. It was especially helpful to take a tour of the plant's north gate section to see the challenges it presents for the security officer stationed there.

Expresses gratitude for an interview and singles out main company feature

As you noted at the interview, my training in surveillance electronics has prepared me to operate the sophisticated equipment Global Tech has installed at the north gate. I was grateful to Ms. Turner for taking time to demonstrate this technology.

Reemphasizes qualifications

I am looking forward to receiving the handbook about Global Tech's employee services. Would you also kindly e-mail me a copy of the newsletter from last quarter that introduced the new security equipment to employees?

Asks for newsletter to express further interest

Thank you, again, for considering me for the position and for the hospitality you showed me. Please let me know if you have any other questions for me. I look forward to hearing from you.

Ends politely by thanking interviewer

Sincerely yours,

Marcia Le Borde

Marcia Le Borde

Revision Checklist

☐ Enhanced my professional image to apply for jobs for which I am qualified.
☐ Prepared a dossier at school placement office, including supporting letters from professors, employers, and community officials.
☐ Identified places where relevant jobs are advertised.
☐ Networked with instructors, friends, relatives, and individuals who work for the companies I want to join, notified them that I am looking for a job.
☐ Researched the companies I am interested in—on the Internet, through printed sources, and by networking with current employees.
☐ Inventoried my strengths carefully to prepare résumé.
☐ Wrote a focused and persuasive career-objective statement.
☐ Determined the most beneficial format of résumé to use—chronological, functional, or both.
☐ Investigated creating a website for my job search-related documents.
☐ Prepared a portfolio/webfolio that included documents demonstrating my professional skills and achievements relevant for the job.
☐ Made résumé attractive and easy to read with logical and persuasive headings and descriptive keywords.
☐ Made sure résumé contains neither too much nor too little information.
☐ Proofread résumé to ensure everything is correct, consistent, and accurate.
☐ Created properly formatted online résumé to send to prospective employers.
☐ Wrote a letter of application that shows how my specific skills and background meet an employer's exact needs.
☐ Prepared short oral presentation about myself and my accomplishments for an interview.
☐ Researched prospective employer's company/organization and salary range(s).
☐ Sent prospective employer a follow-up letter within a few days after interview to show interest in position.

Exercises

1. Using at least four different sources, compile a list of 10 employers for whom you would like to work. Get their names, street and e-mail addresses, phone numbers, and the names of the managers or human resources officers. Then select one company and write a profile about it—locations, services, kinds of products or services offered, number of employees, clients served, awards, contributions to the community or environment, and any other pertinent facts.

2. Which of the following would belong on your résumé? Which would not belong? Why?
   a. student ID number
   b. Social Security number
   c. the ZIP codes of your references' addresses
   d. a list of all your English courses in college
   e. section numbers of the courses in your major
   f. statement that you are recently divorced
   g. subscriptions to journals in your field
   h. the titles of any stories or poems you published in a high school literary magazine or newspaper
   i. your GPA for every year you were in college
   j. foreign languages you studied
   k. years you attended college
   l. the date you were discharged from the service
   m. names of the neighbors you are using as references
   n. your religion
   o. job titles you held
   p. your summer job busing tables
   q. your telephone number
   r. the reason you changed schools
   s. your current status with the National Guard
   t. the URL of your website or blog
   u. your volunteer work for the Red Cross
   v. hours per week you spend reading science fiction
   w. the title of your last term paper in your major
   x. the name of the agency or business where you worked last

3. Indicate what is wrong with the following career objectives, and rewrite them to make them more precise and professional.
   a. Job in a lawyer's office.
   b. Position with a safety emphasis.
   c. Desire growth position in a large department store.
   d. Am looking for entry position in health sciences with an emphasis on caring for older people.
   e. Position in sales with fast promotion rate.
   f. Want a job working with semiconductor circuits.
   g. I would like a position in fashion, especially one working with modern fashion.
   h. Desire a good-paying job, hours: 8–4:30, with time and a half for overtime. Would like to stay in the Omaha area.
   i. Insurance work.
   j. Working with computers.
   k. Personal secretary.
   l. Job with preschoolers.
   m. Full-time position with hospitality chain.
   n. I want a career in nursing.

    o. Police work, particularly in a suburb of a large city.

    p. A job that lets me be me.

    q. Desire fun job selling cosmetics.

    r. Any position for a qualified dietitian.

    s. Although I have not made up my mind about which area of forestry I shall go into, I am looking for a job that offers me training and rewards based upon my potential.

4. As part of a team or on your own, revise the following poor résumé to make it more precise and persuasive. Include additional details where necessary and exclude any details that would hurt the job seeker's chances. Also correct any inconsistencies.

> RÉSUMÉ OF
> Powell T. Harrison
> 8604 So. Kirkpatrick St.
> Ardville, Ohio
> 345 37 8760
> 614 234 4587
> harrison@gem.com

<u>PERSONAL</u>	Confidential
CAREER <u>OBJECTIVE</u>	Seek good paying position with progressive Sunbelt company.
<u>EDUCATION</u>	
2010–2012	Will receive degree from Central Tech. Institute in Arch. St. Earned high average last semester. Took necessary courses for major; interested in systems, plans, and design development.
2008–2010	Attended Ardville High School, Ardville, OH; took all courses required. Served on several student committees.
<u>EXPERIENCE</u>	None, except for numerous part-time jobs and student apprenticeship in the Ardville area. As part of student app. worked with local firm for two months.
<u>HOBBIES</u>	Surfing the Net, playing Nintendo DSi, Member of Junior Achievement.
<u>REFERENCES</u>	Please write for names and addresses.

5. Determine what is wrong with the following sentences in a letter of application. Rewrite them to eliminate any mistakes, to focus on the "you attitude," or to make them more precise.

    a. Even though I have very little actual job experience, I can make up for it in enthusiasm.

b. My qualifications will prove that I am the best person for your job.

c. I would enjoy working with your other employees.

d. This e-mail résumé is my application for any job you now have open or expect to fill in the near future.

e. Next month, my family and I will be moving to Detroit, and I must get a job in the area. Will you have anything open?

f. If you are interested in me, then I hope that we make some type of arrangements to interview each other soon.

g. I have not included a résumé since all pertinent information about me is in this letter.

h. My GPA is only 2.5, but I did make two B's in my last term.

i. I hope to take state boards soon.

j. Your company, or so I have heard through the grapevine, has excellent fringe benefits. That is what I care about most, so I am applying for any position that you may advertise.

k. I am writing to ask you to kindly consider whether I would be a qualified person for the position you announced in the newspaper.

l. I have made plans to further my education.

m. My résumé speaks for itself.

n. I could not possibly accept a position that required weekend work, and night work is out, too.

o. In my own estimation, I am a go-getter—an eager beaver, so to speak.

p. My last employer was dead wrong when he let me go. I think he regrets it now.

q. When you want to arrange an interview time, give me a call. I am home every afternoon after 4:00.

6. Explain why the following letter of application is ineffective. Rewrite it to make it more precise and appropriate.

```
Apartment 32
Jeggler Drive
Talcott, Arizona

Monday

Grandt Corporation
Production Supervisor
Capital City, Arizona


Dear Sir:


I am writing to ask you if your company will consider me for
the position you announced online recently. I believe that
with my education (I have an associate degree) and experi-
ence (I have worked four years as a freight supervisor), I
could fill your job.
```

My schoolwork was done at two junior colleges, and I took
more than enough courses in business management and information technology. In fact, here is a list of some of my
courses: Supervision, Materials Management, Work Experience
in Management, E-commence, Safety Tactics, Introduction
to Software Analysis, Art Design, Contemporary Business
Principles, and Small Business Management. In addition, I
have worked as a loading dock supervisor for the last two
years, and before that I worked in the military in the
Quartermaster Corps.

Please let me know if you are interested in me. I would
like to have an interview with you at the earliest possible
date, since there are some other firms also interested in
me, too.

Eagerly yours,

George D. Milhous

7. From the Sunday edition of your local newspaper or from one of the other sources discussed on pages 155–157, find notices for two or three jobs you believe you are qualified to fill and then write a letter of application for one of them.

8. Write a chronologically organized résumé to accompany the letter you wrote for Exercise 7.

9. Write a functional résumé to accompany your application letter in Exercise 7.

10. Bring the two résumés you prepared for Exercises 8 and 9 to class to be critiqued by a collaborative writing team. After your résumés are reviewed, revise them. Write an e-mail to your instructor about the revisions you made and why they will help you in your job search.

11. Prepare a hard-copy portfolio. Collect appropriate documents and arrange them into the most relevant order. Provide an annotation for each.

12. Write a letter to a local business inquiring about summer employment. Indicate that you can work only for one summer because you will be returning to school by September 1. Include an appropriate résumé.

6 Designing Successful Documents, Visuals, and Websites

The success of your document depends as much on how it looks as on what it says. You will be expected to design professional-looking memos, letters, instructions, and reports; to create websites; and to provide appropriate visuals to support these documents. This chapter gives you practical advice for making your work more reader-friendly and visually appealing. It also surveys the kinds of visuals you will encounter most frequently and shows you how to read, construct, and write about them.

To expand your understanding of designing documents and visuals, take advantage of the Web Links, Additional Activities, and ACE Self-Tests at www.cengage.com/english/kolin/writingatwork concise3e

Characteristics of Effective Design

In designing documents and websites, you need to project a positive, professional image of yourself, your company, and your product or service. A report or website filled with nothing but thick unbroken, long paragraphs crowded to the margins, with no visual clues to break them up or to make information stand out, is sure to intimidate readers and turn them away. They will conclude that your work is too complex and not worth their effort or time. Your company, too, will win or lose points because of your design choices. A visually appealing document or website will enhance a company's reputation and improve its sales. A poorly designed one will not.

Make your documents look user-friendly—clear and logical—by signaling to your audience that your message is

- easy to read
- easy to follow and understand
- easy to recall

You can do all this by breaking material into smaller units that are visually appealing. You can help readers find key points at a glance through **chunking** (using smaller paragraphs) and using lists, boldface type, and bullets. That way they can find your ideas easily the first time through or on a second reading if they have to double back to check or verify a point.

Organizing Information Visually

Today's web-based culture prizes visual thinking when you design all kinds of documents. The way you organize and visualize information can help readers move quickly and clearly through your document. Take a quick look at Figures 6.1 and 6.2 (pages 199–200). The same information is contained in each figure. Which visually appeals to you more? Which do you think would be easier to read? Which is better designed? As the two figures show, design or layout plays a crucial role in an audience's overall acceptance of your work.

Figure 6.2 exemplifies the positive design characteristics (on the left) while Figure 6.1 displays the unfavorable ones (on the right).

Positive Design	Negative Design
■ visual appeal	■ crowded
■ logical organization	■ disorganized
■ clarity	■ hard to follow
■ accessibility	■ difficult to read
■ functionality	■ uneven
■ balanced	■ inconsistent

Study the annotations to these figures to see how the errors in Figure 6.1 are corrected in Figure 6.2. By modeling your written work after the document in Figure 6.2, you can guarantee that your message will be well received.

The ABCs of Print Document Design

The basic elements of effective document design are

- page layout
- typography, or type design, including using color
- graphics, or visuals

The proper arrangement and balance of type, white space, and graphics involve the same level of preparation that you would spend on your research, drafting, revising, and editing. Just as you do research to find information, you have to research and experiment in order to adopt the most effective design for your document.

Page Layout

Each of your pages needs to coordinate space and text pleasingly. Too much or too little of one or the other can jeopardize the reader's acceptance of your message. To design an effective page layout, pay attention to the following elements.

1. White space. White space (blank space), which refers to open areas on a page, such as margins and space around images, is free of text and visuals. It can help you increase the impact and tone of your message. Skimping on white space by packing too much print on the page only distracts the reader from the message you really

A poorly designed document.   Figure 6.1

No title

The results for the recent cholesterol screening at *our company's Health Fair* were distributed to each employee last week. Many employees wanted to know more information about cholesterol in general, the different types of cholesterol, what the results mean, and the foods that are high or low in cholesterol.

Single spacing makes document difficult to read

We hope the information provided below will help employees better answer their questions concerning cholesterol and our cholesterol screening program.

High cholesterol, along with high blood pressure and obesity, is one of the primary risk factors that may contribute to the development of coronary heart disease and may eventually lead to a heart attack or stroke. Cholesterol is a fatty, sticky substance found in the bloodstream. Excessive amounts of the bad type of cholesterol can deposit on the walls of the heart arteries. *This deposit is called plaque and over a long period of time plaque can narrow or even block the blood flow through the arteries.*

Lack of headings makes it hard for readers to organize material

Total cholesterol is divided into three parts—LDL (low-density lipoprotein), or bad cholesterol; HDL (high-density lipoprotein), or good cholesterol; and VLDL (very low-density lipoprotein), a much smaller component of cholesterol you don't have to worry about. Bad (LDL) cholesterol forms on the walls of your arteries and can cause a lot of damage. Good cholesterol, on the other hand, functions like a sponge, mopping up cholesterol and carrying it out of the bloodstream.

You should have received *three cholesterol numbers.* One is for your HDL, or good cholesterol, and the other is for your LDL, or bad cholesterol, reading. These two numbers are added to give you the third, or composite, level of your total cholesterol. You are doing fine.

Frequent and arbitrary font changes are confusing for readers

As you can see, a total cholesterol reading of below 200 is considered safe. Continue what you have been doing. If your reading falls in the moderate risk range of 200–239, you need to modify your diet, get more exercise, and have your cholesterol checked again in six months. *If your reading is above 240, see your doctor.* You may need to take cholesterol-lowering medication, if your doctor prescribes it. Reducing your total cholesterol by even as little as 25% can decrease your risk of a heart attack by 50%.

Uneven presentation of numbers

The Surgeon General recommends that your LDL, or bad, cholesterol should be below 130; and your HDL, or good, cholesterol needs to be at least above 36. Ideally, the ratio between the two numbers should not be greater than 5 to 1. That is, your HDL should be at least 20% of your LDL. The higher your HDL is, the better, of course. So even if you have a high LDL reading, if your HDL is correspondingly high you will be at less risk.

Lack of margins makes document look dense and complex

One of the easiest ways to decrease your cholesterol is to modify your diet. Cholesterol is found in foods that are high in saturated fat. *Saturated fat comes from animal sources and also from certain vegetable sources.* Foods high in bad cholesterol that you should restrict, or avoid, include whole milk, red meat, eggs, cheese, butter, shrimp, oils such as palm and coconut, and avocados. Generally, *food groups low in cholesterol* include fruits, vegetables, and whole grains (wheat breads, oatmeal, and certain cereals), lean meats (fish, chicken), and beans.

Unnecessary italics are confusing

The goal of our cholesterol screening is to help each employee lower his or her cholesterol level and eventually reduce the risk of heart disease. **Besides the advice given above,** you can do the following: get regular aerobic exercise—bicycling, brisk walking, swimming, rowing—for at least 30 minutes 3–4 times a week. But get your doctor's approval first. *Eat foods low in cholesterol* but high in dietary fiber (beans, oatmeal, brown rice). Maintain a healthy weight for your frame to lower your body fat. Minimize stress, which can increase cholesterol. Learn relaxation techniques.

Inconsistent use of italics and boldfacing

Figure 6.2   An effectively designed document with the same text as Figure 6.1.

Title clearly set apart from text by caps, larger font, and boldfacing

Text is double spaced with more ample margins, making it more readable

Page does not look cluttered

Headings divide material into easy-to-follow units for readers

Consistent use of one font

Only key word being defined is boldfaced

Types of cholesterol are helpfully labeled with numbers

Cholesterol Screening

The results for the recent cholesterol screening at our company's Health Fair were distributed to each employee last week. Many employees wanted to know more information about cholesterol in general, the different types of cholesterol, what the results mean, and the foods that are high or low in cholesterol. We hope the information provided below will help employees better answer their questions concerning cholesterol and our cholesterol screening program.

Determining Risk Factors

High cholesterol, along with high blood pressure and obesity, is one of the primary risk factors that may contribute to the development of coronary heart disease and may eventually lead to a heart attack or stroke. Cholesterol is a fatty, sticky substance found in the bloodstream. Excessive amounts of the bad type of cholesterol can deposit on the walls of the heart arteries. This deposit is called **plaque** and over a long period of time plaque can narrow or even block the blood flow through the arteries.

Separating Types of Cholesterol

Total cholesterol is divided into three parts: (1) **LDL** (low-density lipoprotein), or bad cholesterol; (2) **HDL** (high-density lipoprotein), or good cholesterol; and (3) **VLDL** (very low-density lipoprotein), a much smaller component of cholesterol you don't have to worry about. Bad (LDL) cholesterol forms on the walls of your arteries and can cause a lot of damage. Good cholesterol, on the other hand, functions like a sponge, mopping up cholesterol and carrying it out of the bloodstream.

Understanding Your Cholesterol Results

You should have received three cholesterol numbers. One is for your **HDL** (or good cholesterol) and the other is for your **LDL** (or bad cholesterol) reading. These two numbers are added to give you the third, or composite, level of your total cholesterol.

Cholesterol levels can be classified as follows:

Minimal Risk	Moderate Risk	High Risk
below 200	200–239	above 240

As you can see, a total cholesterol reading of below 200 is considered safe. You are doing fine. Continue what you have been doing. If your reading falls in the moderate risk range of 200–239, you need to modify your diet, get more exercise, and have your cholesterol checked again in six months. If your reading is above 240, see your doctor. You may need to take cholesterol-lowering medication, if your doctor prescribes it. Reducing your total cholesterol by even as little as 25% can decrease your risk of a heart attack by 50%.

Relationship Between Bad and Good Cholesterol

The Surgeon General recommends that your **LDL**, or bad, cholesterol should be below 130. And your **HDL**, or good, cholesterol needs to be at least above 36. Ideally, the ratio between the two numbers should not be greater than 5 to 1. That is, your **HDL** should be at least 20% of your **LDL**. The higher your **HDL** is, the better, of course. So even if you have a high **LDL** reading, if your **HDL** is correspondingly high you will be at less risk.

Recognizing Food Sources of Cholesterol

One of the easiest ways to decrease your cholesterol is to modify your diet. Cholesterol is found in foods that are high in saturated fat. Saturated fat comes

2

Concise paragraph provides clear opening for new section

Emphasizes range of risk categories by setting them apart and shading

Paragraph is not crowded with numbers

Includes more space between sections

Paragraphs are neither too long nor too short

Page is numbered

Figure 6.2	**(Continued)**

from animal sources and also from certain vegetable sources. Foods high in bad cholesterol that you should restrict include:

A double-spaced, numbered list helps readers easily identify foods with bad cholesterol

1. whole milk
2. red meat
3. eggs
4. cheese
5. butter
6. shrimp
7. oils such as palm and coconut
8. avocados

Easy-to-follow examples of foods low in cholesterol

Generally, food groups low in cholesterol include fruits, vegetables, and whole grains (wheat breads, oatmeal, and certain cereals), lean meats (fish, chicken), and beans.

Realizing It Is Up to You

Heading signals conclusion

The goals of our cholesterol screening program are to help each employee lower his or her cholesterol level and eventually reduce the risk of heart disease. Besides the advice given above, you can do the following:

- Get regular aerobic exercise—bicycling, brisk walking, swimming, rowing—for at least 30 minutes 3–4 times a week. But get your doctor's approval first.

Bulleted list serves as both summary and plan for future action

- Eat foods low in cholesterol but high in dietary fiber (beans, oatmeal, and brown rice).
- Maintain a healthy weight for your frame to lower your body fat.
- Minimize stress, which can increase cholesterol. Learn relaxation techniques.

3

Source: Thanks to Sgt. Mannie E. Hall of the U.S. Army for his advice in drafting this document.

want to convey. White space, on the other hand, can entice, comfort, and appeal to the reader's "psychology of space" by

- attracting the reader's attention
- assuring the reader that information is presented logically
- announcing that information is easy to follow
- assisting the reader to organize information visually
- allowing the reader to highlight important information

Again, compare Figures 6.1 and 6.2. Which document shows that it was designed by someone who understands the importance of white space?

   2. Margins. Use wide margins, usually 1 to 1 ½ inches, to "frame" your document with white space surrounding text and visuals. Margins prevent your document from looking cluttered or overcrowded. If your document requires binding, you may have to leave a wider left margin (2 inches).

   3. Line length. Most readers find a text line of 10 to 14 words, or 50 to 70 characters (depending on the type size you choose), comfortable and enjoyable to read. Excessively long lines that bump into the margins signal that your work is difficult to read. In the following example, note how the extra-long lines unsettle your reading and tax your eye movement; they signal rough going.

> In order to succeed in the world of business, workers must learn to brush up on their networking skills. The network process has many benefits that you need to be aware of. These benefits range from finding a better job to accomplishing your job more easily and efficiently. Through networking you are able to expand the number of contacts who can help you. Networking means sharing news and opportunities. The Internet is the key to successful networking. It does this through linking texts of all sorts.

On the other hand, do not print a document with too short or extremely uneven lines.

> In order to succeed in the world of
> business, workers must learn
> to brush up on their networking skills. The
> network process has many
> benefits you need to be
> aware of.

Readers will suspect your ideas are incomplete, superficial, or even simple-minded.

   4. Columns. Document text usually is organized in either single-column or multicolumn formats. Memos, letters, and reports are usually formatted without columns, whereas documents that intersperse text and visuals (such as newsletters and magazines) work better in multicolumn formats.

Typography

Typography consists of font (also called typeface), font size, font styles, justification, heads and subheads, lists, and captions.

Font

Readability of your text is crucial. Select a font, therefore, that ensures your text is

- legible
- attractive
- functional
- appropriate for your message
- complementary with accompanying graphics

The most familiar fonts are the following four.

Times New Roman	Arial
Helvetica	Palatino

The font you use can make your document look businesslike or too casual. For instance, avoid using a font that looks like script, and don't mix and switch fonts. The result makes your work look amateurish and disorganized, as in Figure 6.1.

Font Size

Font size options are almost unlimited, depending on your software package and printer capabilities. Font size is measured in units called **points,** 72 points to the inch. The bigger the point size, the larger the type. Never print your letter or report in 6- or 8-point newspaper ad type or in a size larger than 12-point type. Here are some suggestions.

Times New Roman 8 point = footnotes/endnotes

Helvetica 12 point = letters, reports, and e-mails

Palatino 14 point = headings

Arial 22 point = title of your report

Font Styles

Font styles include boldface, italics, shadow, underlining, small caps, and shading.

Boldface
Italics
Shadow
Underlining
Small Caps
Shading

Avoid overusing boldface and italics. Use them only when necessary and not for decoration. Do *not* underline the text unless absolutely necessary. Not only will too many special visual effects make your work harder to read, but you will lose the dramatic impact these features have to distinguish and emphasize key points that rightfully deserve to be set boldface or italic type.

Justification

Sometimes referred to as alignment, justification consists of left, right, full, and center options. Left-justified (also called **unjustified** or **ragged right**) is the preferred method because it allows space between words in lines of text to remain constant and is easier to read.

Our new website offers consumers a mall on the Internet. It gives them access to our products and services and makes buying easy and fun. The new website gives shoppers access to a wide range of	Our new website offers consumers a mall on the Internet. It gives them access to our products and services and makes buying easy and fun. The new website gives shoppers access to a wide range of
Left-justified text	Right-justified text

Heads and Subheads

Heads and subheads are brief descriptive words or phrases that signal starting points and major divisions in your document. They provide helpful road signs for readers charting their course through a document as Figure 6.2 does. Without heads and subheads, your work will look unorganized and cluttered. Heads and subheads should also parallel each other grammatically. Note how the following headings, from a poorly organized proposal from the Acme Company, are not parallel.

- What Is the Problem?
- Describing What Acme Can Do to Solve the Problem
- It's a Matter of Time . . .
- Fees Acme Will Charge
- When You Need to Pay
- Finding Out Who's Who

Revised, the heads are parallel and easier for a reader to understand and follow.

- A Brief History of the Problem
- A Description of Acme Solutions
- A Timetable Acme Can Follow
- A Breakdown of Acme's Fees
- A Payment Plan
- A Listing of Acme's Staff

Heads and subheads immediately attract attention and quickly inform readers about the function, scope, purpose, or contents of the document or section. The space around a heading is like an oasis for the reader, signaling both a rest and a new beginning. In designing a document with heads and subheads, follow these guidelines.

- Use a larger type size for heads than for text; major heads should be larger than subheads. If your text is in 10-point type, your heads can be in 16-point type and your subheads in 12- or 14-point type.

Fourteen-Point Head
Subhead in 12 Point
Use larger type for heads than you do for text; major headings should de larger than subheads. If your text is in 10-point type, your heads may be in 14-point type and your subheads in 12.

- Modify type to differentiate sections. For example, for heads and subheads, use uppercase, bold type, underlining, and changes in font style or type.
- Establish a horizontal position for a head, such as centered or aligned left, and keep the placement of heads consistent throughout the document.
- Do not overuse heads (e.g., inserting them after each paragraph); having too many heads are as bad as having too few.
- Allow additional space above a head to set it off from the preceding section. You should also allow additional space below a head.

Lists
Placing items in a list helps readers by dividing, organizing, and ranking information. Lists emphasize important points and contribute to page design that is easy to read. Lists can be numbered, lettered, or bulleted. Take a look at the memos, reports, and proposals in Chapters 3, 8, and 9 that effectively use lists.

Captions
Used to accompany, explain, highlight, or reference pictures or other graphics (like charts and graphs), captions are titles that help a reader identify a visual and quickly explain the nature of the picture or other graphic. (Chapter 9 discusses using captions with visuals in a document.)

Using Color
Using color in workplace documents is a good way to enhance readability, break up long segments of text, and tie important ideas together. Tastefully done, color can help sell ideas more effectively than black and white alone. You can use color for borders and graphic accents, headings, titles, keywords, Internet addresses, sidebars, rules, and boxes that link related facts, figures, or information. But first

determine if using color serves a functional purpose or if it is merely a decoration. If it is window-dressing, stick with black and white.

Guidelines on Using Color Effectively

Here are some guidelines to follow when you use color in your documents or websites:

- Estimate how the color will look on the page—colors look different on the screen than they do on a sheet of paper. Print a sample page to get a clear view.
- Make sure text colors contrast sharply with background colors in both print and web-based documents.
- Use no more than two or three colors on a page or screen unless there are photographs, illustrations, or graphics.
- Too many bright colors overwhelm the eye, so use them sparingly—only to call attention to important elements.
- Select "cool" colors, such as blue, turquoise, purple, and magenta, for backgrounds. However, avoid light blue text, which is hard to read against a dark background, or yellow text on a light blue background.
- Use colors that respect an international reader's cultural heritage. See pages 236–237.

Successful Document Design: A Wrap-Up

By following these three simple guidelines, you can design effective documents for your readers.

1. Keep it simple. Don't try to impress your reader with visual effects—such as the overuse of italics or boldface (they will lose their purpose), fancy designs and scripts. Also, avoid printing everything in capitals or other large type size usually reserved for headlines.

2. Make it clear. Make your message logical and easy to understand by including heads, bulleted lists, numbered steps, and the like.

3. Have it flow. Help your readers get through and understand your document easily and, if necessary, to go back and find information quickly. Don't cram too much information on a page (e.g., lots of visuals, including photos, or several lists). Skimping on white space also makes your document difficult to digest. Also, steer clear of tight spacing and small type size.

The Purpose of Visuals

Visuals are essential in the world of work. They are vital to the success of newsletters, reports, proposals, instructions, PowerPoint presentations, blogs, and, most certainly, websites. Even your company's logo reveals a great deal about its image and mission. The reason visuals are so important is that they help readers understand your work. Experts estimate that as much as 80 percent of our learning comes through our sense of sight. Here are several reasons why visuals can improve your work; each point is graphically reinforced in Figure 6.3 below.

1. Visuals arouse readers' immediate interest. They catch the reader's eye quickly by setting important information apart and by giving relief from sentences and paragraphs. Note the eye-catching quality of the visual in Figure 6.3.

2. Visuals increase readers' understanding by simplifying concepts. A visual *shows* ideas whereas a verbal description only *tells* about them in the abstract. Visuals are especially important and helpful if you have to explain a technical process to a nonspecialist audience. Visuals also help readers more clearly see percentages, trends, comparisons, and contrasts. Figure 6.3, for example, shows at a glance the growth of market share of online computer sales.

3. Visuals are especially important for non-native speakers of English and multicultural audiences. Given the international audiences for many business documents, visuals will make communicating with them easier and clearer.

4. Visuals emphasize key relationships. Through their arrangement and form, visuals quickly show contrasts, similarities, growth rates, and downward and upward movements, as well as fluctuations in time, money, and space.

| Figure 6.3 | **A line-and bar chart depicting market share of online computer purchases in the southwestern United States compared with those at brick-and-mortar stores.** |

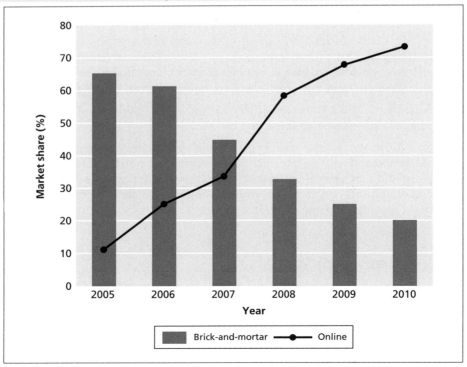

5. **Visuals condense and summarize a large quantity of information into a relatively small space.** A visual allows you to streamline your message by saving words and time. It can record data in far less space than it would take to describe those facts in words alone. Note how in Figure 6.3 the market share of two different types of businesses are concisely expressed and documented.

6. **Visuals are highly persuasive.** Visuals have sales appeal. They can persuade readers to buy your product or service or to accept your point of view and reject your competitor's.

Choosing Effective Visuals

Select your visuals carefully. You'll find that many of the visuals you use will originate from the Internet or from a graphics program on your computer. But you can also scan hard-copy visuals and upload digital photographs. Regardless of the source of your visuals, the following suggestions will help you to choose them effectively

1. **Include visuals only when they are relevant for your purpose and audience.** Never include a visual simply as a decoration. A short report on fire drills, for example, does not need a picture of a fire station. Similarly, avoid any visual that is too technical for your readers or that includes more details than you need to show.

2. **Use visuals in conjunction with—not as a substitute for—written work.** Visuals do not take the place of words. In fact, you may need to explain information contained in a visual. A set of illustrations or a group of tables alone may not satisfy readers looking for a summary or a recommendation. Note how the visual in conjunction with the description of a magnetic resonance imager (MRI) in Figure 6.4 on page 210 makes the procedure easier to understand than if the writer had used only words or only a visual. This visual and verbal description are appropriately included in a brochure teaching patients about MRI procedures.

3. **Experiment with several visuals.** Evaluate a variety of options before you select a particular visual. For instance, software such as PowerPoint allows you to represent statistical data in several ways. Preview a few different versions of a visual or even different types of visuals to determine which one would be best.

4. **Always use easy-to-read and relevant visuals.** If readers have trouble understanding its function and arrangement, your visual probably is not appropriate and you need to change or revise it.

5. **Consider how your visuals will look on the page.** Don't cram visuals onto a page or allow them to spill over text or margins. Also be careful not to resize your image so that it becomes distorted. Images are limited to a certain size by their *resolution*—the number of pixels (or dots) per inch (also known as dpi). If you make a low-resolution image too large, it will become fuzzy and look unprofessional. Many software programs help you move your visuals directly into

Figure 6.4   **A visual used in conjunction with written work.**

A PICTURE FROM THE INSIDE OUT
At the heart of the magnetic resonance imager is a large magnet that is big enough for you to lie inside. Look at the picture below. The **magnet** directs radio signals to surround sections of your body. When the signals pass through your body, they **resonate** (release a signal). Then your body's response is picked up by a receiver and sent to a computer. The computer analyzes the signal and converts it into a visual **image** of your tissues on a video screen.

1. **The MR Imager** surrounds your body with a harmless **magnetic** field and radio signals that safely pass through your body.

2. **A receiver** picks up and measures the radio signals that leave, or **resonate** from, your body.

3. **The radio signals** are turned into a computerized picture—or **image**—of your body's tissues.

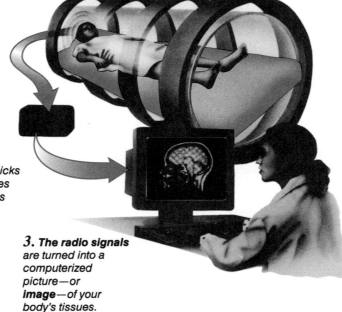

Reprinted by permission of Krames Communications.

your document so you can insert them properly. Review pages 198–203 for advice on effective page layouts.

6. Be prepared to revise and edit your visuals. Just as you draft, revise, and edit your written work to meet your audience's needs, create several versions of your visual to get it right. Expect to change shapes or proportions; experiment with different colors, shadings, labels, and various sizes. Replace any visuals that may distract the reader or contradict your message. Also make sure to replace or revise any visuals that may be unethically distorted or exaggerated. (See "Using Visuals Ethically" on pages 230–235.)

7. **Take advantage of scanners.** Using technology similar to that of a photocopier, a scanner captures an image and creates a digital reproduction that can then be altered or enhanced and used as a visual in a document. Scanners provide an efficient means of incorporating those visuals unavailable in digital form, such as older photographs, sketched diagrams, or pictures from books into your documents.

Generating Your Own Visuals

You can easily generate charts, graphs, or tables by using the templates available in your word processing software, such as Microsoft Word, and by selecting Insert Chart or Insert Table. In fact, some of the visuals in this book, especially those in the long report in Chapter 9, were created using such software. Graph and chart templates allow you to insert your raw numerical data into the appropriate blanks and then add titles, labels, and color-coded keys. Such software helps you to construct different visuals, including pie charts (discussed on pages 218–219) and flow charts (page 221), and also allows you to show a visual from different perspectives and emphases, such as shading, 3-D, exploded view, etc.

Spreadsheet programs, such as Microsoft Excel, also provide easy-to-use templates for creating graphs and charts using the numerical data you have entered into your spreadsheet. When using these software programs, though, avoid the temptation to make your visuals too showy. Although dazzling graphics may attract your reader's immediate attention, you don't want them to take over the report. Computer-generated graphs and charts should work with your words, not overshadow them.

Inserting and Writing About Visuals: Some Guidelines

Using a visual requires more of you as a writer than simply inserting it into your written work. You need to use visuals in conjunction with what you write. The following guidelines will help you to (1) identify, (2) cite, (3) insert, (4) introduce, and (5) interpret visuals for your readers.

Identify Your Visuals

Give each visual a number and caption (title) that indicates the subject or explains what the visual illustrates. An unidentified visual is meaningless. A caption helps your audience to interpret your visual—to see it with your purpose in mind. Tell your readers what you want them to look for by doing the following:

- Use a different typeface (bold) and size in your caption than what you use in the visual itself.
- Include key words about the function and the subject of your visual in a caption.
- Make sure any terms you cite in a caption are consistent with the units of measurement and the scope (years, months, seasons) of your visual.

Tables and figures should be numbered separately throughout the text—Table 1 or Figure 3.5, for example. (In the latter case, Figure 3.5 is the fifth figure to appear in Chapter 3.)

Cite the Source for Your Visuals

If you use a visual that is not your own work, give credit to your source (newspaper, magazine, textbook, company, federal agency, individual, or website). If your paper or report is intended for publication, you must first obtain permission to reproduce copyrighted visuals from the copyright holder.

Insert Your Visuals Appropriately

Since many of the images you use will come from outside sources, especially the Web, you have to incorporate them clearly and in appropriate places in your written work.

Here are some guidelines to help you incorporate the visuals in the most appropriate places for your readers.

- Never introduce a visual *before* a discussion of it; readers will wonder why it is there. Include a sentence or two to introduce your visuals.
- Always mention in the text of your paper or report that you are including a visual. Tell readers where it is found—"below," "on the following page," "to the right," "at the bottom of page 3."
- Place visuals as close as possible to the first mention of them in the text. Try not to put a visual more than one page after the discussion of it. Never wait two or three pages to present it. By inserting a visual near the beginning of your discussion, you help readers better understand your explanation.
- Use an appropriate size for your visual. Don't make it too large or so small it is hard to read.
- If the visual is small enough, yet clear, insert it directly in the text rather than on a separate page. If your visual occupies an entire page, place that page containing your visual on the facing page or immediately after the page on which the first reference to it appears.
- Center your visual and, if necessary, box it. But leave at least 1 inch of white space around it. Squeezing visuals toward the left or right margins looks unprofessional.
- Never collect all your visuals and put them in an appendix. Readers need to see them at those points in your discussion where they are most pertinent.

Introduce Your Visuals

Refer to each visual by its number and, if necessary, mention the title as well. In introducing the visual, though, do not just insert a reference to it, such as "See

Figure 3.4" or "Look at Table 1." Help readers to understand the relationships in your visual. Here are two ways of writing a lead-in sentence for a visual.

> Poor: Our store saw a dramatic rise in the shipment of electric ranges over the five-year period as opposed to the less impressive increase in washing machines. (See Figure 3.)

This sentence does not tie the visual (Figure 3) into the sentence where it belongs. The visual just trails insignificantly behind.

> Better: As Figure 3 shows, our store saw a dramatic rise in the shipment of electric ranges over the five-year period as opposed to the less impressive increase in washing machines.

Mentioning the visual in Figure 3 alerts readers to its presence and function in your work and helps them to more easily understand your message.

Interpret Your Visuals

Give readers help in understanding your visual and in knowing what to look for. Let them know what is most significant about the visual. Do not expect the visual to explain itself. Inform readers what the numbers or images in your visual mean. What types of conclusions can you and your readers logically make after seeing the visual?

In a report on the benefits of vanpooling, the writer supplied the following visual, a table:

TABLE 1   Travel Time (in minutes): Automobile Versus Vanpool

Private Automobile	Vanpool
25	32.5
30	39.0
35	45.5
40	52.0
45	58.5
50	65.0
55	71.5
60	78.0

Source: U.S. Department of Transportation. *Increased Transportation Efficiency Through Ridesharing: The Brokerage Approach* (Washington, D.C., DOT-OS—40096): 45.

Then to interpret the table, the writer called attention to it in the context of the report on transportation efficiency.

Although, as Table 1 above suggests, the travel time in a vanpool may be as much as 30 percent longer than in a private automobile (to allow for pickups), the total trip time for the vanpool user can be about the same as with a private automobile because van-pools eliminate the need to search for parking spaces and to walk to the employment site entrance.[1]

Two Categories of Visuals: Tables and Figures

Visuals can be divided into two categories—tables and figures. A **table** arranges information—numbers and/or words—in parallel columns or rows for easy com-parison of data. Anything that is not a table is considered a figure. **Figures** include graphs, pie charts, bar charts, organizational charts, flow charts, pictographs, maps, photographs, and drawings. Expect to use both in your work.

Tables

Tables are parallel columns or rows of information organized and arranged into categories to show changes in time, distance, cost, employment, or some other dis-tinguishable or quantifiable variable. They allow readers to compare a great deal of information in a compact space. Tables also summarize material for easy recall—causes of wars; provisions of a law; or differences between a common cold, the flu, and pneumonia (see, for example, Table 6.2 on page 237, which provides informa-tion on different cultural meanings for various gestures).

Parts of a Table

To use a table properly, you need to know the parts that constitute it. Refer to Table 6.1, which labels these parts, as you read the following:

- The main **column** is "Amount Needed to Satisfy Minimum Daily Requirement," and the **subcolumns** are the protein sources for which the table gives data.
- The **stub** refers to the first vertical column on the left side. The stub column heading is "Source." The stub lists the foods for which information is broken down in the subcolumns.
- A **rule** (or line) across the top of the table separates the headings from the body of the table.

Guidelines for Using Tables

When you include a table in your work, follow these guidelines.

- Number the tables according to the order in which they are discussed in the text (Table 1, Table 2, Table 3). Tables are numbered separately from figures (charts, graphs, photos) in your text.

---

[1]James A. Devine, "Vanpooling: A New Economic Tool," *AIDC Journal.*

- Include the table on the same or facing page, where it is most appropriate, whenever possible.
- Keep your table on one page; it is difficult for readers to follow a table spread across different pages.
- Give each table a concise and descriptive title to show exactly what is being represented or compared.
- Use words in the **stub** (a list of items about which information is given), but put numbers under column headings. The "Source" column below in Table 6.1 is the stub.
- Supply footnotes (often indicated by small raised letters: a, b) if something in the table needs to be qualified—for example, the number of cups of milk in Table 6.1. Then put that information below the table.
- Arrange the data you want to compare vertically; it is easier to read down than across a series of rows.
- Place tables at the top (preferable) or bottom of the page and center them on the page rather than placing them up against the right or left margin.
- Don't use more than five or six columns (note that Table 6.1 has five), tables wider than that are more difficult for readers to use.
- Round off numbers in your columns to the nearest whole number to assist readers in following and retaining information.
- Always give credit to the source (the supplier of the statistical information) on which your table is based.

TABLE 6.1   Parts of a Table

Table number

TABLE 1 Efficiency of Some Protein Sources in Meeting an Adult's Minimum Daily Requirements				
Source	*Percent of Protein*	*Percent of Amino Acids*	*Amount Needed to Satisfy Minimum Daily Requirement*	
			(grams)	(ounces)
Cheese[a]	27	70	227	7.2
Corn	10	50	860	30.0
Eggs	11	97	403	14.1
Fish[a]	22	80	244	8.5
Kidney beans	23	40	468	16.4
Meat[a]	25	68	253	8.8
Milk	4	82	1,311	45.9[b]
Soybeans	34	60	210	7.3

← Title
← Rule
← Column headings
← Subheadings
Stub {
← Origin of data

Source: From *Biology: The Unity and Diversity of Life,* 4th edition by C. Starr and R. Taggart. Copyright © 1987. Reprinted with permission of Brooks/Cole, a division of Thomson Learning: www.thomsonrights.com. Fax 800-730-2215.

a = Average value
b = Equivalent of 6 cups   }   *Footnotes*

Figures

As we saw, any visual that is not a table is classified as a figure. The types of figures we will look at next are

- graphs
- pie, or circle, charts
- bar charts
- organizational charts
- flow charts

- pictographs
- maps
- photographs
- drawings
- clip art

Graphs

Graphs transform numbers into pictures. They take statistical data presented in tables and put them into rising and falling lines, steep or gentle curves. Two main types of graphs are **simple line graphs** and **multiple-line graphs.**

Functions of Graphs

Graphs vividly portray information that changes, such as

- sales/stocks
- costs
- trends
- distributions

- employment
- energy levels
- temperatures
- crop prices

Graphs not only describe past and current situations but also forecast trends.

Simple Line Graphs

Basically, a simple graph consists of two sides—a **vertical** or **y-axis** and a **horizontal** or **x-axis**—that intersect to form a right angle, as in Figure 6.5. The space between the two axes contains the picture made by the graph—the amount of snowfall in Springfield between November 2010 and April 2011. The vertical line represents the **dependent variable** (the snowfall in inches); the horizontal line, the **independent variable** (time in months). The dependent variable is influenced most directly by the independent variable, which almost always is expressed in terms of time or distance. The vertical axis is read from bottom to top; the horizontal axis from left to right.

Multiple-Line Graphs

The graph in Figure 6.6 contains only one line per category. But a graph can have multiple lines to show how a number of dependent variables (conditions, products) compare with one another.

The six-month sales figures for three salespeople can be seen in the graph in Figure 6.6. The graph contains a separate line for each of the three salespersons. At a glance, readers can see how the three compare and how many dollars each salesperson generated per month. Note how the line representing each person is clearly

A simple line graph showing the amount of snowfall in Springfield from November 2010 to April 2011.

Figure 6.5

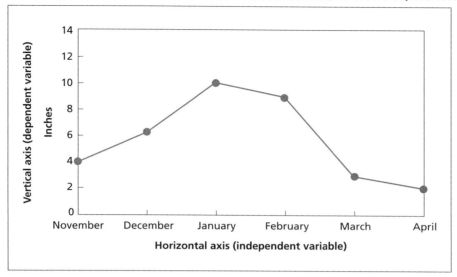

Graph is easy to read and follow

A multiple-line graph showing sales figures for the first six months of 2011 for three salespeople.

Figure 6.6

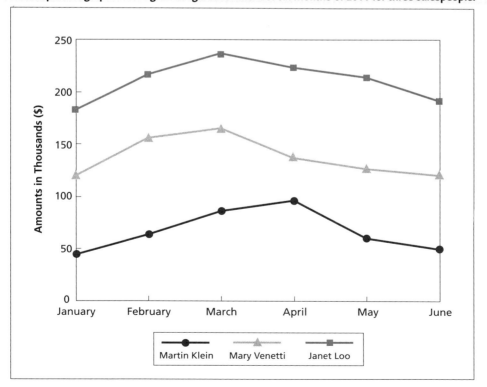

Uses different colors and symbols to represent each salesperson

Legend explains what different symbols and colors represent

differentiated from the others by symbols and colors. Each line is clearly tied to a **legend** (an explanatory key below the graph) specifying the three salespersons.

Guidelines for Creating Graphs

1. Use no more than three lines in a multiple-line graph, so readers can interpret the graph more easily. If the lines run close together, use a legend to identify individual lines.
2. Label each line to identify what it represents for readers. Include a key, or legend, as in Figure 6.6.
3. Keep each line distinct in a multiple-line graph by using different colors, dots or dashes, or other symbols. Note the different symbols in Figure 6.6.
4. Make sure you plot enough points to show a reasonable and ethical range of the data. Using only three or four points may distort the evidence. See page 231 on unethical uses of graphics.
5. Keep the scale consistent and realistic. If you start with hours, do not switch to days or vice versa. If you are recording annual rates or accounts, do not skip a year or two in order to save time or be more concise.

Charts

Among the most frequently used charts are (1) pie, or circle, charts, (2) bar charts, (3) organizational charts, and (4) flow charts.

Pie Charts

Pie charts are also known as **circle charts,** a name that descriptively points to their construction and interpretation. Tables are more technical and detailed than pie charts. Figure 6.7 shows an example of a pie chart used in a government document.

Figure 6.7	**A three-dimensional pie chart showing the breakdown by department of the proposed Palmer City Budget for 2012.**

Puts largest slice first

Does not use more than six slices

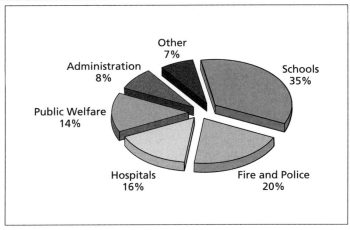

Source: Palmer City Statistics

A table or graph with a more detailed breakdown of, say, a city's budget would be much more appropriate for a technical audience (auditors, budget and city planners).

The full circle, or pie, represents the whole amount (100 percent or 360 degrees) of the data being represented; the entire budget of a company or a family, a population group, an area of land, the resources of an organization or institution. Each slice or wedge represents a percentage or portion of the whole.

A pie chart effectively allows readers to see two things at once: the relationship of the parts to one another and the relationship of the parts to the whole.

Preparing a Pie Chart

Follow these seven rules to create and present your pie chart.

1. **Keep your pie chart simple.** Don't try to illustrate technical statistical data in a pie chart. Pie charts are primarily used for general audiences.

2. **Do not divide a circle, or pie, into too few or too many slices.** If you have only three wedges, use another visual to display them (a bar chart, for example, discussed below). If you have more than seven or eight wedges, you will divide the pie too narrowly, and overcrowding will destroy the dramatic effect. Instead, combine several slices of small percentages (2 percent, 3 percent, 4 percent) into one slice labeled "Other," "Miscellaneous," or "Related Items."

3. **Make sure the individual slices total 100 percent, or 360 degrees. Do the math.**

4. **Put the largest slice first, at the 12 o'clock position, and then move clockwise with proportionately smaller slices.** Schools occupy the largest slice in Figure 6.7 because they receive the biggest share of taxes.

5. **Label each slice of the pie horizontally.** Do not put in a label upside down or slide it in vertically. If the individual slice of the pie is small, draw a connecting line from the slice to a label positioned outside the pie.

6. **Shade, color, or cross-hatch slices of the pie to further separate and distinguish the parts.** Note how Figure 6.7 effectively uses color. But be careful not to obscure labels and percentages; also make certain that adjacent slices can be distinguished readily from each other. Do not use the same color or similar colors for two adjacent slices.

7. **Give percentages for each slice to further assist readers,** as in Figure 6.7.

Bar Charts

A bar chart consists of a series of vertical or horizontal bars that indicate comparisons of statistical data. For instance, in Figure 6.8 (page 220), vertical bars depict increases in the number of working mothers. Figure 6.9 (page 221) uses horizontal bars to depict the nation's top 20 metropolitan areas in 2009, based on building permits issued during that calendar year (at the time Houston lead the nation). The length of the bars is determined according to a scale that your computer software can easily calculate.

Figure 6.8   Vertical bar chart.

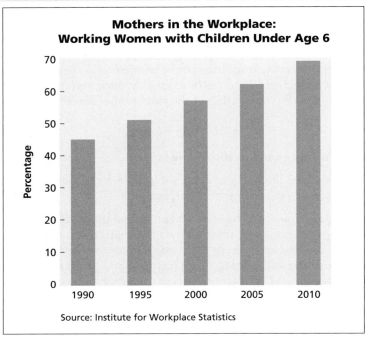

Bars are evenly spaced and clearly labeled

Length of bar determined by the percentages listed on the left hand side of visual

Organizational Charts

An organizational chart denotes the chain of command in a company or agency, with the lines of authority stretching down from the chief executive, manager, or administrator to assistant manager, department heads, or supervisors to the work force of employees. Figure 6.10 (page 222) shows a hospital's organizational chart for its nursing services.

Organizational charts have many advantages:

- informing employees and customers about the makeup of a company
- depicting the various offices, departments, and units
- showing where people work in relationship to each other in a business
- coordinating employee efforts in routing information to appropriate departments

Flow Charts

A **flow chart** displays the stages in which something is manufactured, is accomplished, develops, or operates. Flow charts are highly effective in showing the steps of a procedure. They can also be used to plan the day's or week's activities.

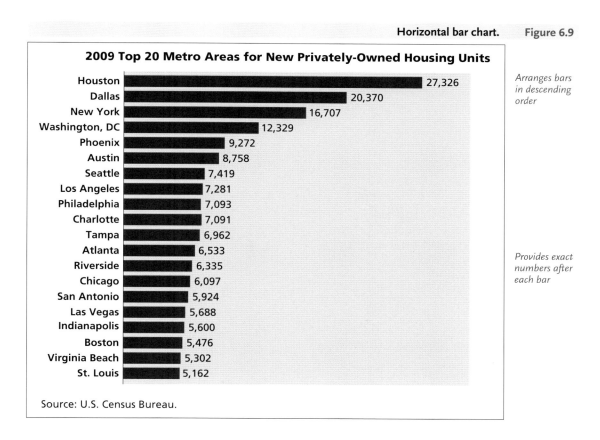

2009 Top 20 Metro Areas for New Privately-Owned Housing Units

Houston	27,326
Dallas	20,370
New York	16,707
Washington, DC	12,329
Phoenix	9,272
Austin	8,758
Seattle	7,419
Los Angeles	7,281
Philadelphia	7,093
Charlotte	7,091
Tampa	6,962
Atlanta	6,533
Riverside	6,335
Chicago	6,097
San Antonio	5,924
Las Vegas	5,688
Indianapolis	5,600
Boston	5,476
Virginia Beach	5,302
St. Louis	5,162

Source: U.S. Census Bureau.

Arranges bars in descending order

Provides exact numbers after each bar

A flow chart tells a story with arrows, boxes, and sometimes pictures. Boxes are connected by arrows to show the stages of a process. Flow charts often proceed from left to right and back again, as in the one below, showing the steps students must take to graduate

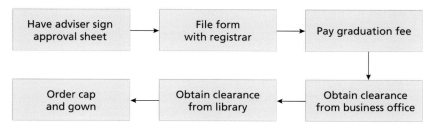

Flow charts can also be constructed to read from top to bottom. Computer programming instructions often are written that way. See, for example, Figure 6.11 (page 233), which like a computer programming chart lists the steps that an employee must follow in writing a research report.

Figure 6.10   An organizational chart representing critical care nursing services at Union General Hospital.

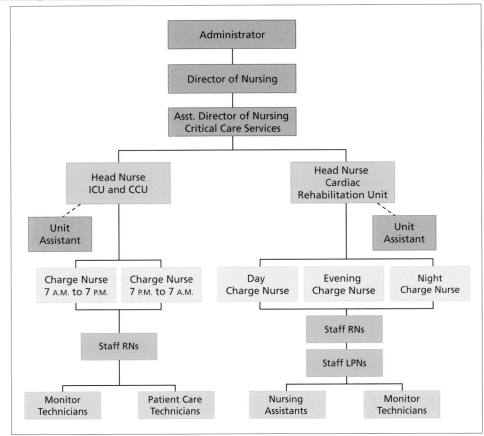

Pictographs

A **pictograph** uses picture symbols (called **pictograms**) to represent differences in statistical data, as in Figure 6.12 (page 234). A pictograph repeats the same symbol or icon to depict the quantity of items being measured. Each symbol or icon stands for a specific number, quantity, or value. Visually appealing and dramatic, pictographs are appropriate for a nontechnical audience trying to understand a process.

When you create a pictograph, follow these three guidelines:

1. Choose an appropriate symbol for the topic—such as a smartphone icon to represent the increase in the number of sales of BlackBerrys.
2. Always indicate the precise quantities involved by placing numbers after the pictures or at the top of the visual.
3. Increase the number of symbols rather than their sizes because differences in size are often difficult to construct accurately and harder for readers to interpret.

A programming flow chart showing steps in writing a research report.   Figure 6.11

Select collabo-
rative writing team

↓

Meet with team

↓

Set up ground rules

↓

State objectives

↓

Prepare working
bibliography

↓

Are there enough   No
relevant sources? ————————→ Gather more
                                    sources
Yes ↓

Take notes on topic

↓

Prepare preliminary
outline

↓

Are there enough   No   Take more
notes on topic? ————————→   notes

Yes ↓

Draft outline

↓

Write drafts

↓

"Cool off" for a
couple of days

↓

Revise and edit draft
Read for content,
organization, style

↓

Choose most effective
layout and visuals

↓

Prepare final copy

↓

Proofread—check for spelling,
punctuation, and grammar

↓

Submit
completed report

↓

Stop

*Process moves
from start to
finish*

*Arrows clearly
tell readers
what steps to
take and when*

*Indicates when
and why a step
may have to
be repeated*

*Each step
included in a
separate box*

Figure 6.12   A pictograph showing the growth of one state's retirement assets.

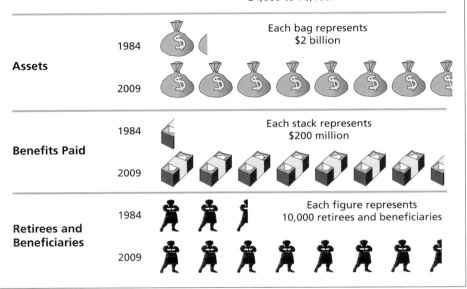

How does the Public Employees' Retirement System (PERS) compare today with 25 years ago?

During that time, PERS experienced spectacular growth, with both assets and benefits paid increasing more than sixfold.

- The market value of assets increased from $2.3 billion to $15.5 billion.
- PERS paid benefits of $94.3 million in FY '92. The total reached $1.525 billion during FY '09.
- Retirees and beneficiaries drawing benefits more than doubled, from 24,000 to 76,100.

Assets

1984

2009

Each bag represents $2 billion

Benefits Paid

1984

2009

Each stack represents $200 million

Retirees and Beneficiaries

1984

2009

Each figure represents 10,000 retirees and beneficiaries

Source: *PERS Member Newsletter*, February 2010. Official Publication of the Public Employees' Retirement System of Mississippi. By permission of Public Employees' Retirement System.

Maps

The maps you use on the job may range from highly sophisticated and detailed geographic tools to simple sketches such as the map in Figure 6.13, which shows the location of a town's water filter plants and pumping stations.

You may have to construct your own map, like the one in Figure 6.13, or scan one in a printed source or on the Internet. If you scan a map, be sure to obtain permission to use it from the copyright holder.

Visit www. cengage.com/ english/kolin/ writingatwork concise3e for an online exercise, "Assessing Online Maps, Photographs, Drawings, and Clip Art."

Guidelines for Creating a Map

Follow these steps when you create a map:

1. Always acknowledge your source if you did not construct the map yourself.
2. Use distinct lines, colors, symbols, and shading to indicate features.
3. If necessary, include a legend, or map key, explaining dotted lines, colors, shading, and symbols, as in Figure 6.13.

A map showing the location of the Smithville Water Department's filter plants and pumping stations. Figure 6.13

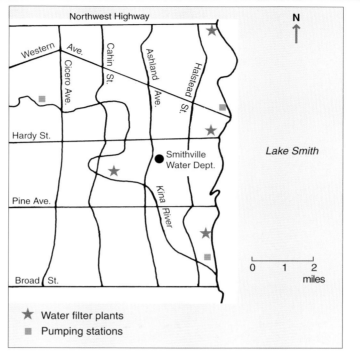

Provides directional indication

Includes only necessary information

Includes a legend explaining symbols

4. Exclude features (rivers, elevations, county seats) that do not directly relate to your topic. For example, a map showing the crops grown in two adjacent counties need not show all the roads and highways in those counties.
5. Indicate direction. Conventionally, maps show north, often by including an arrow and the letter: $N\uparrow$.

Photographs

Correctly taken or scanned, photographs are an extremely helpful addition to job-related writing. A photograph's chief virtues are realism and clarity, as Figure 6.14 (page 226) illustrates. Among its many advantages, a photo can

- show what an object looks like
- demonstrate how to perform a certain procedure
- compare relative sizes and shapes of objects
- compare and contrast scenes or procedures

Digital cameras and camera phones allow you to supply professional-looking, customized photos with your written work. Using them, you can send photographs

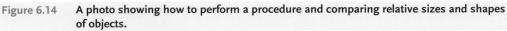

Figure 6.14   **A photo showing how to perform a procedure and comparing relative sizes and shapes of objects.**

Source: ©Corbis

quickly, store them on your computer or on a memory card, or upload them to a website. Moreover, with Photoshop, or other image-manipulation software, you can edit photographs for color, sharpness, contrast, or brightness, and as well as crop them to omit unnecessary details.

Guidelines for Taking Effective Photographs

Whether you use a traditional or digital camera, observe the following guidelines:

1. **Take the photo from the most appropriate distance.** Decide how much foreground and background information your audience needs. For example, if you are photographing a three-story office building, your picture may misleadingly show only one or two stories if you are standing too close to the building.

2. **Select the correct angle.** Take the photo from an angle that makes sense, usually straight on so that the object or person photographed can be viewed in full.

3. **Include only the details that are necessary and relevant for your purpose.** Crop any unnecessary details from the photograph.

4. **Provide a sense of scale.** So that readers understand the size of the object, include a person in the photograph (if the object is very large), a hand (if the object is small), or a ruler.

5. **Make sure you consider lighting and resolution.** Take your photograph in appropriate light, so that the subject of your photograph is clear and crisp. If

you are using a digital camera, take a high-resolution photograph for maximum clarity.

6. **Always ask for permission before you take a photograph of a person, place, or thing, unless you are photographing a public park or building.** Do not take photographs of copyrighted materials, for instance, a work of art, a page from a copyrighted book, or a movie or television screen. Also obtain permission from your boss if you are photographing your company's equipment, sites, or designs.

Drawings

Drawings can show where an object is located, how a tool or machine is put together, or what signals are given or steps taken in a particular situation. A drawing can be simple, such as the one in Figure 6.15, which shows readers exactly where to place smoke detectors depending on the size of their homes.

A simple drawing showing where to place smoke detectors in a house. Figure 6.15

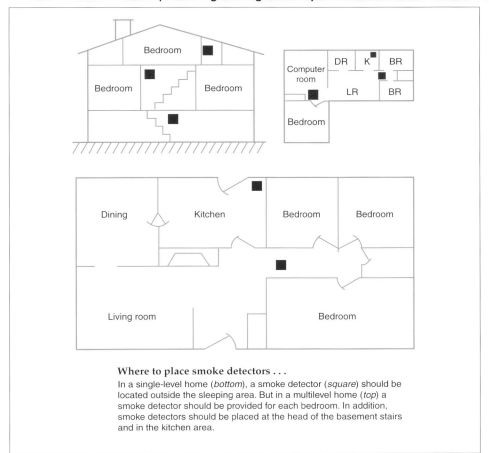

Where to place smoke detectors . . .
In a single-level home (*bottom*), a smoke detector (*square*) should be located outside the sleeping area. But in a multilevel home (*top*) a smoke detector should be provided for each bedroom. In addition, smoke detectors should be placed at the head of the basement stairs and in the kitchen area.

Source: Reprinted by permission of *Southern Building.*

Figure 6.16   Cutaway drawing of an electric car.

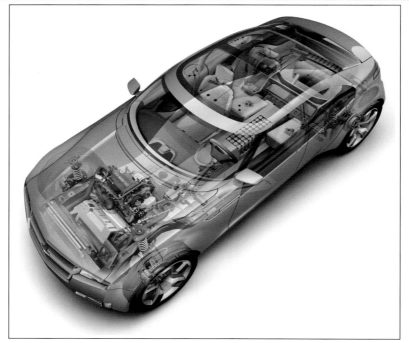

Source: General Motors Corporation. Used with permission, GM Media Archives.

A more detailed drawing can reveal the interior of an object. Such sketches are called **cutaway drawings** because they show internal parts normally concealed from view. Figure 6.16 is a cutaway drawing of an extended-range electric vehicle, the GM Volt. The drawing shows the passenger interior and the battery-powered engine, which allows the car to drive up to 40 miles without the simultaneous use of gasoline.

Another kind of sketch is an **exploded drawing**, which blows the entire object up and apart, as in Figure 6.17, to show how the individual parts are arranged. An exploded drawing comes with most owner's guides to computers and uses **callouts**, or labels, to identify the components. The labels are often attached to the drawing with arrows or lines.

Guidelines for Using Drawings

1. Keep your drawing simple. Include only as much detail as your reader will need to understand what to do, be it to assemble or to operate a mechanism.
2. Clearly label all parts so that your reader can identify and separate them.
3. Decide on the most appropriate view of the object to illustrate—aerial, frontal, lateral, reverse, exterior, interior—and indicate in the title which view it is.
4. Keep the parts of the drawing proportionate unless you are purposely enlarging one section.

Exploded drawing of a laptop computer. Figure 6.17

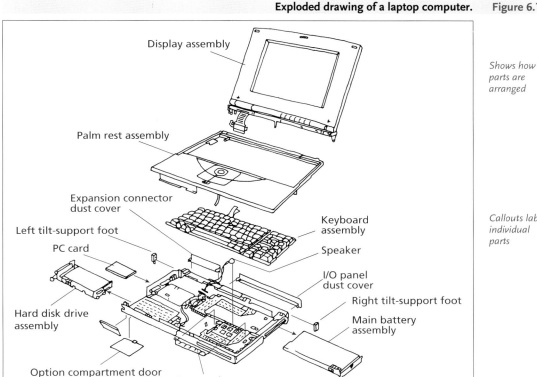

Shows how parts are arranged

Callouts label individual parts

Dell Publishing Co., Inc.

Clip Art

Clip art (or icons) refers to ready-to-use electronic images. These small cartoon-style representations and photographs, such as the ones shown in Figure 6.18 (page 230), depict almost anything you can think of under such headings as energy, government, leisure, health, money, the outdoors, food, technology, and transportation. Free clip art and photo-illustration databases can be found at the following websites (among others): http://www.wpclipart.com, http://www.reusableart.com, and http://freerangestock.com.

When you use clip art, follow these guidelines:

1. **Choose simple, easy-to-understand icons.** Select an image that conveys your idea quickly and directly. Avoid using an icon of an unfamiliar object or of a drawing or silhouette that might confuse your audience, especially a global one.
2. **Use clip art functionally.** Do not insert clip art as decorations. Including too many will make your work look unprofessional.

Figure 6.18   Examples of clip art

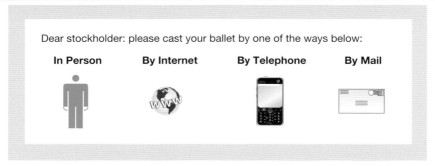

3. **Make sure the clip art is relevant for your audience and your message.** A clip art airplane does not belong in a technical report on fuel capacity or jet engine design.
4. **Make sure your clip art is professional.** Some clip art is humorous, even silly, which may not be appropriate for a professional business report or proposal.

Using Visuals Ethically

Make sure your visuals, whether you create or import them, are ethical. Like your words, your visuals should represent you, your company, and the data truthfully. Ethical visuals convey and interpret statistical information and other types of data, products and equipment, locations, and even individuals without misinterpretation. Your visuals should neither distort events and statistics nor mislead readers. Ethical visuals should be:

- accurate
- honest, fair
- complete
- appropriate
- easy to read
- clearly labeled
- uncluttered
- consistent with conventions

Guidelines for Using Visuals Ethically

To ensure that your visuals are ethical, honest, accurate, and easy to read, avoid the following unethical practices no matter what type of visual you use.

Photos

- Do not distort a photo by omitting key details or by misrepresenting dimensions, angles, sizes, or surroundings or by superimposing one image over another.
- Do not take a photo of your most expensive, top-of-the-line product/model but then place the cost of your lowest-priced product/model under it.
- Do not misrepresent location—for example, taking a photo in a "doctored" or off-site location, studio, or lab and then claiming it as an "actual" location shot.

- Never take a photo of an individual for business purposes (e.g., putting it in a newsletter, ad, or on the web) without his or her permission.
- Don't counterfeit or subtly alter a company's logo to sell, distribute, or promote an imitation as the real thing.

Graphs

- Don't distort a graph by plotting it in misleading or unequal intervals—for example, omitting certain years or dates to hide a decline in profits. Contrast Figure 6.19, and its misleading interpretation, with the ethical revision in Figure 6.20 (page 232).
- Include information in correct chronological sequence along the horizontal axis. Note how Figure 6.19 omits key years.
- Don't switch the type of information usually given along the vertical axis with the horizontal axis.
- Don't project growth or increases without having reliable and valid reasons.
- Don't misrepresent data or trends by making increments along the vertical axis too limited, leaving a much smaller (and incomplete) area to represent. When data are plotted wrongly this way, readers are unethically led to misinterpret the numbers—to read that there was little loss in revenue, or no change in sales, for example. For instance, if the horizontal axis begins at $5 and advances to $6 a share, you leave only an intentionally small and misleading area to measure. If stocks fell below $5 a share, your graph would unethically not represent those declines.

An unethical graph and misleading interpretation.

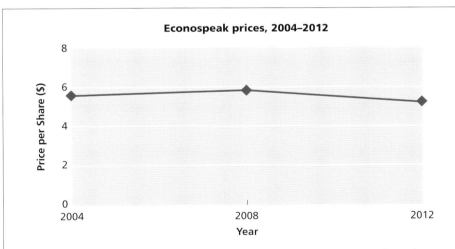

Econospeak's prices during 2004–2012 have been stable, resting securely at about $5.60. The graph above illustrates the stability of Econospeak's stock. Given our steady market, we believe shareholders will be confident in our recent decision to proceed with Econospeak's expansion into global markets.

| Figure 6.20 | Ethical revision of Figure 6.19. |

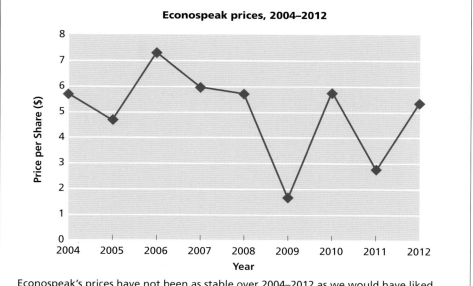

Econospeak's prices have not been as stable over 2004–2012 as we would have liked. The graph above illustrates the challenges the company has faced in the market in the past decade, resulting in fluctuation of prices. We believe, however, that Econospeak's expansion into the global market will be possible by 2013.

Bar Charts

- Don't use color or shading to mislead or distort—for example, shading one bar to make it more prominent than the others.
- Make sure the height and width of each bar truthfully represents the data it purports to. That is, don't make one of the bars larger to maximize the profits, products, or sales in any one year.
- Show bars for every year (or other sales period) covered. Note how the unethical bar chart (and accompanying text) in Figure 6.21 violates this rule, but the chart in Figure 6.22 ethically represents the data.

Pie Charts

- Don't use 3-D to distort the thickness of one slice of the circle and thereby misleadingly deemphasize other slices.
- Avoid concealing negative information (losses, expenses, etc.) by silently including the information in another category, or slice, or lumping it into a category marked "other" or "miscellaneous."
- Make sure percentages match the number and size of the wedges or slices of the pie chart. Study Figures 6.23 and 6.24 on page 234. Note how a larger expense for guest speakers (35% of budget) is unethically misrepresented in Figure 6.23 by using a smaller-sized wedge, while the expenses for venue rental are actually less than for guest speaker expenses but drawn larger to misrepresent costs.

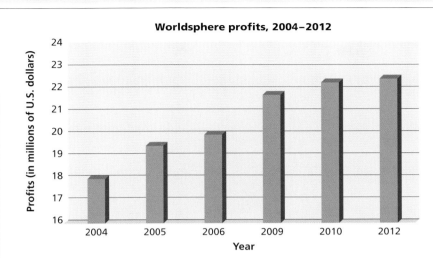

Profits at Worldsphere have seen a healthy increase over recent years. The above bar chart demonstrates the steady increase in profits, which have risen $4.5 million since 2004. Given the profit history of Worldsphere, our investors can be confident of future profits and the security of their stock in our company.

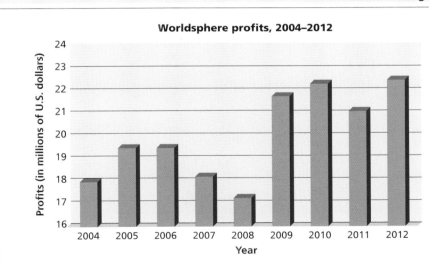

Although profits at Worldsphere have been volatile in the past eight years, we are now at our highest profit margin of the last 8 years. The bar chart above shows the effects ofmarket difficulties for the period 2004–2012, when the economy suffered major cutbacks. However, Worldsphere achieved a successful turnaround in 2009, with profits regaining strength due to advances in research and technology.

Figure 6.23   **An unethical pie chart.**

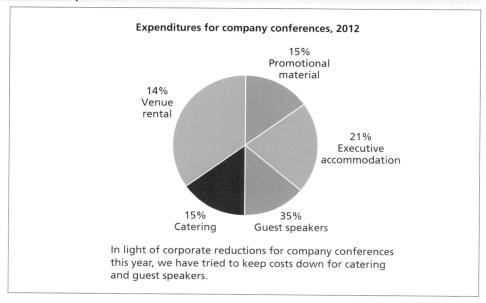

Expenditures for company conferences, 2012

15%
Promotional
material

21%
Executive
accommodation

35%
Guest speakers

15%
Catering

14%
Venue
rental

In light of corporate reductions for company conferences this year, we have tried to keep costs down for catering and guest speakers.

Figure 6.24   **Ethical revision of Figure 6.23.**

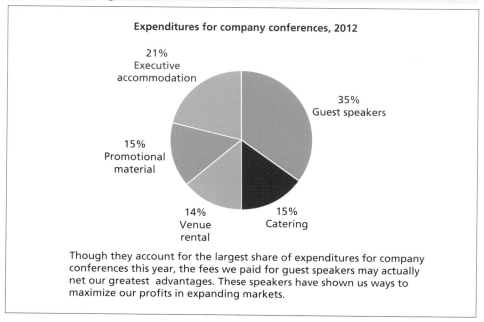

Expenditures for company conferences, 2012

21%
Executive
accommodation

35%
Guest speakers

15%
Promotional
material

14%
Venue
rental

15%
Catering

Though they account for the largest share of expenditures for company conferences this year, the fees we paid for guest speakers may actually net our greatest advantages. These speakers have shown us ways to maximize our profits in expanding markets.

Drawings

- Avoid any clutter that hides features.
- Label all parts correctly.
- Do not omit or shadow any necessary parts.
- Draw an object accurately. Tell readers if your drawing is the actual size of the object or equipment or if it is drawn to scale. Provide a scale.

Using Appropriate Visuals for International Audiences

Whether you are writing for an expanding international business community in South America, China, or for multicultural readers in the United States, you will have to prepare numerous documents that require visuals. These can range from instructions containing warning and caution statements to tables, graphs, charts, and photos for proposals, reports, and online presentations.

Choose your visuals for international readers just as carefully as you do your words. Visuals should be designed to help international audiences understand your message clearly and without bias. Visuals and other graphics must be consistent with your message, not detract from or distort it.

Visuals Do Not Always Translate from One Culture to Another

While there are some internationally recognized icons such as those in Figure 6.25 below, keep in mind that visuals do not automatically transfer from one culture to another. They are not boilerplate images that have the same meaning for all readers across the globe. Like words, visuals—photos, clip art, designs, colors, and other graphic devices—may have one meaning or use in the United States and a radically different one in other countries around the globe.

To avoid confusing or offending an international audience, investigate the meaning and cultural significance of a visual before you include it. Consult a native speaker from your audience's country to see if your visuals are culturally acceptable; then follow up with your collaborative team and any graphic artists assigned to the project to ensure that every visual is appropriate and accurate.

Internationally recognized icons.   Figure 6.25

Guidelines for Using Visuals for International Audiences

To communicate appropriately and respectfully with international readers through visuals and other graphics, follow these guidelines:

1. Do not use any images that ethnically or racially stereotype your readers. A large clothing manufacturer was criticized by the Asian American community when it released T-shirts depicting two men with slanted eyes and rice paddy hats as laundromat owners. Similarly, depicting Native Americans through clip art images of red-faced chiefs is insulting. Rather than using an ethnic or racial pictogram, use neutral stick figures or non-biased clip art.

2. Be respectful of religious symbols and images. A cross to a U.S. audience represents a church, first aid, or a hospital, and, when the cross is a large red one on a white background, the Red Cross. But the symbol for that humanitarian agency in Turkey and in other Muslim countries is the crescent. Portraying a smiling Buddha to sell products is considered disrespectful to residents in Southeast Asia.

3. Avoid using culturally insensitive or objectionable photographs. What is acceptable in photographs in the United States may be taboo elsewhere in the world. Portraying men and women eating together at a business conference is unacceptable in Saudi Arabia. Moreover, depicting individuals seated with one leg crossed over the other is regarded as a disrespectful gesture in many countries in the world. It may be safest not to show photographs for an international audience.

4. Avoid any icons or clip art that international readers would misunderstand. In the United States, an owl can stand for wisdom, thrift, and memory, while in Japan, Romania, and some African countries it is a symbol for death. A software program showing the icon of a mailbox (below) for the designation "You Have Mail" confused readers in other countries who thought the icon represented a birdhouse. A better alternative would be an icon of an envelope.

5. Be cautious about importing images or photos with hand gestures in your work, especially in manuals or other instructional materials. Many gestures are culture-specific; they do not necessarily mean the same thing in other countries that they do in the United States. Table 6.2 lists cultural differences around the globe for some common gestures.

6. Don't offend international readers by using colors that are culturally inappropriate. Emphasizing the color red in a document aimed at Native Americans or the color yellow in a document for Asians or Asian Americans is likely to be seen as offensive. Yet red in China symbolizes happiness while yellow in Saudi Arabia signifies strength.

TABLE 6.2   Different Cultural Meanings of Various Gestures

Gesture	Meaning in the United States	Meaning in Other Countries
OK sign Index finger joined to thumb in a circle	All right; agreement	Sexual insult in Brazil, Germany, Russia; sign for zero, worthlessness in France
Thumbs-up	A winning gesture; good job; approval	Offensive gesture in Muslim countries
Pointing with the index finger	This way; pay attention; turn the page	Rude, insulting in Japan and Venezuela; in Saudi Arabia used only for animals, not people
Motioning with index finger	Signal to "come over here"	Insulting in China—instead, extend arm and wave to ask "come over"

Source: Institute for Study of World Cultures

Always research (on the Internet or by consulting a representative from the audience you want to reach) if the color scheme you have chosen is appropriate for your target audience. For example, purple is the color of death and mourning in Thailand, as is white in China. Although orange is the symbolic color of Northern Ireland, you should avoid it when writing to readers whose culture does not value that color.

7. Be careful using directional signs and shapes. While road signs tend to be fairly recognizable throughout the world—for example, the octagon is the shape for the stop sign—there are country-specific signs and code books. For instance, a pennant on American highways, exclusively signaling a no-passing zone, may not have the same meaning in Nigeria or India. The symbol below points to a railroad crossing for American readers but would baffle an audience in the Czech Republic.

8. Avoid confusing an international audience with punctuation and other writing symbols used in the United States. Not all cultures use a question mark (?) to end a sentence asking a question, to represent the **Help** function in a computer program, or for an **FAQ** link on a website. For instance, ellipses (. . .) and slashes (/) may not be a part of the language your international audience reads and writes.

Writing for and Designing Websites*

In today's global marketplace, you can apply the skills you have just learned in designing documents and visuals to an online environment (websites, blogs, etc.). In fact, companies often ask their employees to write for and prepare visuals for a corporate website, and while you won't be expected to construct such as website on your own, you will likely be part of a team of specialists in information technology, graphics, marketing, and engineering who are responsible for your firm's web presence. To be a helpful member of this team, your employer will expect you to keep up with the latest features of website design and information on how your company can incorporate them. You may also be asked to critique a competitor's website or to assist a client in preparing his/her website. This section of Chapter 6 stresses the principles and guidelines you need to follow to

- help readers navigate your website quickly, moving from one part to another easily
- ensure that your site is current
- make your site user friendly (for global readers as well as English-speaking audiences)
- create a visually attractive site
- keep your design and content ethical

Web Versus Paper Pages

Writing and designing a website requires the same emphasis on planning, drafting, formatting, and testing that printed business documents do. Just as with these paper documents, writers must take into account spatial, rhetorical, and visual considerations when they prepare work for a website. Yet, clearly, a website differs from a hard copy memo, letter, or report in the way content is organized and visually illustrated.

Most important, websites are not fixed on the screen. They can change, be edited, and be available to a far greater audience far more quickly than any print source ever conceivably could be. Every website is really a work in progress. But in order to know how to write for a website, you need to understand how it differs from printed documents.

Web Versus Print Readers

To help design and write for the web, you need to recognize how webpages are read differently from the way print documents are.

1. **Readers do not generally go through a website word for word hunting for information, carefully studying each sentence.** They want to find information at a glance. On the average, Web readers spend 10 to 80 seconds scanning a page. They will not waste their time scrolling in the hopes of finally finding what they need. They'll just click to another site.

---

*I am grateful to Michael Tracey, of Bay St. Louis, Mississippi for his invaluable advice on building and designing websites.

2. **Web readers want articles, news stories, and features to be more condensed, and more strategically arranged** to give only essential information. Unlike a print source, blocks of text are intimidating on the web. Some experts advise using half the word count of a paper document for a website.

3. **A web audience will not necessarily read your entire website.** In fact, they may not have even begun their search at your site. So you can't assume your Web audience, like one for a printed source, will read from the start to the finish of your site. Don't expect to write to build up a climax or to postpone key facts. Web readers may not navigate through your pages in any predetermined order or even read all of its parts. They can start with a home- or later page and then go to any other page on that site.

4. **Text on the Web needs to be shorter because of the size of the screen.** Text displayed on a screen will be harder to read than on a piece of paper. Lots of text, along with clip art and images, on a website squeezed onto a screen, coupled with the problem of glare on the computer screen, can lead to eyestrain and fatigue for a reader more quickly than in a print source.

5. **Web readers will expect navigational cues that readers of a print source do not have.** A web audience will be looking for visual markers such as highlighted keywords to click on or bulleted lists, different colored text, commands such as *search*, *click here*, *go to*, *go back*, *contact us*, arrows and crosses, or hyperlinks of other sites to visit. Note how these cues are incorporated into the website in Figures 6.26 on page 240.

Preparing a Successful Homepage

A homepage has two main functions. First, it has to catch visitors' attention. If it fails to do that, everything else is a waste of time; your visitors will click their mouse and be gone. Second, it has to sell your company's product or service or to introduce your organization. That may seem backward, but unless you attract readers to your website, you cannot promote your business, share news, feature new products, and encourage shopping via your site.

Successful homepages accomplish these objectives clearly and effectively. For example, students applying for financial aid can find it a daunting process, but the design of the FinAid homepage, Figure 6.26, keeps things upbeat and easy to follow. Even the image of the student "jumping for joy" contributes to the site's user-friendliness. The writing style and tone, for example, are appropriately conversational, friendly, and helpful—"Find everything from grants to... tuition payment plans"; "Beware scholarship scams." The FinAid homepage also provides clear links to the various parts of the website.

Designing and Writing for the Web: Eight Guidelines

Designing and writing for a website requires you to follow many of the rules you read about in Chapter 2 (pages 38–47) and earlier in this chapter. By adhering to the following eight guidelines, you can create an effectively designed and written website that will capture your audience's attention.

Figure 6.26 Sample homepage from FinAid.

Left column headers use appropriate icons

Uses concise and clear language

Page is carefully organized, visually balanced, and uncluttered

Links arranged by audience needs

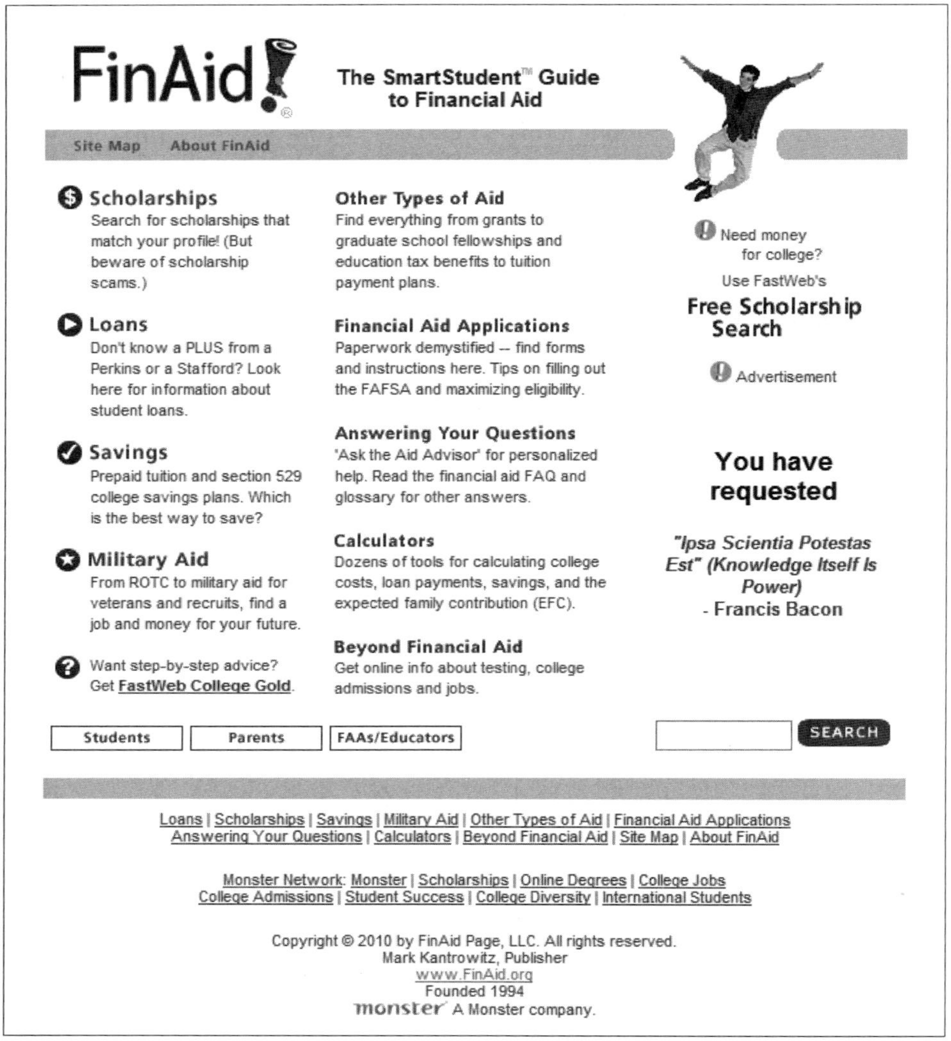

1. **Make your site easy to find.**
   - Obtain a registered domain name for your site. Choose a professional domain name that clearly and quickly tells readers what you do. Avoid cute spellings and fanciful names, e.g., happihouse.com. Instead use toledohousepainters.com. A good place to start is the InterNIC website (*http://www.internic.net*), which provides updated information on Internet domain name registrations.
   - Submit your website to the various search engines and directories (Google, Yahoo!, Search.com, Business.com) to make sure the largest possible audience knows about your site.

■ Make your site search-engine ready. Robin Noble offers this helpful advice, "Since search engines read webpages from top to bottom, include your keywords in the title, description, and tags in the headline. You also want to include keyword-containing text in the first line of your first two or three paragraphs." For instance, you might consider using words such as "Discount," "Latest," or "Improved" in your title to emphasize customer benefits

2. **Make your site easy to download and navigate.**

■ Help your visitors find their way easily through your site with logical and effective navigation tools such as:

hyperlinks	search engines
navigation bars	button links
indexes and menus	rollover icons
site maps and tables of contents	previous, next, and back links

Again note how the FinAid website in Figure 6.26 provides multiple navigation aids: menu buttons that are labeled and illustrated with icons and a site map at the bottom of the page.

■ Test every link to make sure it is current, that it works, and that it connects to related sites. A website that does not identify one of its links or that contains a broken link will drive readers/customers away.

■ Don't overload your page with images so your site look crowded. See again how Figure 6.26 uses a single image, six smaller icons, and boxes at the bottom left.

3. **Make your site informative.**

■ Provide essential information on or through your homepage, including your company's name, address, e-mail, phone numbers, corporate blogs, etc. The more helpful your site is, the more likely it will draw repeat visitors.

■ Tell the visitor what products or services you offer. Figure 6.26 offers tailor-made services to an audience of students, parents, and educators.

■ Indicate what type of information can be obtained through your website, including links to your sales force, customer service, and technical support. See how readers in Figure 6.26 can easily locate information on scholarships, savings, military aid, etc.

■ Offer readers different types of interaction—FAQs, bulletin boards, animated product demonstrations, and free e-mail subscriptions. For example, FinAid in Figure 6.26 offers "Calculators" and "Applications."

4. **Make your site easy to read for both native English speakers and international readers.**

■ Put the most important point first in a seven- to eight-word headline (e.g. "The Smart Student Guide to Financial Aid").

■ Get to the point right away. Don't begin with a question or background information.

- Write short paragraphs (chunks of content)—no more than three to four sentences long.
- Provide headings with attention-grabbing keywords, bulleted lists, and numbered lists to help readers locate information quickly.
- Use plain, concise English—concrete nouns, action verbs. And keep sentences short—use only active voice. Write in a common sentence pattern: (subject-verb-object).
- Respect international readers by using, or converting to, units of measurement they are familiar with (see page 137).
- Include plenty of white space but not after each sentence. Single-sentence paragraphs are harder to process.
- Provide scannable terms and hyperlinks; always highlight them to make them stand out.
- Select fonts that are easy to read—review pages 203–205. Use larger fonts than you would for a print document.
- Select background colors that make your text easy to read. Don't use light yellow text against a white or light blue background or dark green lettering on a black background.

5. **Keep your site updated.**

- New information is vital to sell your product and service on a company website. Otherwise, readers have no reason to make return visits. Feature insider news updates about your business or pre-production information on products services, community projects, greening efforts, etc. Build in hyperlinks to product reviews, conferences, awards, and so on.
- Revise the design of your site if your company offers a new product or service or a new promotional element on its homepage. Clearly, your site does not need a major design overhaul every week, but a new or revised homepage alerts customers to the latest products and services.
- Indicate when your site was last updated, so readers will know your information is kept current.
- Periodically check your external links to make sure they work.

6. **Use images and icons effectively.**

- Arrange images and photos so they do not interfere with text or layout. Proportion is important for achieving a balance between different page elements.
- Choose appropriate icons or images to illustrate menus and page sections. See how Figure 6.26 uses easy-to-recognize icon's for money (costs) of the military.
- Be conservative in using animations or anything that might be viewed as a gimmick because they may distract readers from other content on your page.
- Keep images proportional so that they are not too big or small for the page.

7. **Encourage visitor interaction by soliciting feedback.**

- Ask readers to e-mail you about your product, service, or website. Alert them to any relevant blog sites. And make sure a procedure for handling that feedback is developed within your organization.

- Include a feedback form or survey with specifically targeted questions on your website to encourage visitors to leave useful comments.
- Conduct a usability test. Ask readers what you could do better or what they like about your website. Strive for interaction and connectivity.

8. **Make sure your website is ethical.**

- Never post confidential or proprietary information.
- Avoid posting anything insulting or harassing, or attack a competitor, colleague, other department in your company, or government agency.
- Do not plagiarize from another web (or print) source. If you include any information from another site, including quotations, visuals, or statistics, get permission and acknowledge the source on your site.
- Do not use sexist, racist, and other biased forms of language. Moreover, do not offend an international audience by using terms, names, or visuals that are insulting, stereotypical, or condescending. See pages 236–237.
- Never make false or exaggerated claims. Be honest and accurate. Earn your readers' trust.
- Provide accurate and recent statistics and other data. Let readers know you have done your research.

As this chapter has emphasized, writing requires you to choose effective visuals, layouts, and designs for any print documents or websites you may create or write for, and to make sure they are appropriate for both native-speaking and international audiences.

 Revision Checklist

Printed Documents

☐ Arranged information in the most logical, easy-to-grasp order.
☐ Left adequate, eye-pleasing white space in text and margins to frame document.
☐ Maintained pleasing, easy-to-read line length and spacing.
☐ Chose appropriate typeface for message and type of document.
☐ Did not mix typefaces.
☐ Inserted heads and subheads to organize information for reader.
☐ Used effective type size, neither too small (under 10 points) nor too large (over 12 points) for the body of the text.
☐ Supplied lists, bullets, numbers to divide information for readers.
☐ Chose colors carefully to make sure they look professional.

Visuals

☐ Selected most effective visual (table, chart, graph, drawing, photograph) to represent information the audience needs.
☐ Drafted and edited visual until it met readers' needs.

☐ Selected right amount of detail to include in visual.
☐ Made sure every visual is attractive, clear, complete, and relevant.
☐ Gave each visual a number, a title, and, where necessary, a legend and callouts.
☐ Inserted visual close to the description or commentary it goes with—same or facing page.
☐ Inserted page number where visual can be found.
☐ Introduced and interpreted each visual in appropriate place in report or proposal.
☐ Acknowledged sources for any copyrighted visuals and gave credit to individuals whose statistical data are the basis of a visual.
☐ Used and interpreted visuals ethically and appropriately.
☐ Selected visuals, colors, and images that respect the cultural traditions of my international readers.

Websites
☐ Designed website so that it is easy to find on major search engines.
☐ Made sure navigation is clear and logical, not overly complex.
☐ Provided identification for all pages either with headings or text that explains the purpose of each page.
☐ Ensured that the site is informative and relevant, and that the content is current.
☐ Kept the site current by revising it frequently and including most recent research.
☐ Used headings, subheadings, and white space to break information into readable chunks.
☐ Encouraged visitor interaction by soliciting feedback.
☐ Provided ways for reader to interact with the site, whether via e-mail, a feedback page, or a blog where comments can be posted.
☐ Did not crowd images and text on the same page.
☐ Chose appropriate background colors so text is clear and easy to read.
☐ Strove to make sure the site is ethical.

Exercises

1. Find an ineffectively designed print document—a form, a set of instructions, a brochure, a section of a manual, a catalogue, a newsletter—and assume that you are a document design consultant. Write a sales letter to the company or agency that prepared and distributed the document, offering to redesign it and any other documents they have. Stress your qualifications and include a sample of your work. You will have to be convincing and diplomatic—precisely and professionally persuading your readers that they need your services to improve their corporate image, customer relations, and sales or services.

Additional Activities related to designing documents and visuals are located at www.cengage.com/english/kolin/writingatworkconcise3e

2. Redesign the handwashing document below to make it conform to the guidelines specified in this chapter. Reformat; add headings, spacing, and visual clues; include appropriate clip art; and provide short introduction.

WHY SHOULD YOU WASH YOUR HANDS?

Bacteria and viruses (germs) that cause illnesses are spread when you don't wash your hands.

If you don't wash your hands, you risk acquiring:

The common cold or flu

Gastrointestinal illnesses Shigella or hepatitis A

Respiratory illnesses

Should you wash your hands?

You need to wash your hands several times every day. Some important times to wash your hands are:

BEFORE

Preparing or eating food.

Treating a cut wound.

Tending to someone who is sick.

Inserting or removing contacts

After

Using the bathroom.

Changing a diaper or helping a child use the bathroom (don't forget the child's hands)

Handling raw meats/poultry/eggs

Touching pets, especially reptiles

handling garbage

Sneezing or blowing your nose, or helping a child blow his/her nose

Touching any body fluids like blood or mucus

Being in contact with a sick person

Playing outside or with children and their toys

WHEN SHOULD YOU WASH YOUR HANDS?

There is a right way to wash your hands.

Follow these steps and you will help protect yourself and your family from illness.

Like any good habit, proper hand washing must be taught.

Take the time to teach it to your children and make sure they practice.

3. One government agency supplied statistics on the world production of oranges (including tangerines) in thousands of metric tons for the following countries during the years 2009–2012: Brazil, 2,098, 2,132, 2,760, 2,872; Israel, 909, 1,076, 1,148, 1,221; Italy, 1,669, 1,599, 1,766, 1,604; Japan, 2,424, 2,994, 2,885, 4,070; Mexico, 937, 1,405, 1,114, 1,270; Spain, 2,135, 2,005, 2,179, 2,642; and the United States, 7,658, 7,875, 7,889, 9,245. Prepare a table with that information and then write a paragraph in which you introduce and refer to the table and draw conclusions from it.

4. Write a paragraph introducing and interpreting the following table for a publucation aimed at general readers.

Year	Soft Drink Companies	Bottling Plants	Per Capita Consumption (Gallons)
1945	750	750	10.3
1950	578	611	12.5
1955	457	466	18.6
1960	380	407	17.2
1965	231	292	15.9
1970	171	229	15.4
1975	118	197	16.0
1980	92	154	18.7
1985	54	102	21.1
1990	43	88	23.1
1995	45	82	25.3
2000	37	78	27.6
2005	34	72	30.1
2010	31	70	32.3

5. According to a municipal study in 2012, the distribution of all companies classified in each enterprise industry in that city was as follows: minerals, 0.4%; selected services, 33.3%; retail trade, 36.7%; wholesale trade, 6.5%; manufacturing, 5.3%; and construction, 17.8%. Make a pie chart to represent the distribution, and write a one- or two-paragraph interpretation to accompany (and explain the significance of) your visual.

6. Make an organizational chart for a business or an agency you worked for recently. Include part-time and full-time employees, but indicate their titles or functions with different kinds of shapes or lines. Then write a brief letter to your employer explaining why your organizational chart should be distributed to all employees. Focus on the types of problems that could be avoided if employees had access to such a visual.

7. Prepare a flow chart for one of the following activities:
   a. jumping a "dead" car battery
   b. giving an injection
   c. making a reservation online
   d. using an iPhone to check the status of a flight
   e. checking your credit online
   f. putting out an electrical fire
   g. changing your e-mail password
   h. preparing a visual using a graphics software package
   i. putting your blog online
   j. joining a chat group online
   k. filing for an extension to pay state taxes
   l. any job you do

8. Prepare a drawing of one of the following simple tools, and include appropriate callouts with your visual.

   a. high-definition TV   e. swivel chair
   b. iPod   f. DVD player
   c. pliers   g. ballpoint pen
   d. stethoscope   h. table lamp

9. Prepare appropriate visuals to illustrate the data listed in (a) and (b). In a paragraph immediately after the visual, explain why the type of visual you selected is appropriate for the information.

   a. Life expectancy is increasing in the United States. This growth can be dramatically measured by comparing the number of teenagers with the number of older adults (over age 65) in the United States during the last few years and then projecting those figures. In 1970 there were approximately 28 million teenagers and 20 million older adults. By 1990 the number of teenagers climbed to 30 million, and the number of older adults increased to 25 million. In 2000 there were 27 million teenagers and 31 million older adults. In 2010 the number of teenagers had leveled off to 23 million, but the number of older adults soared to more than 36 million.

   b. Researchers estimate that for every adult in the United States 3,985 cigarettes were purchased in 1990; 4,100 in 1995; 3,875 in 2000; 3,490 in 2005; and 2,910 in 2010.

10. Following are two examples of poorly prepared visuals with brief explanations of how they were intended to be used. Redo one of the visuals to make it easier to read by re-organizing information. Supply a paragraph to introduce your new visual.

    a. To accompany a report on problems that pilots have encountered with a particular model of jet engine.

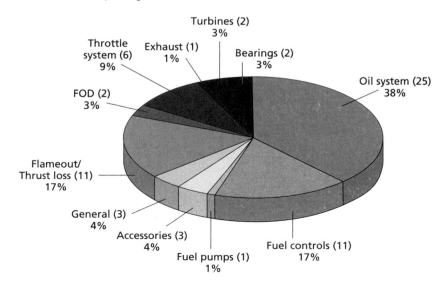

b. To show that a hiking trail is compatible with wheelchair access laws.

11. Explain why the following would be inappropriate in communicating with an international audience and how you would revise the document containing such visuals/graphics symbols:
    a. clip art showing a string tied around an index finger
    b. picture of a man with a sombrero on a website for Pronto Quick Check Cashing Service
    c. clip art/drawing of a light bulb and logo "Smart Ideas" from a CPA firm
    d. clip art showing someone crossing the middle finger over the index finger (wishing sign)
    e. drawing of a cupid figure for a caterer
    f. a sales brochure showing a white and blue flag for a French audience
    g. an advertisement showing a woman's track and field team used to advertise a brand of footware to an Arabic-speaking audience
    h. a satisfied customer making the gesture of OK in an ad aimed at a Japanese audience
    i. an image of a rabbit on a website for an automobile manufacturer to stress how fast its cars can accelerate
    j. a photograph of a roll of Scotch tape to show international readers that your company can solve problems quickly and economically
    k. a photograph of a baseball umpire holding his hands up to ask international readers to repeat a step in a set of instructions
    l. an icon of a white glove to sell home and carpet cleaning

12. Locate two webpages that advertise a similar product, service, industry, or other topic. Analyze some of the webpage elements each one uses, comparing the strengths and weaknes of each site. Write a one-page memo to your instructor explaining which is the more effective site and why

13. Find a webpage that you believe is ineffective. Using the four keys to effective writing (see Chapter 1, pages 5–12), as well as your knowledge of webpage elements, write a one-page assessment of the site, discussing three or four changes you think would make it more effective. Attach a printed hard copy of the webpage with your assessment.

7

Writing Instructions and Procedures

Clear and accurate instructions are essential to the world of work. Instructions tell—and frequently show—how to do something. They indicate how to perform a specific task (draw blood; change the oil in your car); operate a machine (a pH meter; an iPhone); and construct, install, maintain, adjust, or repair a piece of equipment (an incubator; a scanner). Everyone from the consumer to the specialist uses and relies on carefully written and designed instructions.

To expand your understanding of writing instructions and procedures, take advantage of the Web Links, Additional Activities, and ACE Self-Tests at www.cengage.com/english/kolin/writingatwork concise3e

Instructions, Procedures, and Your Job

As part of your job, you may be asked to write instructions, alone or with a group, for your co-workers as well as for the customers who use your company's services or products. Your employer stands to gain or lose much from the quality and the accuracy of the instructions you prepare.

You may also be called on to write a set of procedures as a part of your job. The purpose of writing procedures is very different from preparing instructions. As we saw, a set of instructions explains how to install, repair, or safely use something. Procedures, however, refer to policies, duties, or protocols that a business or organization expects its employees to follow.

This chapter will first show you how to develop, write, illustrate, and design a variety of instructions, and then move into a discussion of writing procedures (pages 272–275) about job-related rules and responsibilities.

Why Instructions Are Important

Perhaps no other type of occupational writing demands more from the writer than do instructions because so much is at stake—for both you and your reader. The reader has to understand what you write and perform the task as well. You cannot afford to be unclear, inaccurate, or incomplete. Instructions are significant for many reasons, including safety, efficiency, and convenience.

Safety

Carefully written instructions get a job done without damage or injury. Poorly written instructions can be directly responsible for an injury to the person trying to follow them and may result in costly damage claims or even lawsuits. Notice how the product labels in your medicine chest inform users when, how, and why to take a medication safely. Without those instructions, consumers would be endangered by taking too much or too little medicine or by not administering it properly. To make sure your instructions are safe, they must be

- accurate
- consistent
- thorough
- clearly written
- carefully organized
- legally proper

Companies have a legal and ethical obligation to inform customers about what constitutes normal use and to warn them about any injuries or malfunctions if the product or procedure is misused. Instructions are also written to protect employees and equipment (see Figure 3.2).

Efficiency

Well-written instructions help a business run smoothly and efficiently. For example, without the instructions on video display terminal safety precautions (see Figure 7.5 on page 260), employees could potentially suffer health problems. Or imagine how inefficient it would be for a business if employees had to stop their work each time they did not have or could not understand a set of instructions. Equally alarming, what if employees made a number of serious mistakes because of confusing directions, costing a business lost sales and increased expenses? Giving readers helpful tips to make their work easier will also increase their efficiency.

Convenience

Clear, easy-to-follow instructions make a customer's job easier and less frustrating. In the customer's view, instructions reflect a product's quality and convenience. How many times have you heard complaints about a company because the instructions that went with its products were difficult to follow? Poorly written and illustrated instructions will cost you business. Instructions are also a vital part of "service after the sale." Owners' manuals, for example, help buyers to avoid a product breakdown (and the headache and expense of starting over) and to keep it in good working order.

The Variety of Instructions: A Brief Overview

Instructions vary in length, complexity, and format. Some instructions are one word long: *stop, lift, rotate, print, erase.* Others are a few sentences long: "Insert blank disk in external disk drive"; "Close tightly after using"; "Store in an upright position."

Instructions can be given in a variety of formats, as Figures 7.1 through 7.3 (pages 251–253) show. They often use numbered lists (Figures 7.1-7.3), and can employ visuals, sometimes to illustrate each step (Figure 7.1) or the entire process (Figure 7.3). They can also be written in paragraphs (Figure 7.5). You will have to determine which format is most appropriate for the kinds of instructions you are to write. For writing that affects policies or regulations, as in Figure 7.7 (see pages 273–274), you will most often use a memo format.

Instructions that supply a visual with each step. **Figure 7.1**

Proper Brushing

Proper brushing is essential for cleaning teeth and gums effectively. Use a toothbrush with soft, nylon, round-ended bristles that will not scratch and irritate teeth or damage gums.

1

Place bristles along the gumline at a 45-degree angle. Bristles should contact both the tooth surface and the gumline.

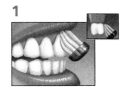

Uses easy-to-follow numbered steps

2

Gently brush the outer tooth surfaces of 2–3 teeth using a vibrating back and forth rolling motion. Move brush to the next group of 2–3 teeth and repeat.

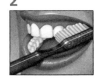

Visuals help readers to follow directions

3

Maintain a 45-degree angle with bristles contacting the tooth surface and gumline. Gently brush, using back, forth, and rolling motion along all of the inner tooth surfaces.

Begins each step with strong, active verbs

4

Tilt brush vertically behind the front teeth. Make several up and down strokes using the front half of the brush.

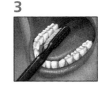

Offers helpful hints

5

Place the brush against the biting surface of the teeth and use a gentle back and forth scrubbing motion. Brush the tongue from back to front to remove odor-producing bacteria.

Explains why a step is important

Source: Reprinted by permission of American Dental Hygienists' Association. Illustrations adapted and used courtesy of the John O. Butler Company, makers of *GUM* Healthcare products.

Figure 7.2   **Instructions given in a numbered list describing a sequence of steps.**

Copying Files to a USB Flash Drive from Your PC or Notebook

1. Insert the USB Flash Drive (see photo below) into a USB portal of your PC or notebook.
2. Find the folder or file to be copied to the USB Flash Drive and right click on it.
   NOTE: The folder or file will be highlighted and a menu with "Open" at the top will appear.
3. Within the menu, move your cursor down to the "Send To" option. Here you will see a list of locations where you may send the selected folder or file.
4. Choose the USB Flash Drive location, and your folder or file will be automatically copied over.
   CAUTION: DO NOT remove the USB Flash Drive at this point, or you will risk damaging it.
5. Go to "My Computer" from the "Start" menu, and double click on the USB Flash Drive. If the folder/files you selected in step #2 are listed here, the copying was successful.
6. Eject the USB Flash Drive before removing it from the computer. To do so, go to "My Computer" again, right click on the USB Flash and select the "Eject" option from the menu.
7. Remove the USB Flash Drive from the USB portal.

Using Word Processing Programs to Design Instructions

To make sure your instructions are easy to read both for native as well as non-native speakers of English, take advantage of the following word processing features that will help you draft, revise, and format:

1. **Brainstorming/clustering to get ideas/steps down.** Find and compile the information you will need to include in the steps of your instructions, and think ahead about possible warnings, cautions, and any helpful hints that users may need to complete the steps (see pages 33–36).
2. **Use the Outline feature.** This element of word processing programs makes it easy to identify, order, and change the steps in a set of instructions, allowing you to try out different options quickly and easily.
3. **Choose a font that is easy to read.** Avoid fonts that may contain unusual letterings. Fonts like Arial or Helvetica provide clean and easy-to-read text for your instructions.
4. **Use numbered and bulleted lists.** Numbering steps helps readers follow the sequence of your instructions and bulleted lists break up text to make it easier to read.

Instructions in a numbered list on how to assemble an outdoor grill.   Figure 7.3

ASSEMBLY INSTRUCTIONS

The instructions shown below are for the basic grill with tubular legs. If you have a pedestal grill, or a grill with accessories, check the separate instruction sheet for details not shown here.

NOTE: Make sure you locate all the parts before discarding any of the packaging material.

TOOLS REQUIRED . . . A standard straight blade screwdriver is the only tool you need to assemble your new Meco grill. If you have a pedestal grill, you will need a 7-16 wrench or a small adjustable wrench.

Starts with helpful note and list of necessary tools

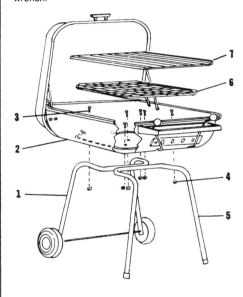

1. Before you start, take time to read through this manual. Inside you will find many helpful hints that will help you get the full potential of enjoyment and service from your new Meco grill.

2. Lay out all the parts.

3. Assemble roller leg (1) to bottom rear of bowl (2) with 1¼" long bolts (3) and nuts (4).

4. Assemble fixed leg (5) to bottom front of bowl (2) with 1¼" long bolts (3) and nuts (4).

5. Place fire grate—ash dump (6) in bottom of bowl (2) between adjusting levers.

6. Place cooking grid (7) on top of adjusting levers. Make sure top grid wires run from front to back of grill.

Exploded drawing shows relations of parts to each other

Easy-to-follow steps for consumer audience

Source: *Meco Assembly Instructions and Owners Manual,* Metals Engineering Corp., P.O. Box 3005, Greenville, TN 37743. Reprinted by permission.

5. **Include boldface sparingly.** Use it to emphasize warnings, cautions, and notes (see pages 204–205) so readers will not overlook the crucial messages they contain.
6. **Be careful about the images you take from a clip art library.** Use only icons that will be immediately recognizable to a broad range of readers. Also, make sure the images are the right size—neither so big that they look frivolous or so small that readers will not see them clearly at once.

Assessing and Meeting Your Audience's Needs

Put yourself in the readers' position. In most instances, you will not be available for readers to ask you questions when they do not understand something. Consequently, they will have to rely only on your written instructions.

Do not assume that members of your audience have performed the process before or have operated the equipment as many times as you have. (If they had, there would be no need for your instructions.) No one who has written a set of instructions ever disappointed readers by making directions too clear or too easy to follow. Remember, too, that your audience will often include non-native speakers of English, a worldwide audience of potential consumers.

Key Questions to Ask About Your Audience

The more you known about your audience, the better your instructions can be. To determine your audience's needs, ask yourself the following questions:

- How and why will my readers use my instructions? (Co-workers, consumers, and expents in the field all have different expectations.)
- What language skills do they possess—is English their first (native) language?
- How much do my readers already know about the product or service?
- How much background information will I have to supply?
- What steps will most likely cause readers trouble?
- How often will they refer to my instructions—every day or just as a refresher?
- Where will my audience most likely be following my instructions—in the workplace, outdoors, in a workshop equipped with tools, or alone in their homes?
- What resources, such as special equipment or power sources, will my readers need to perform my instructions successfully?

Writing Instructions for International Audiences

Your instructions will often be aimed at a worldwide audience of potential customers, many of whom do not use English as their first language and may be following your instructions through a translation of them into their native language. Here are some useful guidelines when writing instructions for this diverse group of readers:

1. **Write in plain, simple language.** Use international English (see pages 131–142).
2. **Use terms and units of measurement that your readers will understand.** Avoid abbreviations, acronyms, and jargon, and don't assume that international readers will use or understand the units of measurement found in the United States.
3. **Make sure all of your visuals are clear and culturally appropriate.** In some instances, your instructions may be given exclusively through visuals. See pages 235–237 for some guidelines.
4. **Be aware that colors can have different meanings.** Red, yellow, and green, for instance, may not convey the same meaning in other cultures that they do in the United States. (Refer to pages 236–237.)
5. **Ask a non-native speaker to review your instructions.** Doing this will help ensure that your instructions are easy to understand because you used clear sentences and appropriate language.

The Process of Writing Instructions

As we saw in Chapter 2, clear and concise writing requires you to follow a process. To make sure your instructions are accurate and easy for your audience to perform, follow these steps.

Plan Your Steps

Before writing, do some research to understand completely the job, process, or procedure that you are asking someone else to perform. Make sure you know

- the reason for doing something
- the parts or tools required
- the steps to follow to get the job done
- the results of the job
- the potential risks or dangers

If you are not absolutely sure about the process, ask an expert for a demonstration. Do some background reading and talk to or e-mail colleagues who may have written or followed similar instructions.

Perform a Trial Run

Actually perform the job (assembling, repairing, conducting, maintaining, testing) yourself or with all your writing team present. Go through a number of trial runs. Take notes as you go along, and be sure to divide the job into simple, distinct steps for readers to follow. Don't give readers too much to do in any one step. Each step should be **complete, sequential, reliable, straightforward,** and **easy** for your audience to identify and perform.

Write and Test Your Draft

Transform your notes into a draft (or drafts) of the instructions you want readers to follow. Test your draft(s) by asking someone from the intended audience (consumers, technicians) who may never have performed the job to follow your instructions as you have written them. Observe where the individual runs into difficulty, that is, cannot complete or seems to miss a step, or gets a result different from yours.

Revise and Edit

Based on your observations and user feedback, revise your draft(s) and edit the final copy of the instructions that you will give to readers. Always consider whether your instructions would be easier to accomplish if you included visuals.

Analyzing the needs and the background of your audience will help you to choose appropriate words and details. A set of instructions accompanying a chemistry set would use different terminology, abbreviations, and level of detail than would a set of instructions a professor gives a class in organic chemistry.

General Audience: Place 8 drops of vinegar in a test tube with a piece of limestone about the size of a pea.

Specialized Audience: Place 8 gtts of CH_3COOH in a test tube and add 1 mg of CO_3.

The instructions for the general audience avoid the technical abbreviations and symbols the specialized audience requires. If your readers are puzzled by your directions, you defeat your reasons for writing them.

Using the Right Style

To write instructions that readers can understand and turn into effective action, observe the following guidelines.

1. Choose verbs in the present tense and imperative mood. Imperatives are commands that have deleted the pronoun *you*. Note how the instructions in Figures 7.1 through 7.4 contain imperatives—"Pull the cord" instead of "You need to pull the cord." In instructions, deleting the *you* is not discourteous, as it would be in a business letter or report. The command tells readers, "These steps work, so do them exactly as stated." Choose action verbs such as those listed in Table 7.1.

Figure 7.4   Exploded drawing showing how to assemble an industrial extension cord.

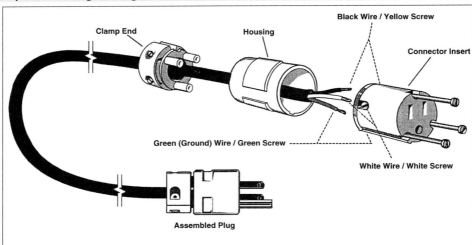

1.   Run the end of the cord through the clamp end and then through the center hole of the housing.
2.   Pull the cord through until it extends 2 inches beyond the housing.
3.   Strip about 1¼ inches of outer insulation from the end of the cord.
4.   Twist the exposed ends to prevent stray strands.

Source: Drawing courtesy of Sally Eddy

TABLE 7.1   Some Helpful Imperative Verbs Used in Instructions

add	eliminate	pass	shift
adjust	enter	paste	shut off
apply	exit	peel	slide
attach	find	pick up	slip
blow	flip	plug	spread
call up	flush	point	squeeze
change	forward	pour	start
check	gather	press	strain
choose	group	prevent	switch
clean	hold	print	tear
click	include	provide	thread
clip	increase	pull	tie
close	insert	push	tighten
connect	inspect	raise	tilt
contact	install	re-boot	trace
copy	lift	release	transect
cover	link on	remove	transfer
create	load	replace	trim
cut	log on	reply	turn
delete	loosen	review	twist
determine	lower	roll	type
dig	lubricate	rotate	unplug
display	maintain	rub	use
double-click	measure	run	ventilate
download	mix	save	verify
drag	mount	scan	wash
drain	move	scroll	weigh
drill	navigate	scrub	wind
drop	notify	select	wipe
ease	oil	send	wire
eject	open	set	wrap

2. Write clear, short sentences in the active voice. Keep sentences short and uncomplicated. Sentences under 20 words (preferably under 15) are easy to read. Note that the sentences in Figures 7.1 through 7.4 are, for the most part, under 15 words. But do not omit "a," "the," or any connective words (such as "and," "but," and "however,") which will make your instructions harder to follow.

3. Use precise terms for measurements, distances, and times. Indefinite, vague directions leave users wondering if they are doing the right thing. The following vague direction is better expressed through precise revision.

> Vague: Turn the distributor cap a little. (*How much is a little?*)
> Precise: Turn the distributor cap one quarter of a rotation.

4. Include connective words as signposts. Connective words specify the exact order in which something is to be done (especially when your instructions are written in paragraphs). Words such as *first, then,* and *before* help readers stay on course, reinforcing the sequence of your steps.

5. Number each step when you present your instructions in a list. Insert plenty of white space between steps to separate them for the reader.

Using Visuals Effectively in Instructions

Readers welcome visuals in almost any set of instructions. Visuals will help your readers your get a job done more quickly and increase their confidence. A visual can help users

- locate parts, areas, and so on
- identify the size and placement of parts
- learn how to assemble parts effectively and easily
- understand the right and wrong way to do something
- identify possible sources of danger, injury, or malfunction
- determine whether a problem is serious or minor
- recognize and follow normal and safe limits/ranges

The number and kinds of visuals you include will, of course, depend on the process or equipment you are explaining and your audience's background and needs. Some instructions may require only one or two visuals. The instructions in Figure 7.2 show users how to copy files using a USB Flash Drive. The illustration clearly shows what a USB Flash Drive looks like. But in Figure 7.1, each step is accompanied by a visual demonstrating a proper technique of brushing teeth.

Another frequently used visual in instructions is an exploded drawing, such as the one in Figure 7.3 that helps consumers see how the various parts of the grill fit together, or the one in Figure 7.4 that labels and shows the relationship of the parts of an industrial extension cord.

Guidelines for Using Visuals in Instructions

Follow these guidelines to use visuals effectively in your instructions:

1. Set visuals off with white space so it is easy to find and read them.
2. Place each visual next to the step it illustrates, not on another page or buried at the bottom of the page.
3. Select a visual that is appropriate for your audience (e.g., a photo showing a person doing a proper stretch is sufficient for a general audience rather than including an elaborate medical illustration of the muscular system).
4. Assign each visual a number (Figure 1, Figure 2) and refer to visuals by figure number in your instructions.
5. Make sure the visual looks like the object the user is trying to assemble, maintain, run, or repair. Don't use a photo of a different model.
6. Always inform readers if a part is missing or reduced in your visual.
7. Where necessary, label or number parts of the visual, as in Figure 7.3.

For further guidelines, refer to chapter 6 on numbering visuals (pages 211–212) and inserting them in your document (pages 212–213).

The Five Parts of Instructions

Except for very short instructions, such as those illustrated in Figures 7.1 through 7.4, or for procedures on policies and regulations (see Figure 7.7 pages 274–275), a set of instructions generally contains five main parts: (1) an introduction; (2) a list of equipment and materials; (3) the actual steps to perform the process; (4) warnings, cautions, and notes; and (5) a conclusion (when necessary).

Visit www. cengage.com/ english/kolin/ writingatwork concise3e for an online exercise, "Rewriting Online Instructions."

Introduction

The function of your introduction is to provide readers with enough *necessary* background information to understand why and how your instructions work. An introduction must make readers feel comfortable and well prepared before they turn to the actual steps.

What to Include in an Introduction

You can do one or all of the following in your introduction. Not every introduction to a set of instructions will contain all five categories of information listed here. Some instructions will require less detail. You will have to judge how much background information to give readers for the specific instructions you write.

1. State why the instructions are useful for a specific audience. Many instructions begin with introductions that stress safety, educational, or occupational benefits. Here is an introduction from a set of safety instructions describing protective lockout of equipment.

> The purpose of these instructions is to provide plant electrical technicians with a uniform method of locking out machinery or equipment. This will prevent the possibility of setting moving parts in motion, energizing electrical lines, or opening valves while repair, setup, or cleaning work is in progress.

2. Indicate how a particular piece of equipment or process works. An introduction can briefly discuss the "theory of operation" to help readers understand why something works the way your instructions say it should. Such a discussion sometimes describes a scientific law or principle. An introduction to instructions on how to run an autoclave briefly begins by explaining the function of the machine: "These instructions will teach you how to operate an autoclave, which is used to sterilize surgical instruments through the live additive-free stream."

3. Point out any safety measures or precautions a reader may need to be aware of. By alerting readers early in your instructions, you help them to perform the process much more safely and efficiently. Note how Cliff Burgess opens his instructions in Figure 7.5 (page 260) with a warning about users' safety.

4. Stress any advantages or benefits the reader will gain by performing the instructions. Make the reader feel good about buying or using the product by explaining how it will make a job easier to perform, save the reader time and money, or allow the reader to accomplish a job with fewer mistakes or false starts. See how Figure 7.5 explains EMFs and health risks.

Figure 7.5   An instructional memo listing safety precautions for a general audience.

BURTON WORLDWIDE SYSTEMS
www.burton.com

TO: All Burton Employees
FROM: Cliff Burgess, Director, Environmental Safety *CB.*
RE: Video Display Terminal (VDT) Safety Precautions
DATE: September 12, 2012

Introduction emphasizes reasons for instructions

You may experience some possible health risks in using your computer video display terminal **(VDT)**. These risks include sleep disorders, behavioral changes, danger to the reproductive system, and cancer.

Nontechnical explanation offers an analogy

The source of any risks comes from the electromagnetic fields **(EMFs)** that surround anything that carries an electric current—for example, copiers, circuit breakers, and especially VDTs. Your computer monitor is a major source of EMFs. Magnetic fields can go through walls as easily as light goes through glass.

Although **EMFs** may affect your health, you can considerably reduce your exposure to these fields by following these three simple steps:

Steps stand out through bullets, bold-face, and spacing

- **Stay at least three feet (an arm's length) away from the front of your VDT.** (The magnetic field is significantly reduced with this amount of distance.)

- **Stay at least four feet away from the sides and back of someone else's VDT.** (The fields are weaker in the front of the **VDT** but much stronger everywhere else.) Refer to the following sketch, which you may want to post in your office or next to your workspace at home.

Simple visual clarifies instructions

- **Switch your VDT/computer off when you are not using it.** (If the computer has to remain on, be sure to switch off the monitor; screen savers do not affect the exposure to **EMFs**.)

Conclusion reassures readers they are acting safely

Our environmental safety team will continue to monitor and investigate any problems. Observing the guidelines above, however, will help you to take all the necessary precautions in order to minimize your exposure to **EMFs**.

Lists helpful hyperlink

To view a demonstration of proper safety techniques, go to www.burton.com/vdtsafety/.

1215 Madison • St. Louis, Missouri 63174
314-555-4300 • FAX 314-555-4311

Source: Reprinted by permission.

Note how the following introduction to a set of instructions for an all-in-one scanner encourages the reader to learn how to operate this system.

> Congratulations on choosing the MegaScan 3000 all-in-one scanner. The MegaScan 3000 will handle all of your electronic document needs and features fast and simple menu options, 1 terabyte of storage capacity, and wireless networking capabilities. You can scan 25 hard copy pages in 1 minute.

5. Provide hyperlinks. When readers will be following your instructions online, provide hyperlinks to any sites or materials they need to know about before starting your instructions, as Cliff Burgess does in Figure 7.5. Similarly, provide relevant cross-references in printed instructions.

List of Equipment and Materials

Clearly, some instructions, as in Figure 7.1, do not need to inform readers of all equipment or materials they will need. But when you do, make your list complete and clear. Do not wait until the readers are actually performing one of the steps to tell them that a certain type of drill or a specific kind of chemical is required. They may have to stop what they are doing to find the equipment or material; moreover, the procedure may fail or present hazards if users do not have the right equipment at the right time. For example, if a Phillips screwdriver is essential to complete one step, specify that type of screwdriver under the heading "Equipment and Materials"; do not list just "screwdriver." See the brief, helpful section "Tools Required" in Figure 7.3.

Steps for Your Instructions

The heart of your instructions will consist of clearly distinguished steps that readers must follow to achieve the desired results. Figure 7.6 (pages 265–271) contains a model set of steps on how to set up a printer. Note how each step is precisely keyed to the visual, further helping readers perform the procedure. Refer to Figure 7.6 as you study this section.

Guidelines for Writing Steps

To help your readers understand your steps, observe the following rules:

1. Put the steps in their correct order and number them. If a step is out of order or is missing, the entire set of instructions can be wrong or, worse yet, dangerous. Double-check every step, and number each step to indicate its correct place in the sequence of tasks you are describing.

2. Include only the right amount of information in each step. Make each step short and simple. Giving readers too little information can be as risky as giving them too much. Keep in mind that each step asks readers to perform a single task in the entire process. In the following example, note how too many steps are combined to access voice mail.

Incorrect:
1. To access your voice mail, make sure you've listened to old messages and then press "1" to obtain your new messages.
2. When each new message is finished, press "7" to delete the message or "8" to store it in the archives. Press "2" to replay the message.
3. To review your saved messages, press "9." To end the call, press "#."

Correct:
1. To access your voice mail, press "1" to obtain your new messages.
2. When each new message is finished, press "7" to delete the message or "8" to store it in the archives. Press 2 to replay the message.
3. To review your saved messages, press "9."
4. To end the call, press "#."

3. Group closely related activities into one step. Sometimes closely related actions do belong in one step to help the reader coordinate activities and to emphasize their being done at the same time, in the same place, or with the same equipment.

Don't divide an action into two steps if it has to be done in one. For example, instructions showing how to light a furnace would not list as two steps actions that must be performed simultaneously to avoid a possible explosion.

Incorrect:
1. Depress the lighting valve.
2. Hold a match to the pilot light.

Correct:
1. Depress the lighting valve while holding a match to the pilot light.

Similarly, do not separate two steps of a computer command that must be performed simultaneously.

Incorrect:
1. Press the CONTROL key.
2. Press the ALT key.

Correct:
1. While holding down the CONTROL key, press the ALT key.

4. Give the reader hints on how best to accomplish the procedure. Obviously, you cannot do that for every step, but if there is a chance that the reader might run into difficulties, you should provide assistance. Particular techniques on how to operate or service equipment also help readers: Readers are told in Figure 7.6 where the best places are to put their new printer, and the worst. Also tell readers it they have a choice of materials or procedures, especially those that would give the best performance: "Several thin coats of paint will give a better finish than one heavy coat."

5. State whether one step directly influences (or jeopardizes) the outcome of another. Because all steps in a set of instructions are interrelated, you could not (and should not have to) tell readers how every step affects another. But stating specific relationships is particularly helpful when dangerous or highly intricate operations are involved. You will save the reader time, and you will stress the need for care. Forewarned is forearmed. Here is an example.

Step 2:   Tighten fan belt. Failure to tighten the fan belt will cause it to loosen and come off when the lever is turned on in step 5.

Do not wait until step 5 to tell readers that you hope they did a good job in tightening the fan belt in step 2. Information that comes after the fact is not helpful.

6. **Where necessary, insert graphics to assist readers in carrying out the step.** See how helpful the different drawings of the printer are in Figure 7.6.
7. **Your instructions might be translated into an international reader's language,** as you can see in the warning statements below.

Warnings, Cautions, and Notes

At appropriate places in the steps of your instructions, you may have to stop the reader to issue a warning, a caution, or a note. Study the examples below as well as those in Figure 7.6, especially for Step 4.

Warnings

A warning ensures a reader's safety. It tells readers that a step, if not prepared for or performed properly, could seriously injure them or even endanger their lives, as the following warning does.

WARNING: UNPLUG MACHINE BEFORE REMOVING PLATEN GLASS.

ADVERTENCIA: DESENCHUFE LA MAQUINA ANTES DE QUITAR EL VIDRIO.

Spanish translation

Cautions

A caution tells a reader how to avoid a mistake that could damage equipment or cause the process to fail, for instance, "Do not force the plug."

Caution: **Formatting erases all data on the disk**

小心： 格式化会删除磁盘的所有资料

Chinese translation

Notes

A note does not relate to the safety of the user or the equipment but does provide clarification or a helpful hint on how to do the step more efficiently.

At 0°C a battery uses only about 68 percent of its power.

Eksi 7 derecede, bir pil enerjisinin yaklasik yuzde 68 ini kullanir.

Turkish translation

Guidelines on Using Warnings, Cautions, and Notes

1. **Do not regard warnings and cautions as optional.** They are vital for legal and safety reasons to protect lives and property. In fact, you and your company

can be sued if you fail to notify the users of your product or service of dangerous conditions that will or could result in injury or death.

2. **Put warnings and cautions immediately before the step to which they pertain.** If you insert a warning or caution statement too early, readers may forget it by the time they come to the step to which it applies, and putting the notification too late exposes the reader, and possibly equipment as well, to risk.

3. **Put warnings and cautions in a distinctive format.** Warnings and cautions should be graphically set apart from the rest of the instructions, as the triangle in the box on the previous page illustrates. Print such statements in capital letters, boldface, different colors. Red is especially effective for warnings and yellow for cautions if your readers are native speakers of English. These colors have different meanings in other countries (see pages 236–237). Also, make sure your audience understands an icon or a symbol, such as a skull and crossbones, an exclamation point inside a triangle, or a traffic stop light, often used to signal a warning, a hazard, or some other unsafe condition.

4. **Include relevant explanations to help readers know what to watch out for and what precautions to take.** Do not just insert the word WARNING or CAUTION. Explain what the dangerous condition is and how to avoid it. Look at the examples of warnings and cautions in Figure 7.6 on pages 266–268.

5. **Do not include a warning or a caution just to emphasize a point.** Putting too many in your instructions will decrease the dramatic impact they should have on readers. Use them sparingly—only when absolutely necessary—so readers will not be tempted to ignore them.

6. **Use notes only when the procedure calls for them and when they help readers,** as in Figure 7.3 and in Figure 7.6 (see pages 269–271).

Conclusion

Not every set of instructions requires a conclusion. For short instructions containing only a few simple steps, such as those in Figures 7.1 through 7.4, no conclusion is necessary. For longer, more involved jobs, a conclusion can provide a succinct wrap-up of what the reader has done or end with a single sentence of congratulations, or reassure readers, as the conclusion in Cliff Burgess's memo (Figure 7.5) does. A conclusion might also tell readers what to expect once a job is finished, describe the results of a test, or explain how a piece of equipment is supposed to look or operate. And always supply contact information and any hyperlinks, should a reader need further information.

Model of Full Set of Instructions

Study Figure 7.6, which is a set of instructions on setting up an Epson printer that includes the parts discussed in this chapter: an introduction; a list of materials; numbered steps; and warnings, cautions, and notes. Pay special attention to how the writer coordinates words with visuals to assist readers.

Complete set of instructions with steps, visuals, cautions, notes, and warnings.   Figure 7.6

Quick Setup

1 Unpack the Printer

Here's how to set up your new EPSON® printer . . .

Remove any packing material from the printer, as described on the Notice Sheet in the box. Save all the packaging so you can use it if you need to transport the printer later. You'll find these items inside:

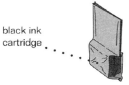

black ink
cartridge

color ink
cartridge

paper
support

Appropriately brief introduction for this straightforward procedure

Each numbered step begins with imperative verb

List (with visuals) of materials

Exploded drawings assist readers to identify/ assemble parts

(continued)

Figure 7.6   (Continued)

Place the printer flat on a stable desk near a grounded outlet. Leave plenty of room in back for the cables and enough room in front for opening the output tray.

Do NOT put the printer:

Bulleted list highlights importance of correct printer placement

- In an area with high temperature or humidity
- In direct sunlight or dusty conditions
- Near sources of heat or electromagnetic interference, such as loudspeakers or cordless telephone base units.

Also, be sure to follow the Safety Instructions in the Introduction of your *User's Guide*.

2 Attach the Paper Support

Insert the paper support in the top slot on the back of the printer.

Groups clearly distinguished steps in the procedure

3 Plug In the Printer

First make sure the power is off. Check the ⏻ power button; it's off when its surface is raised above the printer surface.

Uses appropriate icon to alert reader to possible damage

Caution:
Do not plug the printer into an outlet controlled by a wall switch or timer, or on the same circuit as a large appliance. This may disrupt the power, which can erase memory and damage the power supply.

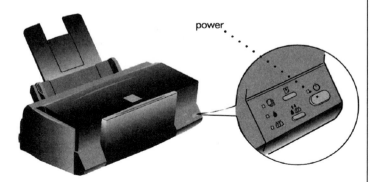

power

Plug the power cord into a properly grounded outlet.

Install the Ink Cartridges

1. Lower the output tray and raise the printer cover.

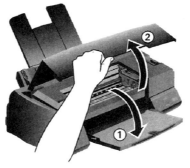

Directional arrows visualize correct movement of parts

2. Press the ⏻ power button to turn on the printer. The ⏻ power light flashes, the ● and ♨ ink out lights come on, and the ink cartridge holders move to the installation position.

3. Pull up the ink cartridge clamps.

Visual clues reinforce right way to perform procedure

Caution:
You must remove the tape seal from the top of the cartridge or you will permanently damage it. Don't remove the tape seal from the bottom or ink will leak.

4. Open the ink cartridge packages. Remove the disposable yellow portion of the tape seal on top.

Warning:
If ink gets on your hands, wash them thoroughly with soap and water. If ink gets in your eyes, flush them immediately with water.

 black ink cartridge

 color ink cartridge

Caution and warning notices inserted in right places with distinctive icons

(continued)

Figure 7.6   (Continued)

5. Lower the ink cartridges into their holders with the labels face up and the arrows pointing toward the back of the printer.

Callouts make parts easy to find

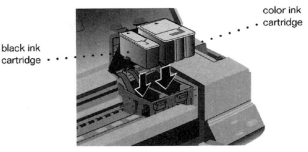

color ink cartridge

black ink cartridge

Generous white space makes steps easy to distinguish and follow

6. Push down the clamps until they lock in place.

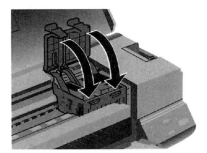

Appropriate icon alerting reader of a possible problem

Caution:
Never turn off the printer when the ⏻ power light is flashing.

7. Press the ●💧 cleaning button to return the print heads to their home position and charge the ink delivery system. Charging can take up to five minutes, with the ⏻ power light flashing until it's finished.

8. Close the printer cover.

5 Load the Paper

1. Slide the left edge guide all the way left and **pull** out the output tray extension.

Includes two visuals, one of them an exploded drawing below, helpfully inserted between the steps to which they apply

2. Fan a stack of plain paper and then even the edges.

3. Load the stack with the printable surface face up. Push the paper against the right edge guide.

Offers clear-cut directions on how to load paper

arrow
mark

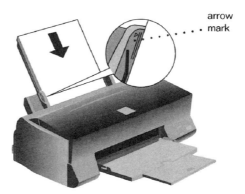

Note:
Don't load paper above the arrow mark inside the left edge guide.

Helpful information in note to ensure best use of equipment

4. Slide the left edge guide back against the stack of **paper**.

(continued)

Figure 7.6   (Continued)

Provides a useful source for further information on performing step correctly

6 Check the Printer

1. Turn off the printer.

2. While holding down the load/eject button, turn on the printer. Then release the buttons.

3. A page prints out showing the ROM version and a nozzle check pattern. When it's finished, turn off the printer. If you have any problems with the test, see Chapter 6 in your *User's Guide* for more information.

Inserts a brief introduction that alerts users to follow one of two procedures, depending on the type of computer they have

7 Connect the Printer to Your Computer

You can connect your EPSON Stylus™ COLOR 600 to either an IBM® compatible PC or an Apple® Macintosh® You'll need a shielded, twisted-pair parallel cable to connect to a PC or an Apple System Peripheral-8 cable to connect to a Macintosh. For a complete list of system requirements, see the Introduction of your *User's Guide*.

Connecting to a PC

1. Turn off the printer and your computer.

2. Connect the cable to the printer's parallel interface; then squeeze the wire clips together until they lock in place. (If your cable has a ground wire, connect it now.)

Chooses informative icon for a note

Note:
The printer is assigned to parallel port LPT1; if you want to use a different port, see your Windows documentation for instructions.

ground wire

Enlarged view of parts helps readers see step

3. Connect the other end of the cable to your computer's parallel port and secure it as necessary.

Connecting to a Macintosh

1. Turn off the printer and your Macintosh.

2. Connect one end of the cable to the serial connector on the back of the printer.

Visual identifies precise location for connecting parts

Note:
If you're using a PowerBook™, connect your printer to the modem port.

3. Connect the other end of the cable to either the modem port ✆ or the printer port 🖶 on your Macintosh.

8 Install the Printer Software

Now you need to install the printer software so you can control printing from your computer.

Functions as both a step—installing software—and a wrap-up

Installing on a PC

You can install the printer software for Windows 95 or Windows 3.1 from the EPSON printer software CD-ROM. If you don't have a CD-ROM drive, you can install the software using the EPSON printer software diskettes.

Installing from the CD-ROM

In addition to the printer driver and utilities, the CD-ROM contains EPSON Answers, a comprehensive online guide that includes:

Provides information for further trouble-shooting and contact with manufacturer

▶ **How To** for step-by-step printer operating instructions
▶ **Color Guide** with practical color printing information
▶ **Problem Solver** to help you fix printer problems
▶ **Test Print** so you can check your print quality

Follow the instructions inside the CD-ROM case to install the software. To run EPSON Answers, click on its icon in the EPSON program group or folder.

EPSON ANSWERS
If you have a CD-ROM drive, you can run EPSON Answers, the on-screen guide to your new printer. It puts you on the right track quickly and easily.

Source: Reprinted by permission of Epson America.

Writing Procedures for Policies and Regulations

Up to this point, we have concentrated primarily on instructions dealing with how to put things together; how to install, repair, or use equipment; and how to alert readers to mechanical or even personal danger(s). But there is another similar type of writing that deals with guidelines for getting things done in the world of work—procedures. These concern policies and regulations found in employee handbooks and other internal corporate communications, such as on websites, in memos, in e-mail messages, or on a company's intranet (a private computer network). Note, however, that some companies do not disseminate policy via e-mail because it is perceived as less formal than hard copy and is deletable. Figure 7.7 shows an example of a company policy on flextime written in a memo format.

Some Examples of Procedures

Procedures deal with a wide range of "how-to" activities within an organization, including the following:

- accessing a company file or database
- preparing for an audit, a transition, a merger
- applying for family or medical leave
- dressing professionally at work or at a job site
- submitting and routing information
- reserving a company vehicle or facility
- taking advantage of telecommuting options
- filing a work-related grievance
- requesting travel expense reimbursement
- seeking safety in case of a fire, flood, spill or other emergency
- fulfilling promotion requirements
- applying for a disability accommodation
- requesting a transfer within a company
- using company e-mail

Policy procedures have a major impact on a company and its workers; they affect schedules, payrolls, acceptable and unacceptable behaviors at work, and a range of protocols governing the way an organization does business internally and externally. Procedures help an organization run smoothly and consistently. Adhering to them, every employee can know and follow the same regulations and standards.

Meeting the Needs of Your Marketplace

As with instructions, you will have to plan carefully when you write a set of procedures. A mistake in business procedures can be as wide ranging and as costly as an error in a set of assembly instructions—perhaps more so since poorly written procedures can land a company and/or its employees in significant financial and legal trouble.

Writing Procedures at Work: An Example

To avoid such difficulties, spell out precisely what is expected of employees—how, when, where, and why they are to perform or adhere to a certain policy. Use the same strategies as for instructions discussed earlier in this chapter. Leave no chance for misunderstanding or ambiguity; be unqualifiedly straightforward and clear-cut. Determine what information employees need in order to comply with the regulations you are specifying.

Many times procedures involve a change in the work environment. Help readers by including, whenever necessary, definitions, headings, some prefatory explanations, and an offer to assist employees with any questions they may have. Always present a copy of the procedures to management to approve or to revise before sending them to employees.

Figure 7.7 (pages 274–275) shows a memo from a human resources director notifying employees how they can take advantage of a new flextime arrangement at work, including what they can and cannot do within the framework of flextime. Note how the writer divides her procedures document into an introduction explaining when flextime will go into effect and what choices employees have about it, a section that clearly and usefully defines and delineates the concept of flextime, and finally a section containing specific guidelines. These guidelines, while not sequential, function as a series of rules that employees have to follow, and ultimately these rules will affect the entire organization.

The various regulations about what employees cannot do in flextime might be seen as the equivalents of warning and caution statements discussed in "Warnings, Cautions, and Notes" (pages 263–264). Observe, too, how the writer does not veer off to discuss benefits to the employer or to examine where and how often flextime has been used elsewhere. Finally, this example of procedural writing protects the employer legally by establishing the policies by which an employee's scheduled work time is clearly defined, delineated, and assessed. There's no room for guesswork.

Some Final Advice

Perhaps the most important piece of advice to leave you with is this: Do not take *anything* for granted when you have to write a set of instructions or procedures. It is wrong and on occasion dangerous to assume that your readers have performed the procedure before, that they will automatically supply missing or "obvious" information, that they will easily anticipate your next step or know what is expected of them at work without being informed. No one ever complained that a set of instructions or procedures was too clear or too easy to follow.

Figure 7.7   **Procedures following a new scheduling policy.**

New Tech, Inc.

www.newtech.com

4300 Ames Boulevard, Gunderson, CO 81230-0999

(303) 555-9721
FAX (303) 555-9876

TO: All Employees
FROM: Tequina Bowers *T.B.*
          Human Resources

DATE: March 20, 2011
SUBJECT: Opting for a Flextime
          Schedule

Notifies readers of new policy and states purpose of the memo

Effective sixty days from now, on May 10, 2011, employees will have the opportunity to go to a flextime schedule or to remain on their current 8-hour fixed schedule. This memo explains the flextime option and establishes the guidelines and rules you must follow if you choose this new schedule option.

Flextime Defined

Spells out precisely how company defines flextime—uses boldface for most important information,

Flextime is based on a certain number of **core hours** and **flexible hours**. Our company will be open twelve hours, from 6:00 a.m. to 6:00 p.m. weekdays, to accommodate both fixed and flextime schedules. During this 12-hour period, all employees on flextime will be expected to work **eight and one-half consecutive hours,** including a half-hour for lunch.

Provides helpful example

Regardless of schedule options, all employees must work a **common core time from 10:00 a.m. to 4:00 p.m.**, but flextime employees will be free to choose their own starting and quitting times. For instance, they might elect to arrive at 8:00 a.m. and leave at 4:30 p.m., or they may want to start at 9:30 a.m. and leave at 6:00 p.m.

Flextime Guidelines and Rules

Stresses employee responsibilities and consequences of violating rules

Employees need to understand their individual responsibilities and adjust their schedules accordingly. All flextime employees must adhere strictly to the following regulations and realize that flextime schedule option privileges will be revoked for violations.

2.

What Flextime Employees Must Do

(1) Be present during core time, arriving and leaving the plant during their flexible work hours.
(2) Observe a minimum unpaid half-hour lunch break each working day.
(3) Cooperate with their supervisors to make sure adequate coverage is provided for their department from 6:00 a.m. to 6:00 p.m.
(4) Notify supervisors at once of absences.
(5) Attend monthly corporate meetings even though such meetings may be outside their chosen flextime schedules.
(6) Adhere to company dress code during any time they are at work, regardless of their flextimes.
(7) Agree to work on a flextime schedule for a minimum 6-month period.

What Flextime Employees Can't Do

(1) Be tardy during core time.
(2) Switch, bank, borrow, or trade flextime hours with other employees without the written approval of an immediate supervisor.
(3) File for overtime without a supervisor's approval.
(4) Self-schedule a vacation or leave by expanding flextime hours.
(5) Switch back and forth between fixed time and flextime.

How Do You Sign Up for Flextime?

If you wish to begin a flextime schedule, first you need to obtain and complete a transfer of hours form from your supervisor. Next, you must bring that signed form to Human Resources (Admin. 201) to participate officially in this program or complete it online at **www.newtech.com/flextime**.

I will be happy to talk to you about this new work option and to answer any of your questions. Please call me at ext. 5121, e-mail me at **tbowers@newtech.com**, or visit my office in Admin. 201. Thank you for your cooperation.

Carefully outlines what is acceptable according to new policy

Numbered points make policy easier to understand, follow, and refer to in the future

Stipulates what new policy will not allow in separate section of memo

Explains steps to begin flextime

Encourages feedback and questions

Revision Checklist

- [ ] Analyzed my intended audience's background, especially why and how they will use my instructions.
- [ ] Tested my instructions to make sure they include all necessary steps in their proper sequence.
- [ ] Made sure all measurements, distances, times, and relationships are precise and correct.
- [ ] Avoided technical terms if my audience is not a group of specialists in my field.
- [ ] Used the imperative mood and wrote clear, short sentences.
- [ ] Made my instructions easy to read and follow for an international audience.
- [ ] Chose effective visuals, labeled them, and placed them next to the step(s) to which they apply.
- [ ] Made my introduction proportionate to the length and complexity of my instructions and suitable for my readers' needs.
- [ ] Included necessary background, safety, and operational information in the introduction.
- [ ] Provided a complete list of tools and materials my audience needs to carry out the instructions.
- [ ] Put instructions in easy-to-follow steps and in the right order.
- [ ] Used numbers or bullets to label the steps and inserted connective words to indicate order.
- [ ] Used warnings, cautions, and notes where necessary and includes icons and color that make them easy to find and to follow.
- [ ] Supplied a conclusion that summarizes what readers should have done or reassures them that they have completed the job satisfactorily.
- [ ] Spelled out clearly policies, responsibilities, restrictions, and consequences of procedures for readers.
- [ ] Defined any terms readers may be unfamiliar with in procedures and, where helpful, provided an example.
- [ ] Gave a copy of procedures to administrators for their approval before distributing to employees.

Exercises

1. Find a set of instructions that does not contain any visuals but that you think should to make the directions clearer. Design those visuals yourself, and indicate where they should be inserted in the instructions.

2. From a technical manual in your field or in an owner's manual, locate a set of instructions that you think is poorly written and illustrated. In a memo to your instructor, explain why the instructions are unclear, confusing, or badly formatted.

Then, revise the instructions to make them easier for the reader to carry out. Submit the original instructions with your revision.

3. Write a short set of instructions in numbered steps (or in paragraph format) on one of the following relatively simple activities.
   a. tying a shoe
   b. using an ATM
   c. unlocking a door with a key
   d. sending a text message from a cell phone
   e. planting a tree or a shrub
   f. sewing a button on a shirt
   g. removing a stain from clothing
   h. pumping gas into a car
   i. creating a blog
   j. checking a book out of the library
   k. polishing a floor
   l. shifting gears in a car
   m. photocopying a page from a book
   n. posting a video to your blog

4. Write an appropriate introduction and conclusion for the set of instructions you wrote for Exercise 3.

5. Write a set of full instructions on one of the following more complex topics. Identify your audience. Include an appropriate introduction; a list of equipment and materials; numbered steps with necessary warnings, cautions, and notes; and an effective conclusion. Also include whatever visuals you think will help your audience.
   a. scanning a document
   b. changing a flat tire
   c. testing chlorine in a swimming pool
   d. shaving a patient for surgery
   e. removing "red eye" from a digital photo
   f. surveying a parcel of land
   g. creating a slideshow of the digital photos you took on a job
   h. jumping a dead car battery
   i. using the Heimlich maneuver to help a choking individual
   j. filleting a fish
   k. creating a logo for a letterhead
   l. taking someone's blood pressure
   m. editing digital video
   n. painting a car
   o. recording a podcast
   p. flossing a patient's teeth after cleaning
   q. backing up your computer files to an external hard drive

6. The following set of instructions is confusing, vague, and out of order. Rewrite the instructions to make them clear, easy to follow, and correct. Make sure that each step follows the guidelines outlined in this chapter.

Reupholstering a Piece of Furniture

(1) Although it might be difficult to match the worn material with the new material, you might as well try.

(2) If you cannot, remove the old material.

(3) Take out the padding.

(4) Take out all of the tacks before removing the old covering. You might want to save the old covering.

(5) Measure the new material with the old, if you are able to.

(6) Check the frame, springs, webbing, and padding.

(7) Put the new material over the old.

(8) Check to see if it matches.

(9) You must have the same size as before.

(10) Look at the padding inside. If it is lumpy, smooth it out.

(11) You will need to tack all the sides down. Space your tacks a good distance apart.

(12) When you spot wrinkles, remove the tacks.

(13) Caution: in step 11 directly above, do not drive your tacks all the way through. Leave some room.

(14) Work from the center to the edge in step 11 above.

(15) Put the new material over the old furniture.

P.S. Use strong cords whenever there are tacks. Put the cords under the nails so that they hold.

---

7. Write a set of procedures on greening a student union or an employee rest area/lounge. This exercise can be done collboratively where each member of the team takes a key area—lighting, heating/cooling, recycling, noise pollution, use of drinking/eating. Include a relevant visual with your procedure.

8. Write a set of procedures, similar to Figure 7.7, on one of the following policies or regulations at your school or job site:
   a. offering quality customer service over the phone or via the web
   b. filing a claim for a personal injury on the job
   c. designing an employee's personal space—what is and is not allowed?
   d. using the Internet at work for personal use
   e. ensuring confidentiality at work (to practice professional ethics in the workplace)
   f. enrolling in mandatory courses to maintain a license or certificate
   g. playing music in the workplace

  h. going through an orientation procedure to begin a new job

  i. following an acceptable dress code

  j. allowing tattoos and body piercings in the workplace

  k. registering a domain name for a sponsored group at work or school

  l. receiving reimbursement for carpooling, taking public transportation, or riding a bicycle to work

m. using a company vehicle

  n. going on a service call to a customer's home

8

Writing Effective Short Reports and Proposals

To expand your understanding of writing short reports and proposals, take advantage of the Web Links, Additional Activities, and ACE Self-Tests at www. cengage.com/ english/kolin/ writingatwork concise3e

This chapter shows you how to write short reports and proposals, which are among the most important and frequent types of business communications you may be called upon to prepare. Short, informal reports give up-to-date information (and sometimes what it means and what should be done about it) to help a company or organization run smoothly, efficiently, and profitably. These reports, which cover a wide range of topics, can help a company fulfill its obligations and plan for its future. Proposals are used to keep current business as well as attract new business, or to make recommendations for changes within a company or organization. Both are crucial to day-to-day operations of a company or organization, and both are designed for an audience of busy decision makers.

Why Short Reports Are Important

A short report can be defined as an organized presentation of relevant data on any topic—money, travel, time, technology, personnel, service equipment, weather, the environment—that a company or agency tracks in its day-to-day operations. Short reports are practical and to the point. They show that work is being done, and they also show your boss that you are alert, professional, and reliable. Short reports are written to co-workers, employers, vendors, and clients. When they are intended for individuals within your organization, short reports are most often sent as memos or as e-mails. But, for clients, usually you will send your reports out as letters.

Businesses cannot function without short written reports. Reports tell whether

- work is being completed
- schedules are being met
- costs have been contained
- sales projections are being met
- unexpected problems have been solved

You may write an occasional report in response to a specific question, or you may be required to write a daily or weekly report about routine activities.

Types of Short Reports

To give you a sense of some of the topics you may be required to write about, here is a list of various types of short reports common in the business world.

appraisal report	incident report	production report
audit report	inventory report	progress/activity report
budget report	investigative report	recommendation report
compliance report	laboratory report	research report
construction report	manager's report	sales report
design report	medicine/treatment	status report
evaluation report	error report	survey report
experiment report	operations report	test report
feasibility report	periodic report	travel report

This chapter concentrates on five of the most common types of reports you are likely to encounter in your professional work.

1. periodic reports
2. sales reports
3. progress reports
4. travel reports
5. incident reports

Although there are many kinds of short reports, they all are written for readers who need factual information so that they can get a job accomplished. Never think of the reports you write as a series of casual notes jotted down for *your* convenience.

Seven Guidelines for Writing Short Reports

Although there are many kinds of short reports, the following seven guidelines will help you write any short report successfully.

1. Anticipate How an Audience Will Use Your Report

Knowing who will read your report and why is crucial to your success as a writer. A co-worker or someone else in your field may be familiar with technical information. But managers, who will constitute the largest audience for your report, may not always understand or be interested in such technical information. Instead they will want bottom-line details about costs and schedules, for example. Similarly, audiences outside of your company (clients, media personnel, community agencies, etc.) will likely not be interested in technical information. Rather, they want information that helps them understand your company, how it will work with, serve, or participate with them

All audiences, however, want clear and concise information. For more information on how to make your reports concise and easy to follow, see #6 on page 284.

2. Do the Necessary Research

An effective short report needs the same careful research that goes into other on-the-job writing. Your research may be as simple as instant messaging, e-mailing, or leaving a voice mail for a colleague or checking a piece of equipment. Or you may have to test or inspect a product or service or assess the relative merits of a group of competing products or services. Some frequent types of research you can expect to do on the job include:

- verifying data in reference manuals or code books
- searching online archives and databases for recent discussions of a problem or procedure
- reading background information in professional and trade journals
- reviewing and updating a client's file
- testing equipment
- performing an experiment or procedure
- conferring with or interviewing colleagues, managers, vendors, or clients
- visiting and describing a site; taking field notes
- attending a conference or workshop

Never trust your memory to keep track of all the details that go into making a successful short report. Take notes, either by hand or on your laptop or handheld. Collect all the relevant data you will need—names, model numbers, costs, places, technology, etc.—and organize this information carefully into an outline, which will help you interpret these facts for your readers.

3. Be Objective and Ethical

Your readers will expect you to report the facts objectively and impartially—locations, costs, sales, weather conditions, eyewitness accounts, observations, statistics, test measurements, and descriptions. Your reports should be truthful, accurate, and complete. Here are some guidelines to follow:

- Avoid *guesswork*. If you don't know or have not yet found out, say so and indicate how, where, and when you'll try to find out.
- Do not substitute *impressions* or *unsupported personal opinions* for careful research.
- Using *biased, skewed*, or *incomplete data* is unethical. Provide a straightforward and honest account; don't exaggerate or minimize. Don't omit key facts.
- Make sure your report is relevant, accurate, and reliable. Double-check your details against other sources, and make sure you have sufficient information to reach your conclusions or provide recommendations.

Review the discussion of ethics in business writing in Chapter 1 (pages 19–27).

4. Organize Carefully

Organizing a short report effectively means that you include the right amount of information in the most appropriate places for your audience. Make your report easy to read and to follow. Many times a simple chronological or sequential organization will be acceptable for your readers. Your headings will show readers how you organized your report. Readers will expect your report to contain information on such topics as the purpose, findings, conclusions, and, in many reports, recommendations, as described in the following sections.

Purpose

Always begin by telling readers why you are writing (your reasons) and by alerting them to what you will discuss and why it is significant. Give your readers a summary of key events and details at the beginning to help them follow the remainder of the report quickly. Essential background information alerts readers to the importance of your report. When you establish the scope (or limits) of your report, you help readers zero in on specific times, costs, places, procedures, or problems.

Findings

This should be the longest part of your report and contain the data (the results) you have collected—facts about prices, personnel, equipment, events, locations, incidents, or tests. Gather the data from your research; interviews and/or conversations with co-workers, employers, or clients. Again, remember to choose only those details that are of most importance, relevance, to your reader.

Conclusion

Your conclusion tells readers what your data mean. A conclusion can summarize what has happened; review what actions were taken; or explain the outcome or results of a test, a visit, or a program.

Recommendations

A recommendation informs readers what specific actions you think your company or client should take—market a new product, hire more staff, institute safety measures, select among alternative plans or procedures, and so on. Recommendations must be based on the data you collected and the conclusions you have reached. They need to show how all the pieces fit together.

Note how the periodic report in Figure 8.1 fails to help readers see and understand the organization and importance of the information. The revised version of the report, Figure 8.2 (pages 286–287), clearly illustrates effective report writing. The reader is provided with information at regularly scheduled intervals—daily, weekly, bimonthly (twice a month).

Note that because of the deadlines executives face on a daily basis, some companies prefer a section on recommendations to come at the beginning of the report (as in Figure 8.5), followed by supporting documentation.

5. Use Reader-Centered Headings, Bullets, Numbering, and Visuals

Help readers locate and focus on key information in your report quickly by using headings and subheadings that clearly demonstrate how you have organized and subdivided your report. See, for example, the headings in Figures 8.2, 8.4, and 8.5. You can also use bulleted and numbered lists where relevant, as in Figure 8.2, to break up chunks of information into easy-to-grasp pieces. Finally, as we discussed in Chapter 6 (pages 209–211), include only the most essential visuals to reinforce and summarize important data for your readers, as the table in Figure 8.2 and the map in Figure 8.5 do.

6. Write Clearly and Concisely

Writing clearly and concisely is essential in all business reports. Ask your boss or co-workers who have done short reports for your company about appropriate style. Also look at previous, similar reports to get a sense of your company or house style. Here are a few guidelines to help you write clearly and concisely.

- **Use an informative title/subject line that gets to the point right away.** "Software Options" is not as clear as "Which Software Options Are Most Economical and Why."
- **Write in plain English.** Make every word count, avoid jargon, and keep your writing simple and straightforward. Prune business clichés such as "at the end of the day" or "to put a fine point on it."
- **For global readers, make sure you use international English.** Keep your sentences short, and write in the active voice. Do not use U.S. idioms, slang, or abbreviations. See pages 136–137.
- **Adopt a professional yet personal tone.** Avoid being overly formal or too casual—strike a balance between these two extremes. Don't sound arrogant by adopting a tone that suggests you alone have the final authority.
- **Keep your report as concise as possible while still giving readers the essential information they need.** Don't burden them with lengthy project histories when all they ask for is a quick update on a project, and don't pad the report with unnecessary details to sound important. If a short report is longer than two to three pages, it may be too long.
- **Begin with the most important information.** Give busy readers what they want as quickly as possible Your first sentence should focus on why the report was written and why it is important—e.g., to update, to summarize, to provide monthly figures to solve a problem. See how all the reports in this chapter supply effective opening paragraphs.
- **Supply sufficient and relevant documentation** about schedules, costs, personnel, sales, materials, the environment, and so forth.

7. Choose the Most Appropriate Format

Depending on your audience, you can send your short report as an e-mail, a memo, or a letter. For routine reports to your boss or others inside your company, you will likely use a memo format, as in Figures 8.2, 8.3, and 8.5. Note that with a memo format your readers will not expect you to include an inside address or formal salutation and complimentary close. Depending on your company's policy, you might also send a short report in the body of an e-mail, as in Figures 2.6, 3.4, and 3.5, or as an attachment to an e-mail. Incident reports, however, are often submitted as hard-copy memos for legal reasons; they can also be written as a memo, as in Figure 8.6, or you may be required to complete a special form. When writing to clients and other readers outside your company or organization, it is best to send your report as a formal letter (including a salutation and complimentary close; see pages 292–294), as in Figure 8.4.

An example of a poorly written, poorly organized, and poorly formatted short report.   Figure 8.1

To Serve and To Protect

Springdale Police Department

Emergency 555-1000   **Administration** 555-1001   **Traffic** 555-1002

TO: Capt. Alice Martin
FROM: Sergeants Daniel Huxley, Jennifer Chavez,
          and Ivor Paz
SUBJECT: Crimes
DATE: July 12, 2011

This report will let you know what happened this quarter as opposed to what happened last quarter as far as crimes are concerned in Springdale. This **report is based on statistics** the department has given us over the quarter.

  Here we'll let the facts speak for themselves. From Jan.–Mar. we saw 126 robberies while from Apr.–June we had 106. Home burglaries for this period: 43; last period: 36. 33 cars were stolen in the period before this one; now we have 40. Interestingly enough, **last year at this time we had only 27** thefts. Four of them involved heirlooms.

Homicides were 9 this time versus 8 last quarter; assault and battery charges were 92 this time, 77 last time. Carrying a concealed weapon 11 (10 last quarter). We had 47 arrests (55 last quarter) for charges of possession of **a controlled substance**. Rape charges were 8, **1 less than last quarter.** 319 citations this time for moving violations: speeding 158/98, and failing to observe the signals 165/102 last quarter. DUIs this quarter—only 45, or 23 fewer than last quarter.

  Misdemeanors this quarter: disturbing the peace the 53; vagrancy/public drunkenness 8; violating leash laws 32; violating city codes 39, including **dumping trash**. Last quarter the figures were 48, 59, 21, 43.

We believe this report is **complete and up to date**. We further hope that this report has given you all the **facts** you will need.

Poor format—some paragraphs indented, some not; boldfacing not helpful

Vague subject line

Introduction doesn't tell reader anything about overall picture

Throws facts out without any sense of reader's needs

Irrelevant data

No analysis or guided commentary— just gives undigested numbers

Hard-to-follow comparisons and contrasts

Conclusion provides no summary or recommen- dations

Figure 8.2   **A well-prepared report, revised from Figure 8.1.**

To Serve and To Protect

Springdale Police Department

Emergency 555-1000   **Administration** 555-1001   **Traffic** 555-1002

Precise subject line

TO: Captain Alice Martin
FROM: Sergeants Daniel Huxley, Jennifer Chavez, and Ivor Paz
SUBJECT: Crime rate for the second quarter of 2011
DATE: July 12, 2011

Begins with concise overview of report

From April 1 to June 30, 852 crimes were committed in Springdale, representing a 5 percent increase over the 815 crimes recorded during the previous quarter.

TYPES OF CRIME

Organizes crimes into categories

The following report, based on the table below, discusses the specific types of crimes, organized into four categories: **robberies and theft, felonies, traffic, and misdemeanors**.

Table 1. Comparison of the 1st and 2nd Quarter Crime Rates in Springdale

Supplies easy-to-follow visual

Table is highlighted, making it easier to read

CATEGORY	1st Quarter	2nd Quarter
ROBBERIES and THEFT		
Commercial	63	75
Domestic	36	43
Auto	33	40
FELONIES		
Homicides	16	13
Assault and battery	77	92
Carrying a concealed weapon	10	11
Poss. of a controlled substance	55	47
Rape	9	8
TRAFFIC		
Speeding	165	158
Failure to observe signals	102	98
DUI	78	45
MISDEMEANORS		
Disturbing the peace	48	53
Vagrancy	40	48
Public drunkenness	19	40
Leash law violations	21	32
Dumping trash	35	37
Other	8	12

Uses clear headings to show organization of report

Robberies and Theft

The greatest increase in crime was in robberies, 20 percent more than last quarter. Downtown merchants reported 75 burglaries, exceeding $985,000. The biggest theft occurred on May 21 at Weisenfarth's Jewelers when three armed robbers stole more than $217,000 in merchandise. (Suspects were

Page 2

apprehended two days later.) Home burglaries accounted for 43 crimes, though the thefts were not confined to any one residential area. We also had 40 car thefts reported and investigated.

Felonies
Homicides decreased slightly from last quarter—from 16 to 13. Charges for battery, however, increased—15 more than we had last quarter. Arrests for carrying a concealed weapon were nearly identical this quarter to last quarter's total. But the 47 arrests for possession of a controlled substance were appreciably down from the first quarter. Arrests for rape for this quarter also were less than last quarter's. Three of those rapes happened within one week (May 6–12) and have been attributed to the same suspect, now in custody.

Provides essential background and statistical information and comparative analyses

Traffic
Traffic violations for this period were lower than last quarter's figures, yet this quarter's citations for moving violations (335) represent a 5 percent increase over last quarter's (322). Most of the citations were issued for speeding (158) or for failing to observe signals (98). Officers issued 45 citations to motorists for DUIs, an impressive decrease over the 78 DUIs issued last quarter. The new state penalty of withholding a driver's license for six months of anyone convicted of driving while under the influence appears to be an effective deterrent.

Easy-to-read sentences

Misdemeanors
The largest number of arrests in this category were for disturbing the peace—53. Compared to last quarter, this is an increase of 10 percent. There were 88 charges for vagrancy and public drunkenness, an increase from the 59 charges made last quarter. We issued 32 citations for violations of leash laws, which represents a sizable increase over last quarter's 21 citations. Thirty-seven citations were issued for dumping trash at the Mason Reservoir.

Draws logical conclusion

Includes only data reader needs

CONCLUSION
Overall, while the crime rate has decreased in traffic (especially DUIs) and possession of controlled substances this quarter, we have seen a marked increase in arrests for robberies and battery.

Summarizes findings of report

RECOMMENDATIONS
To help deter robberies in the downtown area, we recommend the following:

- increasing surveillance units to 15 rather than the 10 now in the area
- offering businesses our workshop on safety and security precautions, as we did during the first quarter

Historically, battery arrests have risen during the second quarter. Our recommendations to counter this trend include:

- continuing to work closely with Neighborhood Watch Groups
- providing more foot and bicycle patrols in the neighborhoods with the highest incidence of battery complaints

Offers specific actions/ changes department should make based on conclusion of report in bulleted lists

Periodic Reports

Periodic reports, as their name signifies, provide readers with information at regularly scheduled intervals—daily, weekly, bimonthly (twice a month), monthly, quarterly. They help a company or agency keep track of the quantity and quality of the services it provides and the amount and types of work done by employees. Information in periodic reports helps managers make schedules; order materials; hire, train, or assign personnel; budget funds; and, generally speaking, determine corporate needs.

Figure 8.2 provides essential police information clearly and concisely. As we saw, it summarizes, organizes, and interprets the data collected over a three-month period from individual officers' activity logs. Because of this report, Captain Alice Martin will be better able to plan future protection for the community and to recommend changes in police services.

Sales Reports

Sales reports provide businesses with a necessary and ongoing record of accounts, online and mail purchases, losses, and profits over a specified period of time. They help businesses assess past performance and plan for the future. In doing that, they fulfill two functions: **financial** and **managerial**. As a financial record, sales reports list costs per unit, discounts or special reductions, and subtotals and totals. Like a spreadsheet, sales reports show gains and losses. They may also provide statistics for comparing two quarters' sales.

Sales reports are also a managerial tool because they help businesses make both short- and long-range plans. The restaurant manager's sales report illustrated in Figure 8.3 (page 289) guides the owners to decide which popular entrées to highlight and which unpopular ones to modify or delete. Note how the recommendations follow logically from the figures Sam Jelinek gives to Gina Smeltzer and Alfonso Zapatta, the owners of The Grill. Note, too, that because the readers are familiar with the subject of the report, the writer did not have to supply background information on the entire offerings.

Progress Reports

A progress report informs readers about the status of an ongoing project. It lets them know how much and what type of work has been done by a particular date, by whom, how well, and how close the entire job is to being completed. A progress report emphasizes whether you are

- specifying what work has been done
- keeping your schedule
- staying within your budget
- using the proper technology/equipment
- making the right assignments
- identifying an unexpected problem
- providing adjustments in schedules, personnel, and so on
- indicating what work remains to be done
- completing the job efficiently, correctly, and according to codes

A sales report to a manager.   Figure 8.3

Thegrill

Dayton, OH 43210 • (813) 555-4000 • (813) 555-4100 fax   www.Thegrill.com

TO: Gina Smeltzer   DATE: June 28, 2012
         Alfonso Zapatta, Owners
FROM: Sam Jelinek S.J.   SUBJECT: Analysis of entrée sales,
         Manager   June 10–16 and June 17–23

As we agreed at our monthly meeting on June 4, here is my analysis of entrée sales for two weeks to assist us in our menu planning. Below is a record of entrée sales for the weeks of June 10–16 and June 17–23 that I have compiled and put in a table for easier comparisons.

Begins with purpose and scope of report readers need

	Portion size	June 10–16		June 17–23		2 weeks combined	
		Amount	Percentage	Amount	Percentage	Amount	Percentage
Cornish Hen	6 oz.	238	17	307	17	545	17
Stuffed Young Turkey	8 oz.	112	8	182	10	294	12
Broiled Salmon Steak	8 oz.	154	11	217	12	371	13
Brook Trout	12 oz.	182	13	252	14	434	9
Prime Rib	10 oz.	168	12	198	11	366	11
Lobster Tails	2–4 oz.	147	10	161	9	308	10
Delmonico Steak	10 oz.	56	4	70	4	126	4
Moroccan Chicken	6 oz.	343	25	413	23	756	24
		1,400	**100**	**1,800**	**100**	**3,200**	**100**

Organizes findings of the report in helpful table

Boldfaces totals

Recommendations

Based on the figures in the table above, I recommend that we do the following:

1. Order at least 100 more pounds of prime rib each two-week period to be eligible for further quantity discounts from the Northern Meat Company.
2. Delete the Delmonico Steak entrée because of its low acceptance.
3. Introduce a new chicken or fish entrée to take the place of the Delmonico Steak; I would suggest grilled lemon chicken to accommodate those patrons increasingly interested in tasty, low-fat, lower-cholesterol entrées.

Please give me your reactions within the next week. It shouldn't take more than a few days to implement these changes.

Offers precise and relevant recommendations

Requests authorization to implement recommendations

Almost any kind of ongoing work can be described in a progress report—research for a paper, construction of an apartment complex, preparation of a website, documentation of a patient's rehabilitation. Progress reports are often prepared at key phases, or milestones, in a project.

Audience for Progress Reports

A progress report is intended for people who generally are not working alongside you but who need a record of your activities to coordinate them with other individuals' efforts and to learn about problems or changes in plans. For example, because supervisors may not be in the field or branch office or at a construction site, they will rely on your progress report for crucial information. Customers, such as a contractor's clients, expect reports on how carefully their money is being spent. That way they can adjust schedules or alter specifications if there is a risk of going over budget.

The length of the progress report will depend on the complexity of the project. Contractor Dale Brandt's assessment of the progress his contractor company is making in renovating Dr. Burke's clinic is given in a two-page letter in Figure 8.4 (pages 291–292). A shorter report, such as one on organizing a workshop on time management, or one providing employees with a status report on a project (as in Figure 3.5), however, might require only a short e-mail.

Frequency of Progress Reports

Progress reports can be written daily, weekly, monthly, quarterly, or annually. Your specific job and your employer's needs will dictate how often you have to keep others informed of your progress. Contractor Brandt determined that three reports, spaced four to six weeks apart, would be necessary to keep Dr. Burke posted. Figure 8.4 is the second of those reports.

Parts of A Progress Report

Progress reports should contain information on (1) the work you have done, (2) the work you are currently doing, and (3) the work you will do.

How to Begin a Progress Report

In a brief introduction, cover the following:

- indicate why you are writing the report
- provide any necessary project titles or codes and specific dates
- help readers recall the job you are doing for them

If you are writing an initial progress report, supply brief background information in the opening. But, if you are submitting a subsequent progress report, your introduction should only remind the reader about where your previous report left off and where the current one begins. Note how Dale Brandt's first paragraph in Figure 8.4 calls attention to the successful continuity of his work.

How to Write the Body of a Progress Report

The body of the report should provide significant details about costs, materials, personnel, and times for the major stages of the project.

Brandt Construction Company

Halsted at Roosevelt, Chicago, Illinois 60608-0999 • 312-555-3700 • Fax: 312-555-1731
http://www.brandt.com

*Professional
looking
letterhead*

April 27, 2012

Dr. Pamela Burke
1439 Grand Avenue
Mount Prospect, IL 60045-1003

Dear Dr. Burke:

Here is my second progress report about the renovation work being done at your new clinic at Hacienda and Donohue. Work proceeded satisfactorily in April according to the plans you had approved in March.

*Begins with key
information:
project status*

Review of Work Completed in March
As I informed you in my first progress report on March 31, we tore down the walls, pulled the old wiring, and removed existing plumbing lines. All the gutting work was finished in March.

*Succinctly
recaps for
background*

Work Completed During April
By April 7, we had laid the new pipes and connected them to the main septic line. We also installed the two commodes, the four standard sinks, and the utility basin. The heating and air-conditioning ducts were installed by April 13. From April 16–20, we erected soundproof walls in the four examination rooms, the reception area, your office, and the laboratory. Throughout we used environmentally safe (green) materials. We also installed solar panels on the roof of the reception area and reduced it by five feet to make the first examination room larger, as you had requested.

*Summarizes
current
accomplish-
ments,
including
attention to
greening the
building*

Problems with the Electrical System
We had difficulty with the electrical work, however. The number of outlets and the generator for the laboratory equipment required extra-duty power lines that had to be approved by both Con Edison and Cook County inspectors. Securing their approval slowed us down by three days. Also, the vendor, Midtown Electric, failed to deliver the recessed lighting fixtures by April 26, as promised. Those fixtures and the generator are now being installed. Moreover, the cost of those fixtures will increase the material budget by **$5,288.00**. But the cost for labor is as we had projected—**$94,550**.

*Identifies
problems,
how they were
solved, and
costs involved*

Work Remaining
The finishing work is scheduled for May. By May 11, the floors in the examination rooms, laboratory, washrooms, and hallways should be tiled and the reception area

*Specifies
what work
remains and
when it will be
completed*

(Continued)

Figure 8.4   (Continued)

Page 2

Projects successful completion of work

and your office carpeted. By May 13, the reception area and your office should be paneled and the rest of the walls painted. If everything stays on schedule, touch-up work is planned for May 14–18. You should be able to move into your new clinic by May 21.

Promises to keep reader informed

You will receive a third and final progress report by May 14. Thank you again for your business and the confidence you have placed in our company.

Sincerely yours,

Dale Brandt

Dale Brandt

- Emphasize completed tasks, not false starts. If you report that the carpentry work or painting is finished, readers do not need an explanation of paint viscosity or geometrical patterns.
- Omit routine or well-known details ("I had to use the library when I wanted to read the back issues of *Safety News* that were not archived on the Internet").
- Describe in the body of your report any snags you encountered that may affect the work in progress (see Dale Brandt's section on electrical problems in Figure 8.4). It is better for the reader to know about trouble early in the project so that appropriate changes or corrections can be made.

How to End a Progress Report

The conclusion should give a timetable for the completion of duties or submission of the next progress report. Give the date by which you expect work to be completed. Be realistic; do not promise to have a job done in less time than you know it will take. Readers will not expect miracles, only informed estimates. Even so, any conclusion must be tentative. Note that the good news Dale Brandt gives Dr. Burke about moving into her new clinic is qualified by the words "If everything stays on schedule." He is also well aware of the "you attitude" by thanking Dr. Burke again for her business.

Travel/Trip Reports

Reporting on the trips you take is an important professional responsibility in the world of work. In documenting what you did and saw, travel reports keep readers informed about your efforts and how they affect ongoing or future business. Moreover, such reports help you to better understand your job and to develop your networking skills. Travel reports also should be written after you attend a conference, convention, or sales meeting or call on customers.

Questions Travel Reports Answer

Specifically, travel reports should answer these questions for your readers.

- Where did you go?
- When did you go?
- Why did you go?
- What did you see or witness?
- Whom did you see?
- What did they tell you?
- What did you do about it?

For a business trip, you are also likely to have to inform readers how much it cost and to supply them with receipts for all of your business expenses.

Common Types of Travel/Trip Reports

Travel reports can cover a wide range of activities and are called by different names to characterize those activities. Most likely, you will encounter the following three types of travel reports.

1. **Site inspection reports**. These reports inform managers about conditions at a branch office or plant, a customer's business, or on the advisability of relocating an office or other facility. After visiting the site, you will determine whether it meets your employer's (or customer's) needs. Site inspection reports can provide information about the physical plant, the environment (air, soil, water, vegetation), or computer or financial operations.

Figure 8.5, which begins with a recommendation, is a report written to a district manager interested in acquiring a new site for a fast-food restaurant.

2. **Field trip reports**. These reports, often assigned in a course, are written after a visit to a laboratory, hospital, detention center, or other location to show what you have learned about the operation of a facility. (Such visits might even be done via "virtual tours" on the Internet.) You will be expected to describe how an institution is organized, the technical procedures and/or equipment it uses, pertinent ecological conditions, or the ratio of one group to another. The emphasis in such reports is on the educational value of the trip. For example, "From my visit to Water Valley, I learned a great deal about the health care delivery system at an extended care facility, which will help me during my internship next term."

3. **Home health or social work visits**. Nurses, social workers, and probation officers, for instance, report routinely on their visits to patients and clients. Their reports describe clients' lifestyles, assess needs, and make recommendations based on a variety of sources—clients, health care professionals, charitable organizations, and the like. These reports are often divided into Purpose of the Visit, Description of the Visit, and Action Taken as a Result of the Visit.

Figure 8.5   **A site inspection report using a map.**

VAIL's
Chicken House

TO: Pretha Bandi   *B.A.R.*   DATE: March 30, 2012
FROM: Beth Armando-Ruiz   SUBJECT: New Site for Vail's #7
          Development Department

Begins with most important details about the writer's recommendation

Recommendation
To follow up on our discussions earlier this month, I think the best location for the new Vail's Chicken House is the vacant Dairy World restaurant at the northeast corner of Smith and Fairfax avenues—1701 Fairfax. I inspected this property on March 19 and 20 and also talked to Kim Shao, the broker at Crescent Realty representing the Dairy World Company—kims@crescent.org; (817) 555-1779. The location, building, and parking facilities at the Dairy World site all present the best opportunity for future growth and increased sales for Vail's.

Gives essential contact information

Provides necessary background details

The Location
Please refer to the map below. Located at the intersection of the two busiest streets on the southeast side, the property will allow us to take advantage of the traffic flow to attract customers. Being only one block west of the Cloverleaf Mall should also help attract business.

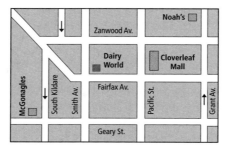

Includes map and traffic flow information essential for reader's purpose

Another benefit is that customers will have easy access to our location. They can enter or exit the Dairy World site from either Smith or Fairfax. Left turns on Smith are prohibited from 7 a.m. to 9 a.m., but since most of our business is done after 11 a.m., the restriction poses few problems.

Denver, CO 87123   (303) 555-7200   http://www.vails.com

(Continued)   Figure 8.5

Pretha Bandi
March 30, 2012
Page 2

Area Competition
Only two other fast-food establishments are in a one-mile vicinity. McGonagles, 1534 South Kildare, specializes in hamburgers; Noah's, 703 Zanwood, serves primarily seafood entrées. Their offerings will not directly compete with ours. The closest fast-food restaurant serving chicken is Johnson's, 1.8 miles away.

Assesses the location, in light of the competition

Parking Facilities
The parking lot has space for 25 cars, and the area at the south end of the property (38 feet × 37 feet) can accommodate 14–15 more vehicles. The driveways and parking lot were paved with asphalt last March and appear to be in excellent condition. We will also be able to make use of the drive-up window on the north side of the building.

Gives only the most essential facts audience needs on parking, seating capacity, and alterations

The Building
The building has 3,993 square feet of heated and cooled space. The air-conditioning units and heating units were installed within the last fifteen months and seem to be in good working order; nine more months of transferable warranty remain on all these units.

Writer has done research on equipment

The only major changes we must make are in the kitchen. To prepare items on the Vail's menu, we need to add three more exhaust fans (there is only one now) and expand the grill and cooking areas. The kitchen also has three relatively new sinks and offers ample storage space in the 16 cabinets.

Writing is clear and concise

Paragraphs are easy to follow with precise topic sentences

The restaurant has a seating capacity of up to 54 persons; 10 booths are covered with red vinyl and are comfortably padded. A color-coordinated serving counter could seat 8 to 10 patrons. The floor does not need to be retiled, but the walls will have to be painted to match Vail's decor.

Does not overwhelm with petty details

How to Gather Information for a Trip/Travel Report
Regardless of the kind of trip report you have to write, your assignment will be easier and your report better organized if you follow these suggestions.

1. **Before you leave on the field trip, site inspection, or visit, be sure you are prepared as follows:**
   a. Obtain all necessary names; street, e-mail, and website addresses; and relevant telephone, cell, and fax numbers.

    b. Check files for previous correspondence, case studies, warranties, or terms of contracts or agreements.

    c. Download work orders, instructions, or other documents pertinent to your visit, for example, websites and ads.

    d. Bring a laptop or notebook with you.

    e. Locate a map of the area and get the directions you'll need beforehand. (Both maps and directions can be obtained easily at *http://www.mapquest.com*, *http://maps.yahoo.com*, and *http://maps.google.com*).

    f. Make a record of appointment times and locations as well as the job titles of the people whom you expect to meet.

    g. Bring a camcorder, tape recorder, iPhone, camera, and/or calculator, if necessary, to record important data.

2. **When you return from your trip, keep the following hints in mind as you compile your report:**

    a. Write your report promptly. If you put it off, you may forget important details.

    b. When a trip takes you to two or more widely separated places, note in your report when you arrived at each place and how long you stayed.

    c. Exclude irrelevant details, such as whether the trip was enjoyable, what you ate, or how delighted you were to meet people.

    d. Double-check to make sure you have listed names and calculated figures correctly. Mistakes in math make you look bad.

Incident Reports

The short reports discussed thus far in this chapter have dealt with routine work. They have described events that were anticipated, planned, or supervised. But every business or agency runs into unexpected trouble that delays routine work. More often than not, these circumstances need to be documented in an incident report. Employers and, on some occasions, government inspectors, law enforcement, insurance agents, and attorneys must be informed about those events that interfere with or threaten normal, safe operations. Incident reports are often submitted as a memo, as in Figure 8.6 (pages 297–298), or on specially prepared forms your employer or government agency expects you to follow.

When to Submit an Incident Report

An incident report is required when there is, for example

- an accident—fire, automobile, physical injury
- a law enforcement offense
- an environmental danger, including a computer virus
- a machine breakdown
- a delivery delay
- a cost overrun
- a production slowdown

THE GREAT HARVESTER RAILROAD
Des Moines, IA 50306-4005
http://www.ghrr.com

TO: Angela O'Brien, District Manager
              James Hwang, Safety Inspector
FROM: Nick Roane, Engineer *Nick Roane*
DATE: August 6, 2011
SUBJECT: Derailment of Train 26 on August 5, 2011

Signs report to verify account of incident

TYPE OF INCIDENT
Two grain cars went off the track while I was driving Engine 457 of Train 26 on August 5, 2011. There were no injuries to the crew.

Begins with most important details

DESCRIPTION OF INCIDENT
At 7:20 a.m. on August 5, 2011, I was traveling north at a speed of 52 miles an hour on the single main-line track four miles east of Ridgeville, Illinois. Weather conditions and visibility were excellent. Suddenly, the last two grain cars, 3022 and 3053, jumped the track. The train automatically went into emergency braking and stopped immediately. But it did not stop before both grain cars turned at a 45° angle. After checking these cars, I found that half the contents of their loads had spilled. The train was not carrying any hazardous chemical shipments.

Gives precise time, location

Describes what happened

I notified Supervisor Bill Purvis at 7:40 a.m., and within 45 minutes he and a section crew arrived at the scene with rerailing equipment. The section crew removed the two grain cars from the track, put in new ties, and made the main-line track passable by 9:25 a.m. At 9:45 a.m. a vacuum car arrived with Engine 372 from Hazlehurst, Illinois, and its crew proceeded with the cleanup operation. By 10:25 a.m. all the spilled grain was loaded onto the cars brought by the Hazlehurst train. Bill Purvis notified Barnwell Granary that their shipment would be at least four hours late.

Explains what was done

CAUSES OF INCIDENT
Supervisor Purvis and I checked the stretch of train track where the cars derailed and found it to be heavily worn. We believe that a fisher joint slipped when the grain cars hit it, and the track broke. You can see the location of the cracked fisher joint in the graphic below.

Determines likely cause

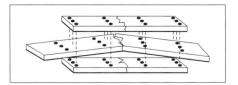

Supplies easy-to-follow exploded visual

Figure 8.6   **(Continued)**

2

Offers precise recommendations to solve or prevent problem from recurring

RECOMMENDATIONS

We made the following recommendations to the switch yard in Hazlehurst to be carried out immediately.

1. Check the section of track for 10 miles on either side of Ridgeville for any signs of defective fisher joints.
2. Repair any defective joints at once.
3. Instruct all engineers to slow down to 5 to 10 mph over this section of the track until the rail check is completed.

The incident report in Figure 8.6 on the train accident was submitted as a memo by the engineer.

Parts of an Incident Report

Include the following information in your incident report. Because it can contain legally sensitive information, for which the reader needs a hard copy and paper trail, an incident report should not be sent as an e-mail. Note how Figure 8.6 includes detailed and accurate information on these parts.

1. Type of incident. Briefly identify the incident—personal injury, fire, burglary, equipment failure. Identify any part(s) of the body precisely. "Eye injury" is not enough; "injury to the right eye, causing bleeding" is better.

2. Identification details. Indicate who was involved. List employee identification numbers. Record titles, department, and employment identification. For customers or victims, record home addresses, phone numbers, and places of employment. Insurance companies will also require policy numbers. Be sure to include any damaged equipment model/serial numbers.

3. Time and location of the incident. Answer key questions of when and where for readers. Include precise calendar dates (not "Thursday") and time (a.m. or p.m.).

4. Description of what happened. Let readers know exactly what happened and why, how it occurred, and what led up to the incident.

5. What was done after the incident. Describe the action you took to correct conditions and to get things back to normal, and what was done to treat the injured, to make the environment safer, to speed a delivery, to repair damaged equipment.

6. **What caused the incident**. Make sure your explanation is consistent with your description of what happened. Pinpoint the trouble. In Figure 8.6, for example, the defective fisher joint is listed under the heading "Causes of Incident."

7. **Recommendations**. Recommendations about preventing the problem from recurring may involve repairing any broken parts, as in Figure 8.6, calling a special safety meeting, asking for further training, adapting existing equipment, doing emergency planning, or modifying schedules.

Protecting Yourself Legally

An incident report can be used as legal evidence; it will then become part of a permanent legal record that can be used by law enforcement and attorneys in court to establish negligence and liability on your and your company's part, including determining a worker's compensation rights. It can also be used by an employer to assess employee responsibility. An incident report frequently concerns the two topics over which powerful legal battles are waged—health and property.

You have to be very careful about collecting and recording details. Make sure your report is not biased, sketchy, or incomplete. To avoid these errors, you may have to interview employees or bystanders; travel to the incident site; check compliance manuals, code books, or other guides; consult safety experts; or research records and archives.

To ensure that what you write is legally proper, follow these guidelines.

1. **Submit your report promptly and sign or initial it**. Any delay might be seen as a cover-up. Send your report to the appropriate parties immediately after you have gathered the necessary information and had it reviewed by your supervisor.
2. **Be accurate, objective, and complete**. Recount precisely what happened in the order it took place. Never omit or distort facts; the information may surface later, and you could be accused of a cover-up. Do not just write "I do not know" for an answer. If you are not sure, state why. Also be careful that there are no discrepancies in your report.
3. **Give facts, not opinions**. Provide a factual account of what actually happened, not a biased interpretation of events or one based on speculation or hearsay. Vague words such as "I guess," "I wonder," "apparently," "perhaps," or "possibly" weaken your objectivity. Stick to details you witnessed or that were seen by eyewitnesses. Identify witnesses or victims by giving names, addresses, places of employment, and so on. Keep in mind that stating what someone else saw is regarded as hearsay and therefore is not admissible in a court of law. State only what you saw or heard. When you describe what happened, avoid drawing uncalled-for conclusions. Consider the following statements of opinion and fact:

   Opinion: The patient seemed confused and caught himself in his IV tubing.
   Fact: The patient caught himself in his IV tubing.

   Opinion: The equipment was defective.
   Fact: The bolt was loose.

Be careful, too, about blaming someone. Statements such as "Baxter was incompetent" or "The company knew of the problem but did nothing about it" are libelous remarks.

4. **Do not exceed your professional responsibilities**. Answer only those questions you are qualified to answer. Do not presume to speak as a detective, an inspector, a physician, a supervisor, or a judge. Do not represent yourself as an attorney or a claims adjuster in writing the report. And don't take sides.

Writing Successful Proposals

A proposal is a detailed plan of action submitted to a reader or group of readers for approval. The readers are usually in a position of authority—supervisors, managers, department heads, company buyers, boards of private foundations, elected officials, military or civic leaders—to endorse or reject the plan. Your proposal must convince these readers that your plan will help them improve their business, save them money, enhance their image, improve customer satisfaction, make the environment safer, or all of these.

Proposals are written for many purposes and many different audiences. You can write an internal proposal, for example, to your boss, seeking authorization to hire staff, change a procedure, or purchase a new piece of equipment. Or you can write a sales proposal to potential customers, offering a product or a service (such as providing training with new, special firefighting gear or selling an office manager a line of ergonomically designed furniture).

Depending on the job, proposals can vary greatly in size and in scope. A formal proposal can be a very long and complex document running into hundreds of pages. A proposal to your employer, however, about redesigning the company website could easily be conveyed in a few pages, the length of a short report. To propose doing a small job for a prospective client—for example, redecorating a waiting room in an accountant's office—a letter with information on costs, materials, and a timetable might suffice. The sales letter in Figure 4.9 (page 115) illustrates a short proposal in letter format. Proposals can be *unsolicited*—that is, they originate with you—or they can be *solicited*, requested by a company or organization, as in Figure 8.9 (pages 311–313).

Proposals Are Persuasive Plans

Proposals, whether large or small, must be highly persuasive to succeed. Without your audience's approval, your plan will never go into effect, however accurate and important you think it is. Your enthusiasm is not enough to persuade readers; you have to supply hard evidence. Your proposals must convince readers that your plan is relevant, practical, based upon careful research, and designed to benefit the reader and his or her company.

Every proposal you write must exhibit a "can do" attitude, putting the reader and his or her company's needs at the center of your work. Show readers how approving your plan will save them time and money, increase productivity, improve employee morale, or attract new business. The tone of your proposal should be "Here is what I can do for you." Yates Engineering has won millions of dollars of business through its reader-centered proposals. Its slogan is "On time... within budget ... to your satisfaction." Time, budget, and your readers' satisfaction and convenience are among the key ingredients of a winning proposal. Customize your proposal by personolizing it. Notice how the advertisement in Figure 8.7 below appeals to customers by offering different security services options: from motion detectors to video recording to supplying well-trained officers. (See also Figure 1.6 on page 17.)

Proposals Frequently Are Collaborative Efforts

Like many other examples of business and technical writing, proposals often are the product of teamwork. Even a short in-house proposal is often researched and put together by more than one individual in the company or agency, as Figure 8.8 illustrates.

Many times, individual employees will pull together information from their separate areas (such as graphics and design, finance, marketing, technology, transportation, and even legal) and put it into a proposal that each member of the team then reads and revises until the team agrees that the document is ready to be released.

An advertisement for a company with a "can-do" attitude.   Figure 8.7

Guidelines for Writing a Successful Proposal

The following guidelines will help you to persuade your audience to approve your plan. Refer to these guidelines and Figures 8.8 (pages 304–307) and 8.9 (pages 311–313) both before and while you formulate your plan. (See also the discussion on ethics and proposal writing on page 308.)

1. **Approach writing a proposal as a problem-solving activity**. Your purpose should reflect your ability to identify and solve problems. Convince your audience you know what their needs are and that you will meet them professionally, safely, and promptly.

2. **Regard your audience as skeptical**. Even though you offer a proposal that you think will benefit your readers, do not be overconfident that they will automatically accept it. Brainstorm, alone or with your collaborative group, to anticipate your readers' questions and objections. To determine whether your proposal is feasible, readers will study it carefully. If your proposal contains errors and inconsistencies, omits important information, or deviates from what your readers are looking for, then your audience is likely to reject it.

3. **Research your proposal topic thoroughly**. A winning proposal is *not* based on a few well-meaning, general suggestions. To provide the detailed information necessary and to convince readers that you know your proposal topic inside and out, you will have to do your homework. Research your topic by studying the latest technology in the field, shopping for the best prices, comparing prices and services with what the competition offers, verifying schedules, visiting customers, making site visits, and interviewing key individuals. Make sure that any technology or equipment you use or sell complies with all codes, specifications, and standards.

4. **Scout out what your competitors are doing**. Become familiar with your competitors' products or services, have a fair idea about their market costs, and be able to show how your company's work is better overall. Provide examples; offer a demonstration. Read competitors' websites and print publications (such as catalogues and marketing brochures) very carefully. Let readers know you have done your homework on their behalf.

5. **Prove that your proposal is workable**. What you propose should be consistent with the organization and capabilities of the company and respect its corporate culture, mission, and capabilities. For instance, recommending that a small company with 25 employees triple its work force to implement your plan would be foolish and risky. Your proposal should not contain statements such as "Let's see what happens."

6. **Be sure your proposal is financially realistic**. "Is it worth the money?" is a bottom-line question you can expect from your readers. For example, recommending that your company spend $20,000 to solve a $2,000 problem is just not feasible. Prepare a cost estimate that is in line with current market conditions, competitors' prices, and your readers' budgets or expenditures. Above all, convince your readers that the benefits are worth the costs. Note how Figure 8.8 saves the bank money and Figure 8.9 stresses that the costs are under control.

7. **Package your proposal attractively**.   Make sure that your proposal is well presented, professional looking, inviting, attractive, and easy to read, and that all visuals are clear and appropriately placed (see Chapter 6, pages 209–313). The visual appearance of your proposal can contribute greatly to whether it is accepted.

Internal Proposals

The primary purpose of an internal proposal, such as the one shown in Figure 8.8, is to offer a realistic and constructive plan to help your company run its business more efficiently and economically.

On your job you may discover a better way of doing something or a more efficient way to correct a problem. You believe that your proposed change will save your employer time, money, or further trouble. (Note how Tina Escobar and Oliver Jabur identified and researched a more effective and less costly way for Community Federal Bank to conduct business and to satisfy its customers in Figure 8.8.) Or your department head, manager, or supervisor may call your attention to a problem and ask you for specific ways to solve it.

Regardless of who identifies the problem, your proposal, generally speaking, will be an informal, in-house message. A brief (one to three-page) memo, as in Figure 8.8, or even a shorter e-mail, should be appropriate.

Some Common Topics for Internal Proposals

An internal proposal can be written about a variety of topics, including the following:

- purchasing new or more advanced equipment to replace obsolete or inefficient computers, appliances, vehicles, and the like, or upgrading equipment technology
- obtaining document security software and offering training sessions to show employees how to use it
- recruiting new employees or retraining current ones on a new technique or process
- eliminating a dangerous condition or reducing an environmental risk to prevent accidents—for employees, customers, or the community at large
- cutting costs—for services, transportation, advertising, etc.
- improving technology/communication within or between departments of a company or agency
- expanding work space or making it greener, more private, ergonomically beneficial to employees, or more inviting to customers

As this bulleted list shows, internal proposals cover almost every activity or policy that can affect the day-to-day operations of a company or agency.

Figure 8.8   An internal unsolicited proposal.

COMMUNITY FEDERAL BANK

http://www.comfedbank.com

EQUAL HOUSING LENDER

POWELL
617-584-5200

MONROE
781-413-6000

LANGSTON
508-796-3009

TO: Michael L. Sappington, Executive Vice President
Dorothy Woo, Langston Regional Manager

FROM: Tina Escobar, Oliver Jabur, ATM Services

DATE: June 10, 2011

RE: A proposal to install an additional ATM at the Mayfield Park branch within the next 60–90 days

Clearly states why proposal is being sent

PURPOSE

We propose a cost-effective solution to what is a growing problem at the Mayfield Park branch in Langston: inefficient servicing of customer needs and rising personnel costs. We recommend that you approve the purchase and installation, within the next two to three months, of another ATM at Mayfield. Such action is consistent with Community's goals of expanding branch banking services and promoting our image as a self-serve yet customer-oriented institution.

Takes into account organization's mission

THE PROBLEM WITH CURRENT SERVICES AT MAYFIELD PARK

Currently, we employ four tellers at Mayfield. But too much is being spent on personnel/salary for routine customer transactions. In fact, as determined by teller activity reports, nearly 25 percent of the four tellers' time each week is devoted to routine activities easily accommodated by the installation of another ATM. Outlined in the table below is a breakdown of teller activity for the month of May:

Identifies problem by giving reader essential background information based on research

Provides easy-to-read table

Teller #	Total Transactions	Routine Transactions
1	6,205	1,551
2	5,989	1,383
3	6,345	1,522
4	6,072	1,518
	24,611	5,974

Page 2

Clearly, we are not taking advantage of our tellers' sales abilities when they are kept busy with routine activities. To compound the problem, we expect business to increase by at least 25 percent at Mayfield in the next few months, as projected by this year's market survey. If we do not install an additional ATM, we will need to hire a fifth teller, at an annual cost of $29,900 ($23,000 base pay plus approximately 30 percent for fringes), for the additional 6,000 transactions we project.

Most important, though, customer needs are not being met efficiently at Mayfield. Recent surveys done for Community Federal by Watson-Perry, Inc. demonstrate that our customers are inconvenienced by not having one more ATM at Mayfield. They are unhappy about long waits both at the ATMs and at the teller windows to do simple banking business, such as deposits, withdrawals, and loan payments. Several discussions we had with manager Rachael Harris-Koyoto at the Mayfield branch confirm customers' complaints.

Ultimately, the lack of another ATM at Mayfield Park hurts Community's image. With plentiful ATMs available to Mayfield residents at local stores and at other banks, our institution risks having customers and potential customers go elsewhere for their banking needs. We not only miss the opportunity of selling them our other services but also risk losing their business entirely.

A SOLUTION TO THE PROBLEM

Purchasing and installing an additional ATM at Mayfield Park will result in significant savings in personnel costs and time. Specifically, we can

- Save money by not having to hire a fifth teller
- Allocate teller duties more efficiently and productively by assisting customers with questions and transactions not handled through an ATM, such as opening a new account; purchasing savings bonds, CDs, and international currency; and providing Internet banking
- Increase time for tellers to cross-sell our services, including our line of nontraditional banking products—annuities, mutual funds, debit cards, and global market accounts

Divides problem into parts— volume, financial, personnel, customer service

Emphasizes the expense if nothing is done

Cites important research

Verifies that problem is widespread

Emphasizes possible future problems

Relates solution to individual parts of the problem

Bulleted list makes recommen- dations easy to follow

Figure 8.8   (Continued)

Page 3

- Service customer retirement options by having tellers track IRAs, 401(k)s, 403(b)s, and Roth IRA's
- Improve customer satisfaction by giving them the option of meeting their banking needs electronically or through a teller
- Ease the stress on tellers at Mayfield Park

Shows problem can be solved and stresses how

It is feasible to install another ATM at Mayfield. This location does not pose the difficulties as at some of our older branches. Mayfield offers ample room to install a drive-up ATM in the stubbed-out fourth drive-up lane. It is away from the heavily congested area in front of the bank, yet it is easily accessible from the main driveway and the side drive facing Commonwealth Avenue, as the photograph below shows.

Photo shows location has room for additional ATM

Photo courtesy Taylor Wilson

Documents that work can be done on time and highlights advantage of doing it now

Judging from our previous experiences, the ATM could be installed and operational within one to two months. Moreover, by authorizing the expenditure at Mayfield within the next month, you will ensure that additional ATM service is available long before the busy Christmas season.

COSTS

The costs of implementing our proposal are as follows:

Itemizes costs

Diebold drive-up ATM	$28,000.00
Installation fee	2,000.00
Maintenance (1 year)	1,500.00
	$31,500.00

Page 4

This $31,500, however, does not truly reflect our annual costs. We would be able to amortize, for tax purposes, the cost of the installation of the ATM over five years. Our annual expenses would, therefore, would actually be

Interprets costs for reader

$$\$30,000 \ (28,000 + 2,000)/5 \text{ years} = \$6,000 \text{ per year}$$
$$+ \$1,500 \text{ (maintenance)}$$
$$\overline{\$7,500 \text{ per year}}$$

Compared with the $29,900 a year the bank would have to expend for a fifth teller position at Mayfield, the annual depreciated cost for the ATM ($7,500) in fact reduces by almost 75 percent ($22,400) the amount of money the bank will have to spend every year for much more efficient customer service.

Proves change is cost-effective, citing specific financial evidence

CONCLUSION

Authorizing another ATM for the Mayfield Park branch is both feasible and cost effective. Endorsement of this proposal will save our bank more than $22,400 in teller services annually, decrease customer complaints, and increase customer satisfaction and approval. We will be happy to discuss this proposal with you anytime at your convenience, and answer any questions you may have.

Ends by stressing benefits for reader and bank as a whole

Thank you for considering our plan.

Ends by thanking reader

Following the Proper Chain of Command

Writing an internal proposal requires you to be sensitive to office politics. It may be wise first to meet with your boss to see if she or he has already identified the problem or has specific suggestions on how to solve it. If given the go-ahead, then you and your team can provide your boss with a draft and ask for revisions or feedback.

But do not assume that your reader(s) will automatically agree that there is a problem or that your plan is the only way to tackle it. To write a successful internal proposal, keep in mind the needs and preferences of your boss and others who may have to sign off on it. Remember that your employer will expect you to be very convincing about both the problem you say exists and the changes you are advocating in the workplace under his or her supervision. Don't rock the corporate boat by going over the boss's head, questioning his or her authority, or suggesting a plan that is too costly; you are not showing respect for your company's mission, chain of command, or its budget.

Ethically Identifying and Resolving Readers' Problems

When you prepare an internal proposal, do not lose sight of the ethical obligations you have and the ways to meet them. Here are three important guidelines:

1. Consider the implications of your plan company-wide. The change you propose (transfers, new schedules or technology, new hires) may have sweeping and potentially disruptive implications in another division of your company.
2. Keep in mind what impact your change may have for co-workers from cultural traditions other than your own. In addition to speaking with your boss, it would be wise, maybe even mandatory, to consult your human resources director.
3. Never submit an internal proposal that offers an idea that you think will work but relies on someone else to supply the specific details on how and when it will work.

Organizing Internal Proposals

A short internal proposal can follow a relatively straightforward plan of organization, from identifying the problem to solving it. Internal proposals usually contain four parts, as shown in Figure 8.8: **purpose, problem, solution**, and **conclusion**. Refer to this figure as you read the following discussion.

The Purpose

Begin your proposal with a brief statement of why you are writing to your supervisor: "I propose that . . ." State right away why you think a specific change is necessary now. Then, succinctly define the problem and emphasize that your plan, if approved by the reader, will solve that problem.

The Problem

Prove that a problem exists. Document its importance for your boss and your company; as a matter of fact, the more you show, with concrete evidence, how the problem affects the boss's work (and area of supervision), the more likely you are to persuade him or her to act. Here are some guidelines for documenting a problem.

- Avoid vague (and unsupported) generalizations such as "We're losing money each day with this procedure (piece of equipment)"; "Costs continue to escalate"; "The trouble occurs frequently in a number of places"; "Numerous complaints have come in"; "If something isn't done soon, more problems will result."
- Provide quantifiable details about the problem, such as the amount of money or time a company is actually losing per day, week, or month. Document the financial trouble so that you can show in the next section how your plan offers an efficient and workable solution. Note how Tina Escobar and Oliver Jabur provide detailed figures about transactions, salary, benefits, and costs of buying and maintaining a new ATM.
- Indicate how many employees (or work hours) are involved or how many customers are inconvenienced or endangered by a procedure or condition. Notice how Escobar and Jabur helpfully include such information in a table in their proposal in Figure 8.8 (page 304).

■ Verify how widespread a problem is or how frequently it occurs by citing specific occasions. Again, see how Escobar and Jabur cite evidence from the Watson-Perry, Inc. survey and from the interviews they conducted with the manager of the Mayfield branch—Rachael Harris-Koyoto

■ Relate the problem to an organization's image, corporate mission, or influence (where appropriate). Pinpoint exactly how and where the problem lessens your company's effectiveness or hurts its standing in the market. Indicate who is affected and how the problem affects your company's business community, as the writers do especially well in the first paragraph of Figure 8.8 (page 304).

The Solution or Plan

In this section describe how you will make the change you propose and want approved, and then demonstrate the likelihood of its succcess. Your reader will again expect to find factual evidence. Be specific. Supply details that answer the following questions: (1) Is the plan workable, is it feasible—can it be accomplished here in our office, plant, or region? (2) Is it cost-effective—will it really save us money in the long run and not lead to even greater expenses?

To get the reader to say "Yes" to these questions, provide the factual evidence you have gathered as a result of your research. For example, if you propose that your firm buy a new piece of equipment, do the necessary homework to locate the most efficient and cost-effective model available, as Tina Escobar and Oliver Jabur did for their proposal in Figure 8.8.

■ Supply the dealer's name, the costs, installation time, types of necessary service and training contracts, and warranties.

■ Describe how your firm could use the equipment to obtain better or quicker results in the future.

■ Document specific tasks the new equipment can perform more efficiently and economically than the models or types you use now.

A **proposal to change a procedure** must address the following questions.

■ How does the new (or revised) procedure work?

■ How many employees or customers will be affected by it?

■ When will it go into operation?

■ How much will it cost the employer to change procedures or to add new equipment, services, or staff.

■ What delays or losses in business might be expected while the company switches from one procedure to another?

■ What employees, equipment, or locations are already available to accomplish the change?

Beyond a doubt, costs will be of utmost importance to your decision-maker reader. Make sure you supply a careful and accurate budget. Moreover, make the costs attractive by emphasizing how inexpensive they are compared with the cost of *not* making the change, as Escobar and Jabur do persuasively in the section labeled "Costs." Link costs to savings and other benefits. And always be sure to double-check your math.

It is also wise to raise alternative solutions, before the reader does, and to discuss their disadvantages. Notice again how Tina Escobar and Oliver Jabur do that by showing why installing an ATM is more feasible than hiring a fifth teller.

The Conclusion

Your conclusion should be short—a paragraph or two at the most. Remind readers that (a) the problem is ongoing and serious, (b) the reason for change is justified and will benefit your organization, and (c) action needs to be taken and by a specified time. Re-emphasize the most important advantages. In Figure 8.8, Escobar and Jabur stress the savings that the bank will see by following their plan as well as the increase in customer satisfaction. Also indicate that you are willing to discuss your plan with the reader and seek his or her input, a necessity in arguing for a corporate change at any level. And always thank your reader(s) for considering your proposal.

Sales Proposals

A sales proposal is the most common type of proposal. Its purpose is to sell your company's products or services for a set fee. The firm or organization you want to sell your product or services to will issue a **request for proposals** or **bids** specifying exactly what it wants done and when, and it will expect you to stay within a clearly defined budget. A short sales proposal is a marketing tool that includes a sales pitch as well as a detailed description of the work you propose to do. Figure 8.9 contains a sales proposal in response to a company's request.

Knowing Your Audience and Meeting Its Needs

Your audience will usually be one or more executives who have the power to approve or reject a proposal. Unlike readers of an internal proposal, your audience for a sales proposal may be even more skeptical since they may not know you or your work. You can increase your chances of success by trying to anticipate their questions such as:

- Does the writer's firm understand our problem?
- Can the writer's firm deliver the services it promises?
- Can the job be completed on time?
- Is the budget reasonable and realistic (neither inflated nor too low)?
- What assurances does the writer offer that the job will be done exactly as proposed?
- What are the qualifications the writer's firm has to get the job done?
- How has the writer demonstrated his or her trustworthiness?

Answer each of these questions by demonstrating how your product or service is tailored to the customer's needs.

Make sure, too, that your proposal has a competitive edge. Readers will compare your plan with those they receive from other proposal writers. Your proposal has to convince readers that the products and the services your company offers are more reliable, economical, efficient, and timely than those of another

A proposal in response to a request from a company.   Figure 8.9

Reynolds Interiors • 250 Commerce Avenue S.W. • Portland, OR 97204-2129

January 21, 2011

Mr. Floyd Tompkins, Manager
General Appliances
Highway 11 South
Portland, OR 97222-1300

Dear Mr. Tompkins:

In response to your request listed on your website for bids for an appropriate floor covering at your new showroom, Reynolds Interiors is pleased to submit the following proposal to meet your specific needs. We appreciated the opportunity to visit your showroom on January 14 in order to submit this proposal.

After carefully reviewing your specifications for a floor covering and inspecting your new facility, we believe that **Armstrong Classic Corlon 900** is the most suitable choice. We are enclosing a few samples of the **Corlon 900** so you can see how carefully they are constructed.

Corlon's Advantages

Guaranteed against defects for a full three years, **Corlon** is one of the finest and most durable floor coverings manufactured by Armstrong. It is a heavy-duty commercial floor 0.085-inch thick for protection. Twenty-five percent of the material consists of interface backing; the other 75 percent is an inlaid wear layer that offers exceptionally high resistance to the heavy, everyday traffic your showroom will see. Traffic tests conducted by the independent Floor Covering Institute have repeatedly proved the superiority of **Corlon's** construction and resistance. You might want to visit the Institute's website (www.fci.org) for a demonstration of how durable and versatile Classic **Corlon** flooring is.

http://www.reynolds.com • 503-555-8733 • Fax: 503-555-1629

Begins with a reference to company's request for bids

Identifies best solution

Describes product's features that will benefit reader

Helpfully refers to an independent source to corroborate the benefits of the product

Distinguishes seller's product from competitors'

(Continued)

Figure 8.9   **(Continued)**

2

Another important feature of **Corlon** is the size of its rolls. Unlike other leading brands of similar commercial flooring—Remington or Treadmaster—**Corlon** comes in 12-foot-wide rather than 6-foot-wide rolls. This extra width will significantly reduce the number of seams on your floor, thus increasing its attractiveness and eliminating the dangers of splitting.

Cites another advantage over competitors' models

Installation Procedures

Explains how job is done professionally

The **Classic Corlon 900** requires that we use the inlaid seaming process, a technical procedure requiring the services of a trained floor mechanic. Herman Goshen, our chief floor mechanic, has more than eighteen years of experience working with the inlaid seam process. His professional work and keen sense of layout and design have been consistently praised by our customers.

Installation Schedule

Gives realistic timetable

We can install the **Classic Corlon 900** on your showroom floor during the first week of March, which fits the timetable specified in your request. The material will take three and one-half days to install but will be ready to walk on immediately. Be assured that your floor will be installed no later than March 7th. We recommend, though, that you not move heavy equipment onto the floor for 24 hours after installation.

Costs

The following costs include the **Classic Corlon** tile, labor, and tax:

Itemizes all costs based on research of market conditions and reader's bid

750 sq. yards of **Classic Corlon** at $23.50/sq. yd.	$ 17,625.00
Labor (28 hrs @ $18.00/hr.)	$ 504.00
Sealing fluid (10 gals. @ $15.00/gal.)	$ 150.00
Total	$ 18,279.00
Sales tax (5 percent)	$ 913.95
GRAND TOTAL	**$ 19,192.95**

(Continued)   Figure 8.9

3

Our costs are at least $300.00 under those you specified in your bid.

Reynolds's Qualifications

Reynolds Interiors has been in business for more than 28 years. In that time, we have installed more than 2,500 commercial floors in Portland and its suburbs. In the last year, we have served more than 60 customers, including the new multipurpose Tech Mart facility in downtown Portland. Our floor designs have also been included in several commercial properties that have won awards from the Portland Architectural Review Board. Reynolds has also consistently received high commendations from our many satisfied customers, and we would be happy to furnish you with a list of our references.

Conclusion

Thank you for the opportunity to submit this proposal to General Appliances. We are confident that you will be pleased with the appearance and durability of an Armstrong Classic **Corlon 900** floor and our installation of it. If we can provide you with any further information about our service or Corlon flooring, please call us at 503-555-8733 or visit us at our website.

Sincerely yours,

Neelow Singh

Neelow Singh
Sales Manager

Jack Rosen

Jack Rosen
Installation Supervisor

Points out proposal comes in under budget—always a major consideration for buyer

Establishes history of service and provides documented evidence of quality work

Encourages reader to accept

company. Wherever relevant, stress that your company offers state-of-the-art technology, exemplary service, and after the sale assistance and warranty. Here is where your homework pays off.

Organizing Sales Proposals

Most sales proposals include the following elements: introduction, description of the proposed product or service, timetable, costs, qualifications of your company, and conclusion.

Introduction

The introduction to your sales proposal can be a single paragraph in a short sales proposal or several pages in a more complex one. Basically, your introduction should persuasively prepare readers for everything that follows in your proposal. The introduction itself may contain the following sections, which sometimes may be combined.

1. **Statement of purpose and subject of proposal.** Tell readers why you are writing, and identify the specific subject of your work. Refer to the request for proposals or bids the reader has issued, as the writers in Figure 8.9 do. Briefly define the solution you propose. Tell readers exactly what you propose to do for them. Be clear about what your plan covers and, if there could be any doubt, what it does not.

2. **Background of the problem you propose to solve.** Show readers that you are familiar with their problem and why it is important. In a solicited proposal like the one in Figure 8.9, this section is usually unnecessary because the potential client has already identified the problem and wants to know how you would address it. In that case, just point out how your company would solve the problem, mentioning your superiority over your competitors (see the fourth paragraph of Figure 8.9 on page 311).

In an unsolicited proposal, you need to describe the problem in convincing detail, identifying the specific trouble areas. Depending on the type of proposal you submit, you may want to focus briefly on the dimensions of the problem—when it was first observed, who/what it most acutely affects, and the specific organizational/community/environmental context in which the problem is most troubling.

Description of the Proposed Product or Service

This section is the heart of your proposal. Before spending their money, customers will demand hard, factual evidence of what you claim can and should be done. Here are some points that your proposal should cover.

1. **Carefully show potential customers that your product or service is right for them.** Stress particular benefits of your product or service most relevant to your reader. Blend sales talk with descriptions of hardware. Note how the proposal in Figure 8.9 references the results of an independent testing agency—the Floor Cover Institute—to stress the benefits of the product it sells.

2. **Describe your work in suitable detail.** Specify what the product looks like; what it does; and how consistently and well it will perform in the readers' office, plant, hospital, or agency. You might include a brochure; picture; diagram; or, as the writers of the proposal in Figure 8.9 do, a few samples of your product for customers to study.

3. **Stress any special features, maintenance advantages, warranties, or service benefits.** Convince readers that your product is the most up-to-date and efficient one they could select. Highlight features that show the quality, consistency, or security of your work. See how Neelow Singh and Jack Rosen in Figure 8.9 show why and how Corlon is the best choice for the heavy traffic of the General Appliances showroom. For a service, emphasize the procedures you use, the terms of the service, and even the kinds of tools you use, especially any state-of-the-art equipment.

Timetable

A carefully planned timetable assures readers that you know your job and that you can accomplish it in the deadline set forth in the call for proposals or bids. Your dates should match any listed in a company's proposal request. Provide specific dates to indicate

- when the work will begin
- how the work will be divided into phases or stages
- when you will be finished
- whether any follow-up visits or services are involved

For proposals offering a service, specify how many times—an hour, a week, a month—customers can expect to receive your help; for example, spraying three times a month if your company offers exterminating services. The proposal writers in Figure 8.9 assure their reader that installation will be done by a specified date.

Costs

Make your budget accurate, complete, and convincing. But give customers more than merely the bottom-line cost. Show exactly what readers are getting for their money so that they can determine if everything they need is included. Itemize costs for

- specific services
- equipment and materials
- labor (by the hour or by the job)
- transportation
- travel
- training

To further persuade readers to accept their proposal, the writers in Figure 8.9 point how their work comes in under the specified budget.

If something is not included or is considered optional, say so—additional hours of training, replacement of parts, upgrades, and the like. If you anticipate a price increase, let the customer know how long current prices will stay in effect. That information may spur them to act favorably now.

Qualifications of Your Company

Emphasize your company's accomplishments and expertise in providing similar services and/or equipment. Mention the names of a few local firms for whom you have worked that would be able to recommend you and cite any awards or commendations, as the writers do in Figure 8.9.

Conclusion of Proposal

This is the "call to action" section of your sales proposal. Encourage your reader to approve your plan by stressing major benefits of your plan. Offer to answer any questions the reader may have. Some proposals end by asking readers to sign and return a copy of the proposal indicating their acceptance, as the proposal in Figure 8.9 does.

Short Reports and Proposals: Some Final Words

To prepare successful short reports and winning proposals, always remember your reader's needs, document your information, and write objectively, persuasively, and ethically.

Revision Checklist

Short Reports

☐ Had a clear sense of how my readers will use my short report.

☐ Provided significant, relevant information about costs, materials, personnel, locations, environmental conditions, and times so readers will know that my work consists of facts, not impressions.

☐ Double-checked all data—names, costs, figures, dates, places, equipment numbers, and so forth.

☐ Kept report concise, to the point, and readable.

☐ Used headings wherever feasible to organize and categorize information.

☐ Began report with statement of purpose that clearly described the scope and significance of my work.

☐ Incorporated tables, maps, graphs, and other pertinent visuals to display data whenever appropriate.

☐ Explained clearly what the data means in a conclusion section.

☐ Determined that recommendations logically follow from the data and are realistic.

☐ Adhered to all ethical and legal requirements in writing an incident report.

Proposals

☐ Identified a realistic problem in my proposal—one that is restricted and relevant to my audience's needs.

☐ Incorporated the scope and importance of the problem.

☐ Effectively convinced audience that the problem exists and that it needs to be solved; documenting about the "you attitude" throughout.

☐ Persuasively emphasized benefits of solving the problem according to the proposal.

☐ Offered a solution that can be realistically implemented—that is, it is both appropriate and feasible, economically and strategically, for audience.

☐ Used specific figures about costs, personnel, technology, and concrete details to show how proposal will save time and money.

☐ *For internal proposals*: Demonstrated how proposal benefits my company and my supervisor; followed the chain of command by discussing proposal with co-workers and/or supervisors who may be affected.

☐ *For sales proposals*: Related my product or service to prospective customer's needs; showed a clear understanding of those needs.

☐ Prepared a comprehensive, realistic, and ethical budget; accounted for all expenses; itemized costs of products and services.

☐ Linked costs to benefits.

☐ Provided a timetable with precise dates for implementing proposal.

Exercises

1. Assume that you are a manager of a large apartment complex (300 units). Write a periodic report based on the following information—26 units are vacant, 38 soon will be vacant, and 27 soon will be leased (by June 1). Also add a section of recommendations to your supervisor (the head of the management company for which you work) on how vacant apartments might be leased more quickly and perhaps at increased rents. Consider important information such as decorating, advertising, installing a new security system, providing Internet access, amenities such as pool, fitness center, clubhouse, and first month discounts.

Additional Activities related to writing short reports and proposals are located at www. cengage.com/ english/kolin/ writingatwork concise3e.

2. Assume you work for a household appliance store. Prepare a sales report based on the information contained in the following table. Include a section on recommendations for your manager.

	Number Sold	
Product	October	November
Kitchen Appliances		
Refrigerators	72	103
Dishwashers	27	14
Freezers	10	36
Electric Ranges	26	26
Gas Ranges	10	3
Microwave Ovens	31	46
Laundry Appliances		
Washers	50	75
Dryers	24	36
Air Treatment		
Room Air Conditioners	41	69
Dehumidifiers	7	2

3. Submit a progress report to your writing instructor on what you have learned in his or her course so far this term, specifying which formatting and writing skills you want to develop in greater detail, and how you propose doing so. Mention specific memos, e-mails, letters, instructions, reports, websites and blogs, or proposals you have written or will soon write.

4. Prepare a site inspection report on any part of the college campus or plant, hospital, office, store, or other facility in which you work that might need remodeling, expansion, rewiring, or new or additional air-conditioning, plumbing, and/or heating work.

5. Write a report to an instructor in your major or to an employer about a trip you have taken recently—to a museum, laboratory, health care agency, correctional

facility, radio or television station, plant or factory, or office. Indicate why you took the trip, name the individuals (with their job titles) you met on the trip, and stress what you learned and how that information will help you in course work or on your job. Include a relevant visual as part of your report.

6. Write an incident report about one of the following problems. Assume that it has happened to you. Supply relevant details and visuals in your report. Identify the audience for whom you are writing and the agency you are representing or trying to reach.
   a. After hydroplaning, your company car hits a tree and has a damaged front fender.
   b. You have been the victim of an electrical shock because an electrical tool was not grounded.
   c. You twist your back lifting a bulky package in the office or plant.
   d. A virus has infected your company's intranet, and it will have to be shut down for 12 hours to debug it.
   e. The crane (or other piece of equipment) you are operating breaks down, and you lose a half-day's work.
   f. The vendor shipped the wrong replacement part for your computer, and you cannot complete a job without buying a more expensive software package.
   g. An electrical storm knocked out your computer; you lost 1,000 mailing label addresses and will have to hire additional help to complete a mandatory mailing by the end of the week.
   h. A scammer has stolen some sensitive files (documents) about a new product your company hoped to launch next month.
   i. An irate customer threatens one of your sales staff, but there was no physical violence though the daily routine of your business was disrupted and several other customers walked out.

7. As a collaborative writing project, prepare a short internal proposal, similar to that in Figure 8.8 (pages 304–307), recommending to a company or a college a specific change in procedure, technology, training, safety, personnel, or policy. Make sure your team provides an appropriate audience (college administrator, department manager, or section chief) with specific evidence about the existence of the problem and your solution to it. Possible topics include:
   a. providing more and safer parking/lighting
   b. offering a job sharing option for mothers
   c. purchasing new office or laboratory equipment or software
   d. hiring more faculty, student workers, or office help
   e. allowing employees to telecommute
   f. changing the lighting/furniture in a student or company lounge to make it more eco-friendly
   g. increasing the number of weekend, night, or online classes in your major
   h. adding more health-conscious offerings to the school or company cafeteria menu
   i. altering the programming on a campus or local commercial radio station
   j. installing a new LCD overhead

8. Write a sales proposal as a collaborative group, similar to the one in Figure 8.9 (pages 311–313), on one of the following services or products you intend to sell or on a topic your instructor approves:
   a. providing exterminating or trash removal service to a store or restaurant
   b. supplying a hospital with rental laptops for patients' rooms
   c. designing websites
   d. obtaining temporary office help or nursing care
   e. supplying landscaping and lawn care work
   f. testing for noise, air, or water pollution in your community or neighborhood
   g. furnishing transportation for students, employees, or members of a community group
   h. offering technical consulting service to save a company money
   i. digging a septic well for a small apartment complex
   j. supplying insurance coverage to a small firm (5 to 10 employees)
   k. cleaning the parking lot and outside walkways at a shopping center
   l. selling a piece of equipment to a business
   m. making a work area safer or greener
   n. preparing a technology seminar or training program for employees
   o. increasing donations to a community or charitable fund
   p. offering discounted or sponsored memberships at a fitness center

9. Write a proposal in letter format (similar to Figure 4.21) for a business you manage to attract new international clients. As part of your letter, stress any new equipment or services you offer and provide as well as any background/history about your business that might appeal to a particular groups' international customers.

9 Writing Careful Long Reports

To expand your understanding of writing long reports, take advantage of the Web Links, Additional Activities, and ACE Self-Tests at www. cengage.com/ english/kolin/ writingatwork concise3e

This chapter introduces you to long reports—and how and why they are written, organized, and documented. It is appropriate to discuss long reports in one of the last chapters of this book because they require you to use and combine many of the writing skills and research strategies you have already learned. In the world of global business, a long report can be the culmination of many weeks or months of hard work on an important company project.

Characteristics of a Long Report

The following sections explain some of the key elements in a long report. You will find a model long report (Figure 9.2) at the end of the chapter (pages 338-354).

Scope

A long report is a major study that provides an in-depth view of a key problem or idea. For example, a long report written for a course assignment may be 8 to 20 pages long; a report for a business or industry may be that long or, more likely, much longer, depending on the scope of the subject. The implications of a long report are wide-ranging for a business—relocating a facility, retraining employees, adding a new line of products or services, or upgrading its technology.

The long report examines a problem in detail, while the short report covers just one part of the problem. Unlike a short report, a long report may discuss not just one or two current and routine events, but rather a continuing history of a problem or idea (and the background information necessary to understand it in perspective).

The titles of some typical long reports further suggest their extensive (and in some cases exhaustive) coverage:

- A Master Plan for the Recreation Needs of Dover Plains, New York
- The Transportation Problems in Kingford, Oregon, and the Use of Monorails
- Promoting More Effective E-Commerce and E-Tailing at TechWorld
- The Use of Virtual Reality Attractions in Theme Parks in Jersey City, New Jersey

- Public Policy Implications of Expanding Health Care Delivery Systems in Tate County
- Internet Medicine in Providing Health Care in Rural Areas: Ways to Serve Southern Montana

Research

A long, comprehensive report requires much more extensive research than a short report does. Information can be gathered over time from primary and secondary research—Internet searches, listservs, books, articles, government documents, laboratory experiments, on-site visits and tests, conferences with your boss and co-workers, interviews, and the writer's own observations. For a course report, you will have to do a great deal of research and possibly interviewing to track down relevant background information and to discover what experts have said about the subject and what they propose should be suggested or have even done.

For a report for class, you will be asked to identify a major problem or topic, while in the business world the topic and even your approach to it will more than likely be dictated to you by your boss and company policy, as the cover letter (see Figure 9.1, page 337) to the long report in Figure 9.2 indicates.

Format

A long report is too detailed and complex to be adequately organized in a memo or letter format. The product of thorough research and analysis, the long report gives readers detailed discussions and interpretations of large quantities of data. To present the information in a logical and orderly fashion, the long report contains various parts, sections, headings, subheadings, documentation, and supplements (appendices) that would never be included in a short report. The long report, as in Figure 9.2, also gives readers a variety of visuals, including charts, a graph, and even a multicultural calendar.

Timetable

A long report is generally commissioned by a company or an agency to explore with extensive documentation a subject involving personnel, locations, technology, costs, safety, or the environment. Many times a long report is required by law—for example, investigating the feasibility of a project that will affect the ecosystem. When you prepare a long report for a class project, select a topic that really interests you and/or the collaborative group you are a part of, because you will spend a good portion of the term working on it. Here is a possible timeline for a long report.

Do research and conduct interviews	Outline or draft	Conferences for revisions and further research	Revise and prepare figures	Proof-read and polish	Submit long report
4 weeks	3 weeks	1 week	3 weeks	1 week	Due Date

Audience

The audience for a long report usually consists of individuals in the top levels of management—presidents, vice presidents, superintendents, directors, heads of departments—who make executive, financial, and organizational decisions. These individuals are responsible for long-range planning, or seeing the big picture, so to speak. A long report written about a campus issue or problem may at first be read by your instructor and then sent to an appropriate decision maker, such as a dean of students, a business manager, a director of athletics, or the head of campus security.

Collaborative Effort

The long report in the world of business may not be the work of one employee. Rather, it may be a collaborative effort—the product of a committee or a group. Your instructor may ask you to work in a group (or alone) in preparing your long report.

The group should estimate a realistic time necessary to complete the various stages of the work—when drafts are due or when editing must be concluded, for example. A project schedule based on that estimate should then be prepared. **But remember: Projects always take longer than initially planned**. Prepare for a possible delay at any one stage. The group may have to submit progress reports (see pages 288–292) to its members as well as to management.

To be successful, a collaborative writing team should also observe the guidelines as well as the procedures for collaboration in the writing process discussed on pages 54–55.

The Process of Writing a Long Report

Because your work will be spread over many weeks, you need to see your report not as a series of static or isolated tasks but as an evolving project. Before you embark on that project, review the information on the writing process in Chapter 2. You may also want to review the flow chart in Figure 6.11 (page 223), which illustrates the different stages in writing a report. The following guidelines will also help you plan and write a long report.

1. **Identify a significant topic.** While in a business setting you won't usually get to choose the topic of a long report, when selecting a topic for a long report for a class, make sure you select one that is important and worth exploring in detail. After all, you can expect to spend a lot of time researching and writing about it. Choose a topic/problem that is relevant for your audience, whether it is a group of college administrators or community leaders, and something that will help them better understand or even solve a problem.

2. **Conduct research.** You'll have to do some preliminary research—widespread reading, online searching, conferring with and interviewing experts—to get an overview of key ideas and individuals involved, and the implications for your company and/or community. Note the kinds of research Terri Smith Ruckel and her

collaborative team did for their long report (see Figure 9.2, pages 338–354). As they did, expect to search a variety of print and Internet sources, to read and evaluate them, to interview experts in the field, and to incorporate this research into your work.

3. Expect to confer regularly with your supervisor(s) and/or team members. In these meetings, be prepared to ask relevant questions based on your research to pin down exactly what your boss wants and how you and your writing team can accomplish this goal.

4. Revise your work often. Be prepared to work on several outlines and drafts. Your revisions may sometimes be extensive, depending on what your boss, instructor, or collaborative team recommends. As you work on the reports you may have to consult new sources and arrive at a new interpretation of those sources.

5. Keep the order flexible at first. Even as you work on your drafts and revisions, remember that a long report is not written in the order in which the parts will finally be assembled. You cannot write in "final" order—abstract to recommendations. Instead, expect to write in "loose" order to reflect the process in which you gathered information and organized it for the final copy of the report. Usually, the body of the report is written first and the introduction later so that the authors can make sure they have not left anything out. The abstract, which appears very early in the report, is always written after all the facts have been recorded and recommendations are made or conclusions drawn.

6. Prepare both a day-to-day calendar and a checklist. Keep both posted where you do your work—above your desk or computer, or use your computer's built-in calendar program, if available—so that you can track your progress. The calendar should mark **milestones**—that is, dates by which each stage of your work must be completed. Match the dates on your calendar with the dates your instructor or employer may have given you to submit an outline, progress report(s), drafts, and then the final copy. Your checklist should list the major parts of the long report. As you complete each section, check it off. Before assembling the final copy of your report, use the checklist to make sure you have not omitted something.

Parts of a Long Report

A long report may include some or all of the following 12 parts, which form three categories: *front matter* (letter of transmittal, title page, abstract, table of contents, list of illustrations), *report text* (introduction, body, conclusion, recommendations), and *back matter* (glossary, references cited, any appendices).

Front Matter

As the name implies, the front matter of a long report consists of everything that precedes the actual text of the report. Such elements introduce, explain, and summarize to help the reader locate various parts of the report. Use lowercase Roman numerals for front matter page numbers, not Arabic numbers.

Letter of Transmittal

This three- or four-paragraph (usually only one-page) letter states the purpose, scope, and major recommendation of the report. It highlights the main points of the report that will most interest your readers. If written to an instructor, the letter should additionally note that the report was done as a course assignment. (See Terri Smith Ruckel's letter on page 337 for a sample letter of transmittal for a business report.)

Title Page

Find out what your boss or instructor prefers. Basically, your title page should contain the full title of your report and how you have restricted it in time, space, or method. Avoid titles that are vague, too short, or too long.

Vague Title:	A Report on the Internet: Some Findings
Too Short:	The Internet
Too Long:	A Report on the Internet: A Study of Dot-com Companies, Their History, Appeal, Scope, Liabilities, and Their Relationship to Ongoing Work Dealing with Consumer Preferences and Identity Protection Within the Past Five Years

Also include the date you submitted the report, and the person(s) for whom you prepared the report. For a report for a class assignment, give your instructor's name and the specific course for which you prepared the report. For most other long reports, the title page needs

- the name(s) of the report writer(s)
- the date of the report
- the name of the firm or individual for whom the report was prepared

Abstract

An abstract summarizes your report. It is usually only one paragraph long and can be included in different parts of your report—on the title page, on a separate page (as in Figure 9.2, page 339), or even on the first page of the report. But the abstract should always be in the front of the report so readers can find it easily. The abstract may be the most important part of your report. Not every member of your audience will read your entire report, but almost every one of them will study the abstract.

There are two types of abstracts—an *informative abstract* and a *descriptive abstract*. An **informative abstract** is more helpful to your readers because it includes the main problems you investigated, the conclusions you reached, and any recommendations you made. See how the informative abstract accompanying Terri Smith Ruckel's report (see page 339) provides all this information for her readers.

To prepare an informative abstract, think of your table of contents as a concise and accurate outline of your report. Guided by your table of contents, you can identify the purpose, scope, and findings of your report—your conclusions and recommendations—that you need to include in your abstract. But keep in mind that your abstract should be neither skimpy nor excessively long. Think of it as a mini version of your report, but even so, avoid including too much detail. Readers want only the most important information in an easy-to-read paragraph that contains connective words and phrases (see page 378). The tone of your abstract should be objective and professional; never use the personal pronoun "I" or "we."

Unlike an informative abstract, a **descriptive abstract** does not give information about results, conclusions, or recommendations. It simply tells the reader what the report is about—what it covers, not how it covers the topic or subject. Busy readers rely on descriptive abstracts to decide whether they need to consult the report itself. Here, for instance, is a descriptive abstract about a report on the use of virtual reality techniques and situations in law enforcement training.

> Virtual reality can be used to teach law enforcement officers firearms training, SWAT team assaults, incident re-creation, and crime scene processing. These training techniques are of interest to law enforcement administrators.

Descriptive abstracts are also useful when you have to summarize the minutes of a conference or give a progress report.

Table of Contents
The table of contents lists the major headings and subheadings of your report and tells readers on which pages they can be found. Essentially, a table of contents shows how you organized your report and emerges from many outlines and drafts. The items on those outlines frequently expand, shrink, combine, and can be moved around until you decide on the formal divisions and subdivisions of your report.

List of Illustrations
This list contains the titles for all of the visuals and indicates where they can be found in your long report.

Text of the Report

The text of a long report is made up of an introduction, the body, conclusions, and sometimes recommendations.

Introduction
The introduction may constitute as much as 10 or 15 percent of your report, but it should not be any longer. The introduction is essential because it tells readers why your report was written and thus helps them understand and interpret everything that follows. Avoid putting your findings, conclusions, or recommendations in your introduction.

Do not regard the introduction as one undivided block of information. It includes related parts, which should be labeled with subheadings. Here are the types of information to include in your introduction.

1. **Background**. To understand why your topic is significant and hence worthy of study, readers need to know about its history. This history may include information on such topics as who was originally involved, when, and where; how someone was affected by the issue; what opinions have been expressed on the issue; and what

the implications of your study are. Note how the long report in Figure 9.2 provides useful background information on when, where, how, and why multinational employees entered the U.S. work force.

2. Problem. Identify the problem or issue that led you to write the report. Because the problem you investigate will determine everything you write about in the report, you need to state it clearly and concisely. Here is a problem statement from a report on how construction designs have not taken into account the requirements of a growing number of seniors and disabled Americans.

> The construction industry in Springfield has not satisfactorily met the needs for accessible workplaces and homes for all age and physical ability groups. The industry has relied on expensive and specialized plans to modify existing structures rather than creating universally designed spaces that are accessible to everyone.

3. Purpose statement. The purpose statement tells readers why you wrote the report and what you hope to accomplish or prove. It expresses the goal of all your research. Like the problem statement, the purpose statement does not have to be long or complex. A sentence or two will suffice. You might begin simply by saying, "The purpose of this report is ..."

4. Scope. This section informs readers about the specific limits—number and type of issues, time, money, locations, personnel, and so forth—you have placed on your investigation. You inform readers about what they will find in your report, or what they won't, through your statement about the scope of your work. The long report in Figure 9.2 concentrates on adapting the U.S. workplace to meet the communication and cultural needs of a work force of multinational employees, not trends in creating international markets—two completely different topics.

Body

This section, also called the *discussion*, is the longest, possibly making up as much as 70 percent of your report. The body contains statistical information, details about the environment, and physical descriptions, as well as the various interpretations and comments of authorities whose work you consulted as part of your research.

The body of your report

- be carefully organized to reveal a coherent and well-defined plan
- separate material into meaningful parts to identify the major issues
- clearly relate the parts to each other
- use headings to help your reader(s) identify major sections more quickly

Your organization should reflect the different headings (and even subheadings) included in your report; use them throughout to make it easy to follow. The headings, of course, will be included in the table of contents and help you prepare your abstract. (Note how the discussion section in Figure 9.2 is carefully organized into three sections—Providing Equal Workplace Opportunities for Multinational Employees; Promoting and Incorporating Cultural Awareness within the company, Making Business Communication More Understandable.)

In addition to including headings at the beginning of each major section of the body, tell readers what they will find in that section and why. The report in Figure 9.2 effectively provides such internal summaries.

Conclusion(s)

The conclusion should tie everything together for readers by presenting the findings of your report. For a research report based on a study of sources located through various reference searches, the conclusion should summarize the main viewpoints of the authorities whose works you have cited. Perhaps your instructor will ask you to assess in your conclusion which resource materials were most thorough and helpful and why. For a marketing report done for a business or organization, you must spell out the implications of your research for readers in terms of costs, personnel, products, location, and so forth.

Regardless of the type of research you do, your conclusions should do the following:

- be based on the information and documentation in the body of the report
- corroborate the evidence/information you gave in the body of your report
- grow out of the work you describe in the body of the report
- keep to the areas that your report covers, and not stay into areas it did not

Recommendations

The recommendation(s) section tells readers what should be done about the findings recorded in the conclusion. Your recommendation tells readers how you want them to solve the problem your report has focused on. Readers will expect you to advise them on a specific course of action—what new technology to purchase, when and where to expand a market, how to improve and safeguard a web presence, or who to recruit, hire, train, and retain for your company, as in Figure 9.2.

Back Matter

Included in the back matter of the report are all of the supporting data that, if included in the text of the report, would bog the reader down in details and cloud the main points the report makes.

Glossary

The glossary is an alphabetical list of the specialized vocabulary used in a long report and the definitions. A glossary might be unnecessary if your report does not use a highly technical vocabulary, as in Figure 9.2, or if *all* members of your audience are familiar with the specialized terms you do use.

References Cited

Any sources cited in your report—websites, books, articles, television programs, interviews, reviews, blogs, audiovisuals—are usually listed in this section. Always ask your instructor or employer how he or she wants information to be documented, that is, what method of documentation to follow (see page 329).

Appendix

An appendix contains supporting materials for the report—tables and charts too long to include in the discussion, sample questionnaires, budgets and cost estimates, correspondence about the preparation of the report, case histories, transcripts of telephone conversations, copies of relevant letters and documents upon which the report is based, etc.

Documenting Sources

Documentation is at the heart of all the research you will do on the job. To document means to furnish readers with information about the print and electronic sources, as well as interviews, you have used for the factual support of your statements, including books, journals, newspapers, manuals, government and corporate reports, websites, and other resources such as blogs, site visits, or surveys.

Documentation is an important component of any long report for at least three reasons.

1. It demonstrates that you have done your homework by consulting experts on the subject and are relying on the most current and authoritative sources to build your case persuasively. Accordingly, do not rely on just one or two sources, and do not repeatedly cite the same source.
2. It gives proper credit to those sources. Citing works by name is not a simple act of courtesy; it is an ethical requirement and, because so much of the material is protected by copyright, a point of law.
3. It informs readers about specific books, articles, or websites you used so they can find your source and verify your facts or quotations.

The Ethics of Documentation: What Must Be Cited

To ensure that your report avoids any type of plagiarism (see page 24) and maintains high ethical standards, follow these guidelines.

- If you use a source and take something from it, document it. Document any direct quotation(s), even a single phrase or keyword.
- Stay away from **patchworking**—using bits and pieces of information and passing them off as your own—which is also an act of plagiarism. Always put quotations marks around anything you take verbatim and document it.
- If any opinions, interpretations, and conclusions expressed verbally or in writing are not your own (e.g., you could not have reached them without the help of another source), you must document them.
- Even if you do not use an author's exact words but still get an idea, concept, or point of view from a source, still document that work in your report.
- Never alter any original material to have it suit your argument. Changing any information—names, dates, times, test results—is a serious offense.
- If you use statistical data you have not compiled yourself, document them.
- Always document any visuals—photographs, graphs, tables, charts, images downloaded from the Internet (and if you construct a visual based on someone else's data, you must acknowledge that source, too).

- Never submit the same research paper for one course that you wrote for another course without first obtaining permission from the second instructor.
- Do not delete an author's name when you are citing or forwarding an Internet document. You are obligated to give the Internet author full credit.

What Does Not Need to Be Cited

Be careful not to distract readers with unnecessary citations. That only demonstrates your lack of understanding of the documentation process and can undercut the professionalism of your report. There is no need to cite the following:

- Common-knowledge scientific facts and formulas, such as "The normal human body temperature is 98.6 degrees Fahrenheit" or "H_2O is the chemical formula of water."
- Readily available geographical data, such as elevation of mountains; depths of lakes, rivers, etc.; population; mileage between two places; and so on.
- Well-known dates, such as the date of the first moon landing in 1969.
- Factual historical information, such as "George W. Bush was the 43rd president of the United States."
- Proverbs from folklore, such as "The hand is quicker than the eye."
- Well-known quotations, such as "We hold these truths to be self-evident …," although it may be helpful to the reader if you mention the name of the person being quoted.
- The Bible, Koran, or other religious texts, but provide a parenthetical reference to the text and to the portion of the text quoted (for instance, *New Jerusalem Bible*, Exod. 2.3).
- Classic literary works, but again provide a parenthetical reference to the original author and the name of the work—for instance, Twain, *The Adventures of Huckleberry Finn*, Chapter 4, or Shakespeare, *The Merchant of Venice*, 5.3.15. Indicate, though, from which edition you took the quotation.

Parenthetical Documentation

Two frequently used systems of parenthetical documentation are found in the *MLA Handbook for Writers of Research Papers* and the *Publication Manual of the American Psychological Association*. The Modern Language Association (MLA) system is used primarily in the humanities and other related disciplines; the APA system is used in psychology, nursing and allied health disciplines, the social sciences, and in some business and technological fields. Both MLA and APA use parenthetical, or in-text, documentation. That is, the writer tells readers directly in the text of the paper, at the moment the acknowledgment is necessary, what reference is being cited.

MLA "Creating an effective website was among the top three priorities businesses have had over the last two years" (Morgan 203).

APA "Creating an effective website was among the top three priorities businesses have had over the last two years" (Morgan, 2011, p. 203).

The MLA citation "(Morgan 203)" or the APA "(Morgan, 2011, p. 203)" informs readers that the writer has borrowed information from a work by Morgan, specifically from page 203. APA also includes the year Morgan's work was published. Such a source (author's last name, year, and page number) obviously does not supply complete documentation. Instead, the parenthetical reference points readers to an alphabetical list of works that appears at the end of the report. The list, called "Works Cited" in MLA or "References" in APA, contains full bibliographic data—titles, dates, web addresses, publishers, page numbers, and so on—about each source cited in your report. Every work that appears in your report must be listed in your references section. (The only exception is that personal communication or well-known works like the Bible do not have to appear in an APA-style References section.) To provide accurate parenthetical documentation for your readers, first carefully prepare your Works Cited or References list (see below) so that you know which sources you are going to cite in the right form and at the right place in your text.

Keep your documentation brief and to the point so that you do not interrupt the reader's train of thought. In most cases, all you will need to include is the author's last name, date, and appropriate page number(s) in parentheses, usually at the end of sentences. When you mention the author's name in your sentence, though, MLA and APA both advise that you do not redundantly cite it again parenthetically; for example:

> MLA Moscovi claims that "tourism has increased 21 percent this quarter" (76).
> APA Moscovi (2011) claims that "tourism has increased by 21 percent this quarter" (p. 76).

For unsigned articles, use a shortened title in place of an author's name parenthetically.

> MLA Shrewd bosses know that "chain-of-command meetings provide the opportunity to pass information up as well as down the administrative ladder" ("Working Smarter" 33).
> APA Shrewd bosses know that "chain-of-command meetings provide the opportunity to pass information up as well as down the administrative ladder" ("Working Smarter," 2010, p. 33).

Similarly, if you list the title of a reference work in the text of your paper, do not repeat it in your documentation.

> MLA According to the *Encyclopedia Britannica*, Cecil B. DeMille's *King of Kings* was seen by nearly 800,000,000 individuals (3: 458).
> APA According to the *Encyclopedia Britannica* (2008), Cecil B. DeMille's *King of Kings* was seen by nearly 800,000,000 individuals (3, p. 458).

The first number in parentheses in both versions refers to the volume number of the *Encyclopedia Britannica*; the second is the page number in that volume.

Works Cited or Reference Pages

Whether you follow MLA or APA, you will need to list your sources at the end of your report, on a new page, under the title of "Works Cited" or "References" at the top and then arrange the list alphabetically by authors' last names (except when no author is listed). But, as Table 9.1 points out, there are major differences between MLA and APA guidelines on the placement of information, punctuation, the use of italics and quotation marks, and capitalization. Below are examples of some of the entries you are most likely to include.

TABLE 9.1   Basic Differences Between Preparing an MLA Works Cited List and APA References List

	MLA	*APA*
Author	• List author's last name first, followed by a comma, and then first name and (if applicable) middle name or initial. • For two or three authors, invert only the first author's name (e.g., Smith, John, and Jose Alvarez), and connect the last two authors' names by and. • For more than three authors, cite just the first author listed on the work (Smith, John) and add *et al.* ("and others"), or you can provide all names in full in the order in which they appear on the title page or byline.	• List author's last name first, followed by a comma, and then cite only the first initial and middle initial (if known). • For multiple authors, invert all authors' names, and separate the last two names with an ampersand (&). • For more than seven authors, invert the first six authors' names, insert an ellipsis (…) and then list the name of the last author (also inverted). • Place the date of publication in parentheses immediately after the author's or authors' names. Then add a period after the parens
Title	• Italicize the full title of the book, newspaper, or journal/magazine, including any subtitles. • Capitalize all words in the title except for prepositions and articles unless the book or journal begins with one of these. • Titles of articles in journals, newspapers, and magazines are set in double quotation marks. • Capitalize all words in the journal/ newspaper/magazine article title except for prepositions and articles.	• For a book, italicize the full title and capitalize only the first word of the title (and also capitalize proper names). If there is a subtitle, place it after the main title, which is followed by a colon, and capitalize only the first word of the subtitle. • For newspapers, journals, or magazines, italicize the full title; capitalize all words in the title except for prepositions and articles. • Titles of articles in newspapers, journals, and magazines should not be in quotation marks. • Capitalize only the first word of the article title (even if it is a preposition) and any proper nouns.

	MLA	*APA*
Volume and Page Numbers	• For articles, cite the volume and the issue number (separated by a period), followed by the year in parentheses—52.1 (2012); for newspapers and magazines, use only the date—12 Aug. 2010. Then include page numbers—91–100-without a "p" or "pp."	Put the volume number of the journal/magazine (in italics) with the issue number (not in italics) in parentheses immediately following. Then, insert a comma and include page numbers—*12*(3), 87–102.
Publication	• For books, give the city of publication, followed by the two-letter abbreviation for the state. Then, after a colon, the publisher's name followed by a comma, the year of publication, and a period. • Include the publication medium for all entries (e.g., Print, Web, PDF, etc.) at the end of the publication information.	• For books, provide the city and two-letter abbreviation for the state, and then include the publisher's name after a colon, for example, Detroit, MI: MegaPress.
Web Sources (websites, blogs, etc.)	• You do not need to include URLs. But indicate the website name, sponsor or publisher, date of publication, and medium of publication, followed by the date of access.	• Insert URLs in place of page numbers with the following designation: Retrieved from (and then list the URL).
Personal Interview	• Provide the name of the person interviewed (last name first), followed by the type of interview that was conducted and the date of the interview.	Interviews, conversations, and presentations are not included in APA reference lists, but you must still cite them within your paper.

Book by One Author

MLA   Spraggins, Marianne. *Getting Ahead: A Survival Guide for Black Women in Business*. Indianapolis: Wiley, 2009. Print.

APA   Spraggins, M. (2009). *Getting ahead: A survival guide for black women in business*. Indianapolis, IN: Wiley.

Book by Two Authors

MLA   Wu, Melody, and Trent Tucker. *China's Role in the Global Economy*. Denver: Tradevision Press, 2011. Print.

APA   Wu, M. & Tucker, T. (2011). *China's Role in the Global Economy*. Denver, CO: Tradevision Press.

Book by Three Authors

MLA   Mallahi, Kamel, Kevin Morrell, and Geoffrey Wood. *The Ethical Business: Challenges and Controversies*. New York: Macmillan, 2010. Print.

APA   Mallahi, K., Morrell, K., & Wood, G. (2010). *The ethical business: Challenges and controversies*. New York, NY: Macmillan.

Book by Four or More Authors (MLA)

MLA   Berkowitz, Harry A., et al. *Collaborating Effectively and Efficiently: A Case Study*. Los Angeles: Collaborative Technology, 2010. Print.

Book by Eight or More Authors (APA)

APA   Berkowitz, H. A., Barner, P. L., Choi, D. G., Osler, T. O., Ruiz, J., Rowell, C. F., ... Emmons, W. D. (2010). *Collaborating effectively and efficiently: A case study*. Los Angeles, CA: Collaborative Technology.

Book Published Online

MLA   Marcus, Bonnie. *Advancing Women's Leadership*. Head over Heels: Women's Business Radio, 2009. Web. 10 Mar. 2011.

APA   Marcus, B. (2009). *Advancing women's leadership* [Adobe Digital Editions version]. Retrieved from http://womenssuccesscoaching.com/wp-content/uploads/2010/01/HeadOver_eBook1-Mar10F.pdf

Edited Collection of Essays

MLA   Chavez, Lisle, and Ted Nowaki, eds. *Urban Planning and People Oriented Space*. Boston: Academic P, 2011. Print.

APA   Chavez, L., & Nowaki, T. (Eds.). (2011). *Urban planning and people oriented space*. Boston, MA: Academic Press.

Work Included in a Collection of Essays

MLA   Papademos, Lucas. "The Effects of Globalization on Inflation, Liquidity, and Monetary Policy." *International Dimensions of Monetary Policy*. Ed. Jordi Gali and Mark Gertler. Chicago: U of Chicago P, 2010. 593–608. Print.

APA   Papademos, L. (2010). The effects of globalization on inflation, liquidity, and monetary policy. In J. Gali & M. Gertler (Eds.), *International dimensions of monetary policy* (pp. 593–608). Chicago, IL: University of Chicago Press.

Book by a Corporate Author

MLA   Computer Literacy Foundation. *PCs in the Classroom*. 3rd ed. New York: Technology P, 2008. Print.

APA   Computer Literacy Foundation. (2008). *PCs in the classroom* (3rd ed.). New York, NY: Technology Press.

Article in a Professional Journal

MLA   Fieseler, Christian, Matthes Fleck, and Miriam Meckel. "Corporate Social Responsibility in the Blogosphere." *Journal of Business Ethics* 91.4 (2010): 599–614. Print.

APA   Fieseler, C., Fleck, M., & Meckel, M. (2010). Corporate social responsibility in the blogosphere. *Journal of Business Ethics, 91*(4), 599–614.

Article in a Print Magazine

MLA   Kurowska, Teresa. "Is the Boss Watching Every Keystroke You Make?" *Today's Workplace* Oct. 2011: 47+. Print.

APA   Kurowska, T. (2011, October). Is the boss watching every keystroke you make? *Today's Workplace, 47,* 72–73.

Article in a Professional Online Journal

MLA   Mayer, Gloria, and Michael Vallaire. "Enhancing Written Communication to Address Health Literacy." *Online Journal of Issues in Nursing* 14.3 (2009). Web. 30 Sept. 2009.

APA   Mayer, G., & Vallaire, M. (2009, September 30). Enhancing written communication to address health literacy. *Online Journal of Issues in Nursing, 14*(3). doi:10.3912/OJIN.Vol14No02ManOS

Article in a Print Newspaper

MLA   Korkki, Phyllis. "Finding a Job by Starting a Business." *New York Times* 31 Jan. 2010: BU2. Print.

APA   Korkki, P. (2010, January 31). Finding a job by starting a business. *New York Times,* p. BU2.

Online Encyclopedia Article

MLA   Kling, Arnold. "International Trade." *Concise Encyclopedia of Economics.* 2nd ed. Library of Economics and Liberty, 2008. Web. 27 Mar. 2010.

APA   Kling, A. (2008). International trade. In *Concise encyclopedia of economics* (2nd ed.). Retrieved from http://www.econlib.org/library/CEE.html

Online Unsigned Encyclopedia Article

MLA   "Inflation." *Encyclopedia Britannica Online.* Encyclopedia Britannica, 2010. Web. 27 Mar. 2010.

APA   Inflation. (2010). In *Encyclopedia Britannica online.* Retrieved from http://www.britannica.com/

Unsigned Article in a Print Magazine or Newspaper

MLA   "The Green Machine." *Economist* 13 Mar. 2010: 7–8. Print.

APA   The green machine. (2010, March 13). *Economist,* 7–8.

Article in an Online Newspaper or Magazine

MLA Zakaria Fareed. "'Swing for the Fences': Energy Secretary Steven Chu on Boosting Technology." *The Washington Post*. The Washington Post Company, 29 Mar. 2010. Web. 31 Mar. 2010.

APA Zakar, F. (2010, March 29). 'Swing for the fences': Energy secretary Steven Chu on boosting technology. *The Washington Post*. Retrieved from http://www.washingtonpost.com/wp-dyn/content/article/2010/03/29/AR2010032901892.html

Government Documents

MLA United States Middle Class Task Force. *Green Jobs: A Pathway to a Strong Middle Class*. Washington: GPO, 2009. Print.

APA United States Middle Class Task Force. (2009). *Green jobs: A pathway to a strong middle class* (Publication No. 0851-K-02). Washington, DC: U.S. Government Printing Office.

Website

MLA Home page. Natl. Council of La Raza, 2010. Web. 28 Mar. 2010.

APA When referencing an *entire* website, APA style is to provide the URL in the body of the text and not list it in the References section.

Radio

MLA *Serious Money*. Hosts Renee Janson and Jason Ayala. Financial News Radio. KFNN, Phoenix, 7 Apr. 2010. Radio.

APA Hanson, R. (Host), & Ayala, J (Host). (2010, April 7). *Serious Money*. Phoeniz, AZ: KFNN Financial News Radio.

Television

MLA "No Frills Business Travel." Prod. Jeff Nathenson. *Business Traveler*. CNN. 14 Apr. 2010. Television.

APA Nathenson, J. (Producer). (2010, April 14). No frills business travel [Television series episode]. In *Business Traveler*. Atlanta, GA: CNN.

Podcast

MLA Gunther, Marc, prod. "Coca-Cola's New PlantBottle Sows Path to Greener Packaging." *Greenbiz Radio*. Greener World Media, 1 Dec. 2009. Web. 28 Mar. 2010.

APA Gunther, M. (Producer). (2009, December 1). *Coca-Cola's New PlantBottle Sows Path to Greener Packaging* [Audio podcast]. Retrieved from http://www.greenbiz.com/podcast/2009/12/01/coca-cola-new-plantbottle-sows-path-greener-packaging

Business Blog

MLA   Musgrove, Marc. "The Evolution of Cisco's Smart + Connected Communities to Colorado." *The Platform: Opinions and Insights from Cisco.* Cisco, 24 Mar. 2010. Web. 31 Mar. 2010.

APA   Musgrove, M. (2010, March 24). The evolution of Cisco's smart + connected communities to Colorado [Web log post]. Retrieved from http://blogs.cisco.com/news/comments/the_evolution_of_ciscos_smartconnected_communities_tocolorado/

Personal Interview

MLA   Alvarez, José. E-mail interview. 15 Oct. 2011.

APA   Interviews, conversations, and presentations are not included in APA reference lists, but you must still cite them within your paper as follows: (J. Alvarez, personal communication, October 15, 2011).

E-Mail Correspondence

MLA   Teke, Cho-Martin. "Re: New Security Cameras." Message to the author. 3 July 2011. E-mail.

APA   Treat as unpublished interview and cite parenthetically in the text; see "Personal Interview" above.

A Model Long Report

The following long report in Figure 9.2 was written by a senior training specialist, Terri Smith Ruckel, and her staff for the vice president of human resources, who had commissioned it. As often happens in the world of work, only Ruckel's name as the senior member or head of a department appears on the report. The main task facing Ruckel and her collaborative team was to demonstrate what RPM Technologies should do recruit and retain multinational workers and thus promote diversity in the workplace. She gathered relevant data from both **primary research** (interviews, direct observations, site visits, and tests) and **secondary research** (consulting and commenting at times on sources already available, such as books, websites, journal articles, reference works, government documents, and even in-house publications from newspapers and company reports on the RPM intranet).

Figure 9.2 contains all the parts of a long report discussed in this chapter except a glossary and an appendix. Intended for a decision maker interested in learning more about the problems multinational workers face, the report does not contain the technical terms and data that would require a glossary or an appendix. Note how the cover letter (Figure 9.1) introduces the report and its significance for RPM Technologies and how the information abstract succinctly identifies the main points of the report.

Transmittal letter for a long report.   Figure 9.1

 RPM *Technologies*

4500 Florissant Drive   St. Louis, MO 63174

314.555.2121   www.rpmtech.com

May 5, 2010

Jesse Butler
Vice President, Human Resources
RPM Technologies

Dear Vice President Butler:

With this letter I am enclosing the report my staff and I prepared on effective ways to recruit and retain a multinational work force for RPM Technologies, which you requested we submit by early May. The report argues for the necessity of adapting the RPM workplace to meet the needs of multinational employees, including promoting cultural sensitivity and making our written business communications more understandable.

Multinational workers undoubtedly play a major role in U.S. businesses and at RPM as well. With their technical skills and homeland contacts, these employees can help RPM Technologies successfully compete in today's global marketplace.

But businesses like RPM need to recruit qualified multinational workers more aggressively and then provide equal opportunities for them in the workplace. We must also be sensitive to cultural diversity and communication demands of such a diverse audience. By including cross-cultural training—for native and non-native English-speaking employees alike— RPM can more effectively promote cultural sensitivity. In-house language programs and plain-English or translated versions of key corporate documents can further improve the workplace environment for our multinational employees.

I hope you will find this report helpful in recruiting and retaining additional multinational employees for RPM Technologies. If you have any questions or want to discuss any of our recommendations or research findings, please call me at extension 5406 or e-mail me at the address below. I look forward to receiving your suggestions.

Sincerely yours,

Terri Smith Ruckel

Terri Smith Ruckel
Senior Training Specialist
truckel@rpmtech.com

Enclosure: Report

Begins with major recommendation of report

Presents findings of report

Alerts reader to major issues covered in the report issue

Offers to answer questions

Enclosure notation specifies report is attached

Figure 9.2   A Long Report

Title page is carefully formatted and uses boldface

Adapting the RPM Workplace for Multinational Employees

Identifies writer and job title

Terri Smith Ruckel

Ruckel presents report from entire staff—writing for another's signature

**Senior Training Specialist
RPM Technologies**

Prepared for

RPM executive who assigned the report

**Jesse Butler
Vice President, Human Resources**

Date submitted

May 5, 2010

2

Abstract

This report investigates how U.S. businesses such as RPM must gain a competitive advantage in today's global marketplace by recruiting and retaining a multinational work force. This current wave of immigrants is in great demand for their technical skills and economic ties to their homeland. Yet many companies like ours still operate by policies designed for native speakers of English. Instead, we need to adapt RPM's company policies and workplace environment to meet the cultural, religious, social, and communication needs of these multinational workers. To do this, we need to promote cultural sensitivity training, both for multinationals and employees who are native speakers of English. Additionally, as other U.S. firms have done, RPM should adapt vacation schedules and daycare facilities for an increasing multicultural work force. Equally important, RPM needs to ensure, either through translations or plain-English versions, that all documents can be easily understood by multinational workers. RPM might also offer non-native speakers of English in-house language instruction while providing foreign language training for employees who are native speakers of English.

Concise, informative abstract that states purpose of report and why it is important for audience

Uses helpful transitional words, e.g., "Additionally," "Equally Important"

While APA does
not recommend
including a table
of contents,
individual
employers such
as RPM may
require one for
their reports

Major divisions
of report in all
capital letters
and boldface

Subheadings
indicated
by different
typeface,
indentations,
and use of
italics

Page numbers
included for
major sections
of report

No subsections
needed here

3

Table of Contents

4

List of Illustrations

*Identifies each
figure by title
and page
number*

*A title is
provided for
each visual.*

*Note the
variety of
visuals used in
the report.*

5

Introduction

Background

The U.S. work force has been undergoing a remarkable revolution. The U. S. Bureau of Labor Statistics predicts that by 2012 the labor force in the United States will comprise 162 million workers who must fill 167 million jobs (2009). According to the U.S. Chamber of Commerce, a shortage of skilled workers is sure to increase "even more heavily in the future when many of the baby boomers begin to retire" (2010).

The most dramatic effect of filling this labor shortage will be in hiring greater numbers of highly skilled multinational employees, including those joining RPM. Currently, "one of every five computer specialists [and] one of every six persons in engineering or science occupations . . . is foreign born" (Kaushal & Fix, 2006). This new wave of immigrants will make up 37 percent of the labor force by 2015 and continue to soar afterward. By 2025 the number of international residents in the United States will rise from 26 million to 42 million, according to the U.S. Chamber of Commerce (2010). The Congressional Budget Office predicts that "over the next decade net immigration will average between 500,000 and 1.5 million people annually" (2010). As Alexa Quincy aptly put it, "The United States is becoming the most multiculturally diverse country in the global economy" (2010).

The following commissioned report explores the impact that this new multinational work force will have on RPM Technologies, and what we must do to recruit and to retain these workers.

Global Immigration

Unlike earlier generations, today's immigrants actively maintain ties with their native countries. These new immigrants travel back and forth so regularly they have become global citizens, exercising an enormous influence on the success of a business like RPM. Five years ago, demographers Crane and Boaz claimed, "Immigration [will] give America an economic edge in the global economy ... most notably in the Silicon Valley and other high-tech centers. Figure 1 reveals the major ethnic groups that have immigrated to New York City in just 2009 alone. They provide business contacts with other markets, enhancing [a company's] ability to trade and invest profitably abroad" (2005).

Today's High-Tech Immigrants

Undeniably, many immigrants today often possess advanced levels of technical expertise. A report by the Kaiser Foundation found that California's Silicon Valley had significantly benefited from the immigrants who have arrived with much needed technical training. Asian, Indian, Pakistani, and Middle Eastern scientists and engineers, who have relocated from a number of countries, now hold more than 40 percent of the region's technical positions ("Immigrants find," 2010). Figure 2 indicates the leading countries of origin for Silicon Valley's immigrants in 2009 and records the percentage for each nationality.

6

Figure 1
Major Ethnic Groups
Immigrating to New York City in 2009

Numbers and
titles figure

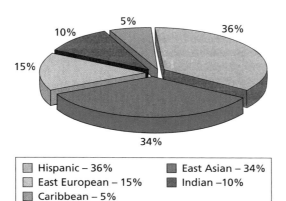

Hispanic – 36%	East Asian – 34%
East European – 15%	Indian –10%
Caribbean – 5%	

Pie chart
reveals differ-
ences in
immigrant
workers

Source: Brown, P. (2010, February). *History of U.S. immigration.*
Retrieved from http://immigration.ucn.edu

Cites source for
visual

Figure 2
Immigration to Silicon Valley in 2009

Bar chart
identifies and
quantifies
major groups
of immigrants

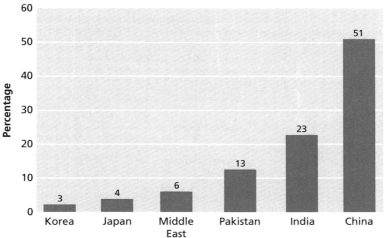

Relevant visual
in appropriate
place in text

Source: Immigrants find the American dream in California's Silicon Valley. (2010, March).
Silicon Valley News, p. 37.

7

Information technology (IT) employers are always enthusiastically searching for skilled international workers. Murali Krishna Devarakonda, president of the Immigrants Support Network in Silicon Valley, states why: "We contribute significantly to the research and development of a company's products and services" (personal communication, April 3, 2010).

E-mail not included in APA Reference list

Problem

Identifies a major problem and why it exists

Two separate, corroborating sources

RPM, like other e-companies, is experiencing a critical talent shortage of IT professionals, making the recruitment of a multinational work force a vital concern for us. Meeting the cultural and communication demands of these workers, however, poses serious challenges for RPM. The traditional workplace has to be transformed to respect the ways multinational employees communicate about business. Native English-speaking employees will also have to be better prepared to understand and to appreciate their international co-workers. Unfortunately, many corporate policies and programs at RPM, and at other U.S. companies as well, have been created for native-born, English-speaking employees (Morales, 2009; Reynolds, 2010). Rather than rewarding multinational workers, such policies unintentionally punish them.

Purpose

Concisely states why the report was written

The purpose of this report is to show that because of the need to increase the numbers of multicultural employees in the workplace, RPM must adapt its business environment to recruit and retain this essential and diverse labor force.

Scope

Informs reader that report will focus directly on RPM's needs

This report explores cultural diversity in the current U.S. workplace and suggests ways for RPM to compete successfully in the global village by providing equal employment opportunities for multinational workers. Doing so, we will foster cross-cultural literacy, and improve training in intercultural communication at our firm.

Discussion

Discussion will be organized into three main headings, each with subheadings

Providing Equal Workplace Opportunities for Multinational Employees

Aggressive Recruitment of IT Professionals from Diverse Cultures

A multilingual work force is essential if RPM wants to compete in a culturally diverse global market. But firms such as RPM must be prepared to adapt or modify hiring policies and procedures to attract these multinational employees, beginning with rethinking our recruitment and retention policies. Routine visits to U.S. campuses by company recruiters or "specialized international recruiters" can help us to identify and to hire highly qualified multinational job candidates (Hamilton, 2010).

8

Moreover RPM should consider visiting universities abroad with distinguished technical programs to attract talented multinational employees. We should encourage students and recent graduates from these universities to apply for a 1-J visa to learn more about RPM through an internship program here. As Catherine Bolgar reported, "Boeing went to Russia for specialist software engineers it couldn't find in the U.S." (2007). These searches should be combined with articles on our websites and executive blogs to emphasize RPM's commitment to globalization. Lobbying more actively to increase the number of H1-B visas for skilled workers will also help RPM. "In 2008 alone, US companies submitted 163,000 applications for the 65,000 H1-B visa slots. Google, for instance, applied for 300 of them, but 90 were denied" (Richtel, 2009).

Emphasizes recruiting multinational workers and suggests how to do so

Capitalizing on a diverse work force, RPM can more effectively increase its multi-cultural customer base worldwide. Logically, customers buy from individuals they can relate to culturally. RPM might take a lead from Union Bank of California, a business that effectively serves a diverse West Coast population, especially its Asian and Hispanic customers. The bank has a successful recruitment history of hiring employees with language skills in Hindi, Vietnamese, Korean, and Spanish. In fact, Union Bank is ranked fourth overall as an employer of minorities (Union Bank of California, 2008). Figure 3 below charts the increase in the percentage of multinational employees from these language groups hired by Union Bank over an 8-year period.

Identifies specific benefits for RPM

Another highly competitive business, Darden Restaurants, Inc., selected Richard Rivera, a Hispanic, to serve as president of Red Lobster, the nation's largest full-service seafood chain. Overseeing 680 restaurants nationwide, Rivera was the "most powerful minority CEO in the restaurant industry" (R. Jackson, personal communication, January 15, 2010).

Personal communication supports the need for recruitment but is not cited in references

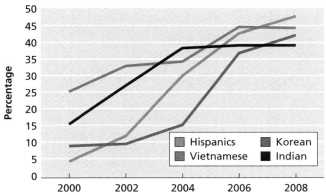

Figure 3
Union Bank of California:
Growth in Percentage of Multinational Employees

Tracks key information in a clear and concise graph

Supplies necessary legend

Source: Hamilton, B. E. (2010, April 15). Diversity is the answer for today's work force. *The Journal of Business Diversity*, *10*(38), 35–37.

9

Under Rivera's leadership, Red Lobster established an impressive history of hiring more international employees—totaling more than 35 percent of its work force—than it had in previous years. Rivera believes that management cannot properly respond to customers from different ethnic backgrounds if the majority of its employees is limited to native speakers of English. Many top Fortune 500 companies, such as Cisco and Intel, can also claim that 40% or more of their work force is comprised of multinationals ("100 Best Companies," 2010). Closer to RPM in St. Louis, Whitney Abernathy— manager of Netshop, Inc.—found that contracts from Indonesia increased by 17 percent after she hired Jakarta native Safja Jacoef (personal communication, April 2, 2010).

Commitment to Ethnic Representation

Many companies today have strong mission statements on diversity and multinational employees in the workplace. The most progressive of these businesses—G.E., American Airlines, IBM, and Wal-Mart—promote multinationals as mentors and interpreters. Kodak, which is committed to all-inclusive workplace, has eight employee cultural network groups, including the Hispanic Organization for leadership and Advocacy, or HOLA, which "helps all Hispanic employees reach their full potential while developing a rewarding career" ("Careers at Kodak," 2008). Such a proactive program, which we might incorporate at RPM, recognizes the leadership abilities of multinational employees. Moreover, "glass ceilings," which in the past have prevented women and ethnic employees from moving up the corporate ladder, are being shattered. The recruitment and promotion of non-native speakers of English is vital for corporate success. Tesfaye Aklilu, vice president at United Technologies, astutely observes:

> In a global business environment, diversity is a business imperative. Diversity of cultures, ideas, perspectives, and values is the norm of today's international companies. The exchange of ideas from different cultural perspectives gives a business additional, valuable information. Every employee can see his/her position from a global vantage point. (Aklilu, 2010)

RPM would also do well to follow the lead of one of our chief competitors, Ablex Plastics, which created the position of Vice President for Diversity two years ago and for which Ablex won an award from the International Business Foundation for hiring more Asian American women managers ("Ablex Appoints," 2008).

Promoting and Incorporating Cultural Awareness Within the Company

Cross-Cultural Training

Many of RPM's competitors are creating cultural awareness programs for international employees as well as native speakers. Committed to diversity, Aetna offers online courses on ethnicity (e.g.,"A Bridge to Asia") to "promote an atmosphere of openness and trust" (Aetna, 2008) while Johnson & Johnson conducts Diversity University "to help employees understand and value differences and the benefits of working collaboratively... to meet business goals" ("Programs and activities," 2010). Employees find it easier to work with someone whose values and beliefs they understand while employers benefit from effective on-the-job collaboration. Such a program could have prevented the problem RPM confronted last quarter when one of our non-native English-speaking employees was offended by a cultural misunderstanding (RPM first quarter, 2010). We may want to model our cultural sharing programs after those at American Express, which has a work force representing 40 nations, or those at Extel Communications with

Personal interview not included in APA References list

Stresses other business precedents that encourage recruiting these employees

Indent quotation of 40 or more words

Second major section

Cites business incentive to adapt as competitors did

References internal document from company intranet

10

its large percentage of Hispanic and Vietnamese employees. United Parcel Service (UPS) profitably pairs a native English-speaking employee with someone from another cultural group to improve on-the-job problem solving and communication. Jamie Allen, a UPS employee since 1998, for instance, found her work with Lekha Nfara-Kahn to be one of the most rewarding experiences of her job (Johnson, 2008).

Although they need to encourage cultural sensitivity training, U.S. firms like RPM also need to be cautious about severing international workers' cultural ties—a delicate balance. When management actively promotes bonds among employees from similar cultures, workers are less fearful about losing their identity and becoming "token" employees. The Lorenzo Lleras law firm, for example, recommends that new multinationals join a "buddy system" to assist their transition to American culture while retaining their cultural identity (U.S.Visa News, 2010). Encouraging such contacts, Globe Citizens Bank has long mobilized culturally similar groups by asking workers of shared ethnic heritages to network with each other (Gordon & Rao, 2009). Employees of Turkish ancestry from Globe's main New York office go to lunch twice a month with Turkish-born employees from the Newark branches. Globe hosts these luncheons and in return receives a bimonthly evaluation of the bank's Turkish and Middle Eastern policies (Hamilton, 2010).

Cultural education must go both ways, though. Both sides have things to learn about doing business in a global environment. The U.S. business culture has conventions, too, and few international employees would want to ignore them, but they need to know what those conventions are (Johnson, 2008). A frequent problem with U.S. corporations such as RPM is that we assume everyone knows how we do things and how we think—it never occurs to us to explain ourselves. Problems often stem from easy-to-correct misunderstandings about business etiquette. For example, native speakers of English are typically comfortable within a space of 1.5 to 2 feet for general personal interactions in business. But workers from Taiwan or Japan, who prefer a greater conversational distance, feel uncomfortable if their desks are less than a few feet away from another employee's workspace (Johnson, 2008; "Taiwanese Business Culture," 2008).

Promotion of Cultural Sensitivity

Corporate efforts to validate different cultures might also include the recognition of an ethnic group's holidays or other memorable historical occassions. RPM has just begun to do this by hosting cultural events, including Cinco de Mayo and Chinese New Year celebrations. Google throws a "Brazilian Carnival-themed edition of a weekly company meeting" (Sanghani & Shaikh, 2009). Emphasizing their "long-standing relationship with the Chinese community," Wells Fargo participates "every year in San Francisco's Chinese New Year parade" by featuring "a 200-foot-long Golden Dragon" (Babal, 2008). Techsure, Inc., an Illinois software firm, allows Muslim employees to alter their work schedules during Ramadan, Islam's holy month of fasting (M. Saradayan, personal communication, February 28, 2010).

Transition to new subdivision—networking of employees with similar cultural backgrounds

Relevant source on topic of immigration

Offers two examples RPM could follow

Another clear transitional sentence

Identifies key RPM problem

Gives cultural example

Gives precise ways RPM can incorporate cultural sensitivity into the workplace

APA lists blogs in references

Includes valuable information from official corporate blogs

11

Includes
appropriate
calendar of
ethnic holidays

Many companies honor National Hispanic Heritage Month in September, which coincides with the independence celebrations of five Latin American countries (U.S. Equal Employment Opportunity Commission [EEOC], 2008). GRT Systems in Ohio sends New Year's greetings at Waisak (the Buddhist Day of Enlightenment) to its Chinese employees and to Indian workers at Rama Dipawli. Chemeka Taylor, GRT's operations manager, wisely points out: "We send native-born employees Christmas cards; why shouldn't we honor our international work force, too?" (personal communication, April 10, 2010). Figure 4 (see page 12) provides a helpful multicultural calendar that RPM needs to follow in developing cultural sensitivity policies.

Identifies
current RPM
programs
and how they
could be easily
modified
to assist
multinalional
workers

Successful U.S. firms have been sensitive to the needs of their English-speaking employees for decades. Flexible scheduling, telecommuting options, daycare, and preventive health programs have become part of corporate benefit plans to take care of employees. Many of these options and benefits have already been in place at RPM. But an international work force presents additional cultural opportunities for RPM management to respond to with sensitivity. For example, our company cafeterias might easily accommodate the dietary restrictions of vegetarian workers or those who abstain from certain foods, such as dairy products or meat. At Globtech, soybean and fish entrees are always available (Reynolds, 2010). Adding ethnic items at RPM would express our cultural awareness and respect for multinational employees.

Cites company
publication
showing
research
within the
organization

Daycare remains a key issue in hiring and retaining skilled employees, whether they are native or non-native speakers of English. RPM's childcare facilities at our high-tech park in San Luis Obispo have brought us much positive publicity over the past eight years ("RPM Daycare Facilities," 2010). But by modifying child care that reflects our workers' culturally diverse needs, RPM can give a multinational work force greater peace of mind and better enable them to do their jobs. A pacesetter in this field is DEJ Computers, which insists that at least two or three of its daycare workers must be fluent in Korean or Hindi (Parker, 2009). Another culturally sensitive employer, ITCorp assists its Hispanic work force by hiring bilingual daycare workers and by serving foods the children customarily eat at home (Gordon & Rao, 2009).

Making Business Communication More Understandable for Multinational Employees

Translation of Written Communications

Third major
section

Turns to writ-
ten communi-
cation and
multinational
workers

Offers practical
solution

All employees must be able to understand business communications affecting them. Among the essential documents causing trouble for multicultural readers are company handbooks, insurance and health care obligations, policy changes, and OSHA and EPA regulations (Hamilton, 2010). To ensure maximum understanding of these documents by a multinational work force, RPM needs to provide a translation or at least a plain-English version of them. To do this, RPM could solicit the help of employees who are fluent in the non-native English speakers' languages as well as contract with professional translators to prepare appropriate work-related documents.

12

Figure 4
A Multicultural Calendar

December 2008

	1	2	3	4	5	6
						Feast of St. Nicholas (some European countries)
7 Advent begins (Christian)	**8** Bodhi Day (Rohatsu-Buddhist)	**9** Eid al-Adha (Islam) Gita Jayanti (Hindu)	**10**	**11**	**12** Datta Jayanti (Hindu); Feast Day—Our Lady of Guadalupe (Catholic)	**13**
14	**15**	**16** Posada begins (Hispanic) (ends Dec. 24)	**17**	**18**	**19**	**20**
21	**22** Ekadashi (Hindu); Hanukkah begins (Jewish) (ends Dec. 29)	**23**	**24** Nocha Buena (Hispanic)	**25** Christmas (Christian)	**26** Kwanzaa begins (Interfaith) (ends Jan. 1)	**27**
28	**29** Muharram (Islam)	**30**	**31** New Year's Eve (Western)			

Some Information About December's Holidays

Bodhi Day: Buddhists celebrate Prince Gautama's vow to attain enlightenment.

Christmas: Christians celebrate the birth of Jesus Christ.

Datta Jayanti: Celebration of the birth of Lord Datta.

Eid al-Adha: Feast of Sacrifice, commemorating the prophet Abraham's sacrifice of his son Ishmael to Allah.

Ekadashi: Twice monthly purifying fast.

Feast of Our Lady of Guadalupe: (Mexico) Roman Catholic holiday honoring the appearance of the Virgin Mary near Mexico City in 1531.

Feast of Saint Nicholas: Christmas celebration in Austria, Belgium, France, Germany, the Netherlands, Czech Republic, Poland, Russia, and Switzerland.

Gitta Jayanti: Celebration of the anniversary of the Bhagavad Gita, the Hindu scripture.

Hanukkah: Eight-day celebration commemorating the rededication of the Temple of Jerusalem.

Kwanzaa: African American and Pan-African celebration of family, community, and culture. Seven life virtues are presented.

Muharram: Islamic New Year.

Nocha Buena: The final day of the Posada celebration.

Posada: Nine-day Latin American Christian celebration, symbolizing Joseph and Mary's journey to find an inn.

Source: Johnson, V. M. (2008). Growing multinational diversity in business sparks changes. *Business Across the Nation, 23*(7), 43–48.

Major visual with a great deal of detail merits a full page to make it readable

Pays attention to major world holidays

Visual helps to convince RPM to adopt similar policy

Provides descriptions of cultural significance of dates

Lists source

13

Workplace signs in particular, especially safety messages, must consider the language needs of international workers. In the best interest of corporate safety, RPM needs to have these signs translated into the languages represented by multinationals in the workplace and/or to post signs that use global symbols. Unquestionably, we need to avoid signs that workers might find hard or even impossible to decipher. For example, a capital **P** for "parking" or an **H** for "hospital" might be unfamiliar to non-native speakers of English (Parker, 2009).

Exchange of Language Learning: Some Options

Providing English language instruction for employees also has obvious advantages for the RPM workplace. Should RPM Technologies elect not to provide in-house language training, the company should consider reimbursing employees for necessary courses, books, tapes, and software. Moreover, we might make available an audio library of specific vocabulary and phrases commonly used on the job as well as multilingual dictionaries with relevant technical terms. Hiromi Naguchi, an e-marketing analyst at PowerUsers Networking, praised the English language training she received on her job. Although she had ten years of English in the Osaka (Japan) schools, Naguchi developed her listening and speaking skills on the job, benefiting her employers, co-workers, and customers. As a result of her diligence, Naguchi was recognized as her company's "Most Valuable Employee for 2009" for her interpersonal skills (H. Naguchi, personal communication, January 30, 2010).

A recent international survey of executive recruiters showed that being bilingual is critical to success in business ("Developing Foreign Language Skills," 2008). According to this survey, 85 percent of European recruiters, 88 percent of Asian recruiters, and 95 percent of those in Latin America stressed the importance of speaking at least two languages in today's global workplace. As Jennifer Torres, CEO of Pacifica Corp., emphasized: "Although English remains the dominant language of international business, multilingual executives clearly have the competitive advantage. This advantage will only increase with the continued globalization of commerce" (as cited in "Developing Foreign Language Skills," 2008). Stockard Plastics, one of RPM's major customers, boasts that its employees collectively know more than 70 languages, everything from Albanian to Yoruba.

Language Training Must Be Reciprocal

But language training has to be reciprocal—for native as well as non-native speakers—if communication is to succeed. Sadly, even though foreign language instruction is on the rise, "fewer than 1 in 8 students at U.S. colleges major in [a] foreign language" (Cicorone, 2010). Unfortunately, this is the case with RPM's native-speaking employees. However, many of the international workers RPM needs to recruit are bi- or even trilingual. In India, Israel, or South Africa, for example, the average worker speaks two or more languages every day to conduct business.

Since RPM needs to recruit such workers, we have to learn more about the cultures and languages of these global employees. RPM management should consider contracting with one of the companies specializing in language instruction for

14

business people found at *http://www.selfgrowth.com/foreignlanguage.html*. We also need to network with international employee groups to solicit their help and advice.

Conclusion

To compete in the global marketplace, RPM must emphasize cultural diversity much more in its corporate mission and throughout the workplace. Through its policies and programs, RPM must aggressively recruit and retain an increasing number of technologically educated and experienced multinational workers who are in great demand today, and will be even more so over the next ten to twenty years. These workers can help RPM increase our international customer base and advance the state of our technology. But we need to ensure that the workplace is sensitive to their cultural, religious, and communication needs. Providing equal opportunities, diversity training and networking, and easy-to-understand business documents will keep RPM globally competitive in recruiting and retaining these essential employees. We can expand our markets only by making a commitment to learn about and respect other cultures and their ways of doing business.

Recommendations

By implementing the following recommendations, based on the conclusions reached in this report, RPM Technologies can succeed in hiring and promoting the IT multinational professionals our company needs for future success in today's global economy.

1. Recruit multinational workers more effectively through our website, international hiring specialists, and visits to college and university campuses here and abroad while working more closely with the Immigration and Naturalization Service (INS) to retain multinationals.
2. Establish a mentoring program to identify and foster leadership abilities in multinational employees, resulting in retaining and promoting these workers.
3. Promote cultural sensitivity and networking groups comprising both multinationals and native English-speaking employees.
4. Foster a group's cultural ties by actively supporting such work-related organizations as the Hispanic Organization for Advocacy and Leadership (HOLA).
5. Develop educational materials for employees who are native speakers of English about the cultural traditions of their multinational co-workers.
6. Reassess and adapt RPM's daycare facilities to more effectively meet the needs of children of multinational employees.
7. Supply relevant translations and plain-English versions of company handbooks, manuals, new regulations, insurance policies, safety codes, and other human resource documents.
8. Develop foreign language training programs to enhance communication and collaboration between multinational and native speaker employees at RPM.

Following the body, conclusion concisely summarizes the highlights of the report without repeating the documentation

Forecasts continuing benefits for RPM

Numbered list format is easy to read for busy executives

Provides specific, relevant recommendations, based on conclusions, to solve the problem at RPM

Uses strong persuasive verbs to introduce such recommendation

15

REFERENCES

Label list of sources actually cited

100 best companies to work for 2010. (2010, February 8). *Fortune 500*. Retrieved from http://money.cnn.com/magazines/fortune/bestcompanies/2010/minorities/

Double spaces between entries

Ablex appoints its first vice president for diversity. (2008, February). Retrieved from http://www.ablexinter.org

All the entries listed by author's last name or (If no author) by the first word of title, e.g., 100 best

Aetna, Inc. (2008, March). *Diversity annual report: The strength of diversity*. Retrieved from http://www.aetna.com/about/aetna/diversity/data/AetnaEnglish_2008.pdf

Aklilu, T. (2010). Diversity at UTC. Retrieved from http://www.utc.com/careers/diversity/index4.htm

Babal, M. (2008, January 26). The year of the ox [Web log post]. Retrieved from http://blog.wellsfargo.com/wachovia/2009/01/the_year_of_the_ox.html

Bolgar, C. (2007, June 18). Corporations need a global mindset to succeed in today's multipolar business world. *Wall Street Journal Online*. Retrieved from http://www.accenture.com/global/highperformancebusiness_business/multipolar businessworld.htm

Date of publication included for every entry after author's name or title (if author's name not given)

Brown, P. (2010, February). *History of U.S. immigration*. Retrieved from http://immigration.ucn.edu

Careers at Kodak. (2008, February). Retrieved from www.kodak.com/US/employment/corp/career/whyemploymentworks.jhtm

Cicorone, M. (2010, November). The importance of foreign languages to success in the world of work. *Language Instruction, 51*(3), 18–27

Congressional Budget Office. (2010, May). *Projections of net migration to the United States*. Retrieved from http://www.cbo.gov/doc.cfm?index=7249.

Only first word and proper nouns capitalized in title

Crane, E. H., & Boaz, D. (Eds.). (2005). *Cato handbook for policymakers* (6th ed.). Washington, DC: Cato.

Developing foreign language skills is good business. (2008, February 3). *Business World*. Retrieved from http://www.businessworld.ca/article.cfm/newsID/7109.cfm

Page numbers provided for print sources

Gordon, T., & Rao, P. (2009, April). Challenges ahead for American companies. *National Economics Review, 11*(3), 38–42, 56.

Hamilton, B. E., (2010, April 15). Diversity is the answer for today's work force. *Journal of Business Diversity, 10*(38), 35–37.

16

Immigrants find the American dream in California's Silicon Valley. (2010, March). *Silicon Valley News*, p. 37.

Johnson & Johnson. (2010). Programs and activities. Retrieved from http://www.jnj.com/connect/about-jnj/diversity/programs

Johnson, V. M. (2008). Growing multinational diversity in business sparks changes. *Business Across the Nation*, *23*(7), 43–48.

Kaushal, N., & Fix, M. The contributions of high-skilled immigrants. In M. Fix (Ed.), *Contributions of immigrants to the United States*. Also published as *Insight*, July 2006, No. 16. Migration Policy Institute: Washington DC.

Morales, J. (2009). *Immigration News & Notes*. Retrieved from http://www.immigrationnewsandnotes.com

Parker, M. (2009, May 26). Multinational Hires—Advice and Advocacy [Web log post]. Retrieved from http://parkeronimmigration.blogspot.com/2009/05/26/multinational_hires-advice_and_advocacy.php

Quincy, A. (2010, March 3). Multiculturalism makes for a good business. *Workforce, Inc., 14*(2), 5–8.

Reynolds, P. (2010, February 3). Serving up culture [Web log post]. Retrieved from http://www.culture.org/2010/02/serving_up_culture/

Richtel, M. (2009, April 12). Tech recruiting clashes with immigration rules. *New York Times*. Retrieved from http://www.nytimes.com/2009/04/12/business/12immig.html

RPM daycare facilities rated high. (2010, January 15). *RPM News*. Retrieved from http://www.rpm.com/rpmnews/01_15_2010/rpm_daycare_facilities_rated_high

RPM first quarter activity report (2010). *RPM Internal Reports*. Retrieved from http://www.rpm.com/internalreports/2010_firstquarter

Sanghani, P., & Shaikh, M. (2009, April 10). Mosaic: Bringing diverse perspectives together [Web log post]. Retrieved from http://googleblog.blogspot.com/2009/04/mosaic-bringing diverseperspectives.html

Taiwanese business culture. (2008). *Executive Planet*. Retrieved from http://www.executiveplanet.com/business-culture-in/132438266669.html

Union Bank of California. (2008, June 23). Federation magazine ranks Union Bank one of the best companies for minorities. Retrieved from http://www.uboc.com/about/main/0,3250,2485_11256_502261585,00.html

Works without author alphabetized by first word of title; first letters of proper nouns are capitalized

Second and subsequent lines indented either five-spaces or 1/2 inch

References blog posts with proper APA citations

Full web addresses are given for verification and to make source easy to find

Cites material available on company intranet

17

Title of government report is italicized

U.S. Bureau of Labor Statistics. (2009). *Overview of the 2008–18 projections.* Retrieved from http://www.bls.gov/oco/oco2003.htm

U.S. Chamber of Commerce. (2010). *Immigration issues.* Retrieved from http://www.uschamber.com/issues/index/immigration/default

U.S. Citizenship and Immigration Service. (2009, December). USCIS reaches FY 2010 H-1B cap. *USCIS.gov.* Retrieved from http://www.uscis.gov/ pressroom

U.S. Equal Employment Opportunity Commission, Federal Hispanic Work Group. (2008). Report on the *Hispanic employment challenge in the federal government.* Retrieved from http://www.eeoc.gov/federal/reports/hwg.

U.S. Visa News. (2010). *Lorenzo M. Lleras immigration attorney.* Retrieved from http://www.usvisanews.com

Final Words of Advice About Long Reports

Perhaps no piece of writing you do on the job carries more weight than the long report. You can simplify your job and increase your chances for success by following these guidelines for scheduling, researching, and collaborating:

1. Plan and work early—do not postpone work until a deadline draws near.

2. Do a thorough search among Internet, print, and other resources.

3. Divide your workload into meaningful units—reassure yourself that you do not have to write the report or even an entire section of a report in a day or two.

4. Set up mini-deadlines for each phase of your work and then meet them.

5. Confer often and carefully with others in your group office.

Revision Checklist

- [ ] Concentrated on a major problem—one with significant implications for my major, community, environment, or employer.
- [ ] Identified, justified, and described the significance of the major problem as opposed to focusing on a minor or side issue.
- [ ] Did sufficient research—in print sources in the library, on the Internet, through interviewing, from personal observation and/or testing—to convince readers that I am knowledgeable about this problem, its scope and effects, and solution(s).
- [ ] Made sure I understand what my employer/instructor wants me to include.
- [ ] Followed company's/instructor's guidelines for style, visuals, and format.
- [ ] Consistently used one method of documentation that my company/instructor wanted me to follow, e.g., APA, MLA, or another model.
- [ ] Adhered to company's/instructor's schedule for completing various parts of the long report.
- [ ] Divided and labeled the parts of long report to make it easy for readers to follow and to show a careful plan of organization.
- [ ] Wrote a one-page letter of transmittal, or cover letter, informing readers why the report was written and describing its scope and findings.
- [ ] Supplied an informational abstract that leaves no doubt in readers' minds about what the report deals with and why.
- [ ] Designed attractive title page that contains all the basic information—title, date, for whom the report is written, my name (or that of my collaborative team)—readers require.
- [ ] Gave readers all the necessary introductory information about background, problem, purpose of report, and scope.
- [ ] Included in body of report the results of all my research—the facts, statistics, expert opinions and descriptions—that my readers need to know that I have done my homework on the topic well.
- [ ] Wrapped up report in a succinct conclusion telling readers what the findings of my research are and accurately interpreted all data.
- [ ] Supplied a recommendations section (if required) that spells out concretely how readers can address/solve the problem. Recommendations made sense—are realistic and practical and related directly to the research supplied in the report. Ensured that recommendations are persuasive.
- [ ] Included a list of references that strictly followed MLA, APA, or another style of documentation my employer or instructor specified.
- [ ] Included in the final copy of report all the parts listed in table of contents.

Exercises

Additional Activities related to writing long reports are located at www. cengage.com/ english/kolin/ writingatwork concise3e

1. What kinds of research did the employee do to write the long report in Figure 9.2?

2. Study Figure 9.2 and answer the following questions based on it.
   a. How has Ruckel and her team successfully limited the scope of the report?
   b. Where and how do they use visuals especially well?
   c. Where and how have they adapted they technical information for their audience (a general reader)?
   d. What visual devices do they use to separate parts of the report and divisions within each part?
   e. How do they introduce, summarize, and draw conclusions from the expert opinions they cite in order to substantiate the main points?
   f. What are the ways in which Ruckel and her team document information they have gathered?
   g. What functions does the conclusion serve for readers? Cite specific examples from the report.
   h. How and why do the recommendations follow from the material presented in the report?

3. As a team or alone, come to class prepared to discuss at least two major problems that would be suitable topics for a long report. Consider an important community problem—traffic, crime, air and water pollution, housing, transportation—or a problem at your college. Then, write a letter to a consulting firm, or other appropriate agency or business, requesting a study of the problem and a report.

4. Write a report outline for one of the problems you decided on in Exercise 3. Use major headings and include the kinds of information discussed in the "Front Matter" section of this chapter (pages 323–325).

5. Have your instructor look at and approve the outline you prepared for Exercise 4. Then, write a long report based on the outline, either individually or as part of a collaborative writing team.

10 Making Successful Presentations at Work

Almost every job requires employees to have and to use carefully developed speaking skills. In fact, to get hired, you have to be a persuasive speaker at your job interview. And to advance up the corporate ladder, you will have to continue to be a confident, well-prepared, and persuasive speaker. The goal of this chapter is to help you be a more successful speaker.

Visit www.
cengage.com/
english/kolin/
writingatwork
concise3e for
this chapter's
online exercises,
ACE quizzes,
and Web links.

Types of Presentations

On the job you will have numerous presentation responsibilities that will vary in the amount of preparation they require, the time they last, and the audience and the occasion for which they are intended. Here are some frequent types of presentations you can expect to make as part of your job:

- sales appeals to prospective customers
- evaluations of products or policies
- progress reports to your boss and clients
- reports to employers about your job accomplishments
- justification of your position or even your department
- appeals and/or explanations before elected officials
- presentations at professional conferences
- explanation of a procedure, decision, or plan before a community/civic group

Whatever type of presentation you are asked to deliver, this chapter gives you practical advice on how to become a better, more assured communicator in both informal briefings and formal presentations.

Informal Briefings

If you have ever given a book report or explained laboratory results in front of a class, you have given an informal briefing. Such semiformal reports are a routine part of many jobs. Here are some of the typical informal briefings you may be asked to deliver at work:

- a status report on your current job or project
- an update or end-of-shift report, like those nurses and police officers give
- an explanation of a policy to co-workers
- a report on a conference you attended or on an environmental issue
- a demonstration of a new procedure or piece of equipment/software
- a follow-up session on equipment or procedures
- a summary of a meeting you attended

Such presentations are usually short (one to seven minutes, perhaps), and you won't always be given advance notice. When the boss tells you to "Say a few words about the new website" (or the new programming procedure), you will not be expected to give a lengthy formal speech but one that fits the occasion.

Guidelines for Preparing Informal Briefings

Follow these guidelines when you have to make an informal briefing:

- Make your comments brief and to the point.
- Write down a few bulleted items you plan to cover.
- Highlight key phrases and terms you need to stress.
- Include in your notes only the major points you want to mention.
- Arrange your points in chronological order or from cause to effect.
- Double-check your facts—names, dates, places, costs, models, and so forth.

Figure 10.1 is an informal outline with key facts used by an employee who is introducing Diana J. Rizzo, a visiting speaker, to a monthly meeting of safety directors.

| Figure 10.1 | **Some notes for an informal briefing to introduce an engineer to a group of safety directors.** |

- Diana J. Rizzo, Chief Engineer of the Rhode Island State Highway Department for 12 years

- Experienced as both a civil engineer and safety expert

- Consultant to Secretary Norman Habib, Department of Transportation

- Member of the National Safety Council and author of *Field Test Procedures in Highway Safety Construction* (2010)

- Instrumental in revising state safety commission's regulations

- Received "Award for Excellence" from the Northeastern Association of Traffic Engineers in May 2011

Formal Presentations

Whereas an informal briefing is likely to be short, generally conversational, and intended for a limited number of people, a formal presentation is much longer, far less conversational, and may be intended for a wider audience. It involves much more preparation.

Avoid speaking "off the cuff." The worst way to make a presentation is to speak without any preparation whatsoever. You only fool yourself if you think you will have all the necessary details and explanations in the back of your head. Without preparation, you are likely to confuse important points or forget them entirely. Mark Twain's advice is apt here: "It takes three weeks to prepare a good impromptu speech."

Expect to spend several days preparing your presentation. You cannot just dash it off. You will need time to

- research the subject
- interview key individuals
- prepare, time, and sequence visuals
- coordinate your presentation with co-workers and/or your boss
- rehearse your presentation

Many of us are uncomfortable in front of an audience because we feel frightened or embarrassed. Yet much of that anxiety can be eased if you know what to expect. The two areas you should investigate thoroughly before you begin to prepare your presentation are (1) who will be in your audience and (2) why they are there.

Analyzing Your Audience

The more you learn about your audience, the better prepared you will be to give them what they need. Just as you do for your written work, for your oral presentation you will need to do some research about the audience, emphasizing the "you attitude" and establishing your own credibility.

Consider Your Audience as a Group of Listeners, Not Readers

While audience analysis pertains both to readers of your work and to listeners of your presentation, there are several fundamental differences between these two groups. Unlike a reader of your report, the audience for your presentation

- may have only one chance to get your message
- has less time to digest what you say
- could have a shorter attention span
- can't always go back to review what you said or jump ahead to get a preview
- is more easily distracted—by interruptions, chairs being moved, people coughing, traffic noise, and so on
- cannot absorb as many of the technical details as you would include in a written report

Take all of these differences into account as you plan your presentation and assess who constitutes your audience.

Guidelines for Analyzing Your Audience

Here are five key questions to ask when analyzing your audience for your talk:

1. How much do they know about my topic?
   - consumers with little or no technical knowledge
   - technical individuals who understand terms, jargon, and background
   - business managers looking only for bottom line information

2. What unites them as a group?
   - members of the same profession
   - customers using the same products
   - employees of the company you work for

3. What is their interest level or stake in your topic?
   - highly motivated
   - neutral—waiting to be informed, entertained, or persuaded
   - uninterested in your topic—only there because attendance is mandatory
   - uncooperative and antagonistic, likely to challenge you

4. What do you want them to do after hearing your presentation?
   - buy a product or service
   - adopt a plan
   - change a schedule
   - understand and accept a new policy (e.g., safety procedures)

5. What questions are they likely to raise?
   - about money
   - about personnel
   - about locations
   - about schedules
   - about technology

Special Considerations for a Multinational Audience

Given the international makeup of audiences at many business presentations, you may have to address a group of listeners whose native language is not English or even make a presentation before individuals in a country other than your own. Consider your audience's particular cultural/communication taboos and protocols. Do they accept your looking at them directly, or do they regard eye contact as an invasion of their privacy?

As you prepare a talk before a multinational audience, keep the following points in mind:

1. Brush up on your audience's culture, especially accepted ways they communicate with one another (see pages 134–135).
2. Find out what constitutes an appropriate length for a talk before your audience.
3. Be especially careful about introducing humor—avoid anything that is based on nationality, dialect, religion, or race.
4. Think twice about injecting anything autobiographical into your speech. Some cultures regard such intimacy as highly inappropriate.
5. Steer clear of politics; you risk losing your audience's confidence.
6. Choose visuals with universally understood icons (see pages 235–237).

The Parts of Formal Presentations

As you read this section, refer to Marilyn Claire Ford's PowerPoint presentation in Figure 10.2. Note how effectively she used the PowerPoint format to persuade a potential client, GTP Systems, to purchase a service contract provided by World Tech, her employer. Her talk consists of seven slides that each contain relevant images and concise text.

The Introduction

The most important part of a presentation is your introduction, which should capture the audience's attention by answering these questions: (1) Who are you? (2) What are your qualifications? (3) What specific topic are you speaking about? and (4) How is the topic relevant to us?

A sample PowerPoint presentation.   Figure 10.2

Switching to Videoconferencing: A Wise Choice for GTP

Marilyn Claire Ford
World Tech
Desktop Videoconferencing
November 15, 2011

Introduces speaker, provides reason for presentation

Type size and font are clear

What We Offer

- User-friendly desktop videoconferencing
- Cutting-edge communication technology
- Flexible low-cost networking
- Single network for data, voice, and video

Chooses short title for each slide

Succinctly lists benefits in short, easy-to-read bulleted points

continued

Figure 10.2   (Continued)

*First of four
slides that
make up the
body of the
presentation*

*Develops
first key sales
feature*

*Relevant visual
does not mask
text*

Easy to Use

With a simple telephone call, you can
- Arrange a meeting with colleagues
  around the globe
- Use your computer to access World
  Tech's conferencing system
- See, hear, and talk with all participants

*Second point
gives only
essential facts*

*Does not
overwhelm
listeners with
numbers*

*Appropriate
icon
emphasizing
"cutting" costs*

Cost Effective

- Dramatically reduces costs for travel
- Upgrades current computer system for less
  than the expense of an existing computer
- Cuts data-processing expenses by 60%

*Third sales
feature
appropriately
mentions
specific
technology*

*Leaves
generous
margins*

*Icon shows
bringing staff
together*

Improves Staff Efficiency

- Brings people together at the right time
- Enhances communication when employees
  see and hear each other
- Whiteboard technology aids collaboration

(Continued)   Figure 10.2

Advantages in the Global Marketplace

- Connects all locations to one virtual office
- Increases sales worldwide
- Strengthens global networking

Addresses the worldwide benefits of choosing speaker's company, as last, most emphatic, point

Conclusion

- Recap of technology benefits
- Please sign up this week—it is easy
- Just log on to www.worldtech.com
- Thank you for considering WorldTech
- Any questions?

mcf@worldtech.com
1-800-271-5555

Conclusion

Issues a call to action; makes contact easy through website, e-mail, and telephone

Your first and most immediate goal is to establish rapport with your audience, win their confidence, and elicit their cooperation. Because your audience is probably at their most attentive during the first few minutes of your presentation, they will pay close attention to everything about you and what you say. Seize the moment and build momentum.

An effective introduction should be proportional to the length of your presentation. A 10-minute speech requires no more than a 60-second introduction; a 20-minute speech needs no more than a two- or three-minute introduction. Note how Marilyn Claire Ford in slides 1 and 2 introduces herself, her company, and its benefits for GTP.

How to Begin

You can begin by introducing yourself, emphasizing your professional qualifications and interests. (A self-introduction is unnecessary if someone else has introduced you or if you know everyone in the room.)

Give Listeners a Road Map

Give listeners a "road map" at the beginning of your presentation so that they will know where you are, where you are going, and what they have to look forward to or to recall. Indicate what your topic is and how you have organized what you have to say about it.

> My presentation today on fraud identification software will last about 20 minutes and is divided into three parts. First, I will outline briefly recent software identity protection changes. Second, I will give a detailed review of how those changes directly affect our company. Third, I will show how our company can profitably implement those changes. At the end of my presentation, please give me your questions and comments.

The most informative presentations are the easiest to follow. Restrict your topic to ensure that you will be able to organize it carefully and sensibly—for example, a tasty diet under 1,000 calories a day or a course in learning Java or another software package.

Capture the Audience's Attention

Use any of the following strategies to get your audience to "bite the hook":

- Ask a question. "Did you realize that every 30 minutes a foreign-owned business opens in China?" or "Do you know what's in your bottled water besides water?"
- Start with a quotation. Winston Churchill said, "We get things to make a living but we give things to have a life." (Consult *Bartlett's Familiar Quotations* online at *http://www.bartleby.com*.)
- Give an interesting statistic. "In 2011, two million heart attack victims will live to tell about it." (Go to the *World Almanac and Book of Facts* online at *http://www.worldalmanac.com* to find information relevant to your presentation topic.)
- Relate an anecdote or story. Be sure it is relevant and in good taste; make your audience feel at ease and friendly toward you by establishing a bond with them.

Be careful about using humor in a business talk. It could backfire—the audience may not get the point or may even be offended by it.

The Body

The body, or discussion, is the longest part of your presentation, just as it is in a long report. Make it persuasive and relevant to your audience by (1) explaining a process, (2) describing a condition, (3) solving a problem, (4) arguing a case, or (5) doing all of these. See how the body of Marilyn Claire Ford's presentation in Figure 10.2 is organized around the customer benefits of GTP's switching to desktop videoconferencing. In slides 3–6 she outlines how easy, economical, and efficient such technology is to use in a global marketplace.

To get the right perspective, recall your own experiences as a member of an audience. How often did you feel bored or angry because a speaker tried to overload you with details or could not concisely stick to the point?

Ways to Organize the Body

Here are a few helpful ways you can present and organize information in the body of your presentation. For a written report, you need to design your document to help readers visually—supplying headings using underscoring and bullets, including necessary white space, and inserting headers and footers. In a presentation, switch from those purely visual devices to such aural ones as the following:

1. **Give signals (directions) to show where you are going or where you have been**. Enumerate your points: *first, second, third*. Emphasize cause-and-effect relationships with *subsequently, therefore, furthermore*. When you tell a story, follow a chronological sequence and fill your speech with signposts: *before, following, next, then* to help your listeners.

2. **Comment on your own material**. Tell the audience if some point is especially significant, memorable, or relevant. "This next fact is the most important thing I'll say today."

3. **Provide internal summaries**. Spending a few seconds to recap what you have just covered will reassure your audience and you as well.

   We have already discussed the difficulties in establishing a menu repertory, or the list of items that the food service manager wants to appear on the menu. Now we will turn to ways of determining which items should appear on a menu and why.

4. **Anticipate any objections or qualifications your audience is likely to have**. Address the issues with relevant facts about costs, personnel, the environment, and/or technology in your presentation.

The Conclusion

Plan your conclusion as carefully as you do your introduction. Stopping with a screeching halt is as bad as trailing off in a fading monotone. An effective conclusion should leave the audience feeling that you and they have come full circle and accomplished what you promised. Let readers know you are near the end of your presentation.

What to Put in a Conclusion

A conclusion should contain something memorable. Never introduce a new subject or simply repeat your introduction. A conclusion can contain the following:

- a fresh restatement of your three or four main points
- a call to action, just as in a sales letter—to buy, to note, to agree, to volunteer
- a final emphasis on a key statistic (For example, "Heating and cooling office space are responsible for almost 40% of carbon dioxide emissions in the U.S. and gobble more than 70% of our total electricity usage"[1]).

---

[1]*Source*: Cullen, L. T. (2007, June 7). Going green at the office. *Time*. Retrieved from http://www.time.com/time/magazine/article/0,9171,1630552,00.html

End your presentation, as Marilyn Claire Ford does in slide 7, with a concise summary of the main points and urge listeners to invest company money in your product or service—for example, World Tech Desktop Videoconferencing system.

Mean It When You Say "Finally"

When you tell your audience you are concluding, make sure you mean it. Saying "In conclusion" and then talking for another 10 minutes frustrates listeners and will makes them less receptive to your message.

Visit www. cengage.com/ english/kolin/ writingatwork concise3e for two online exercises, "Creating Presentations Using Presentation Software" and "Tailoring Presentation Software Presentations for International Audiences."

Presentation Software

As Marilyn Claire Ford's presentation in Figure 10.2 demonstrates, business presentations very frequently rely on PowerPoint, whiteboards, or other software to deliver information quickly and accurately. (A whiteboard—or smartboard—looks like a chalkboard but is really an interactive screen with computer capabilities.) Using these kinds of technology is a crucial skill your employer will expect you to master. These software packages enable you to create a presentation with concise text and carefully chosen visuals. They allow you to plan, write, and add visuals to your slides all at once.

Organize the Presentation

Map your presentation before you actually create your slides. One way is to prepare an outline (see pages 34–35 in Chapter 2). First, identify the large topics you want to cover, as Marilyn Claire Ford did in presenting information to save GTP time and money, increase staff efficiency, and help the company compete in a global economy. Divide your presentation into major subjects that best accomplish your objective and determine only those points that relate directly to your topic and to your audience's needs. But don't overwhelm audiences with too much information. Note that Marilyn Claire Ford used only seven slides in Figure 10.2. Also, choose your visuals as carefully as your words. Resist the temptation to dazzle your audience with electronic special effects.

Test the Technology

Find and eliminate any bugs at the rehearsal stage. Call in advance to confirm the room and any technology you may need. Save and preview your presentation in the format in which you will be giving it. Bring your own notebook to the meeting and set up the projector and your notebook together in advance. Make sure your software is compatible with the equipment you will use. Also, be sure any websites you use are (1) valid and (2) not broken links.

Prepare Handouts

It's always wise to have backup handouts with you in case your computer malfunctions—before or during your presentation. Handouts of your slides (make sure you keep a copy for yourself) will allow you to continue with your presentation, and they will also help your audience follow your presentation and take notes about it while you speak.

Guidelines on Using Presentation Software Effectively

Here are some tips to ensure that the design, organization, and delivery of your presentation go smoothly. Again, refer to Figure 10.2 as you study them.

Readability

- Make sure each slide is easy to read—clear, concise, and uncluttered.
- Use a type size that is crisp and easy to see, even from a distance. For a small presentation on your notebook use 24-point type or larger. Increase your type size for headings, as in Figure 10.2.
- Keep your type style and size consistent. Don't switch from one font to another.
- Avoid ornate or script type, and do not put everything in boldface or in all capital letters. See how Marilyn Claire Ford boldfaces only her headings in Figure 10.2.

Text

- Keep text short and simple. Use easy-to-recall names, words, and phrases. Your audience will not have the time to read long, complex messages.
- Use bulleted lists instead of unbroken paragraphs. But put no more than five bulleted lines on a slide, and limit each line to seven or eight words. Don't squeeze words on a line. Include no more than 40 words per slide.
- Double-space between bulleted items, and leave generous margins on all sides—do not have text running to the edges of the slide.
- Title each slide—using a question, a statement, or a key name or phrase as in Figure 10.2 (e.g., "Easy to Use," "Cost Effective").

Sequencing Slides

- Keep your slides in the appropriate order in which you need to show them.
- Retain the same transition (cover left or straight right) from slide to slide to avoid visual confusion.
- Spend about two to three minutes per slide, but don't read each slide verbatim. Summarize main ideas or concisely expand them while looking at your audience, not the slide.
- Time your slides so your audience can read them. Never continue to show a slide after you have moved on to a new topic.
- If you invite audience participation and interaction, build in extra time between your slides.

Background/Color

- Find a pleasant contrasting background to make your text easy to read. Avoid extremely light or dark backgrounds that make your slides hard to read. Stay away from stark backgrounds, such as using cold white images on a black screen.
- Use the same background for each slide.
- Avoid shadowing your text for "decorative" visual effect.
- Use color sparingly, and make sure it is professionally appropriate. Don't turn each slide into a sizzling neon sign.

Graphics

- Keep graphics clear, simple, and positioned appropriately on the slide (see discussion regarding effective use of visuals in Chapter 6, pages 209–214).
- Be sure your visuals do not cover or shadow text.
- Show only those visuals that support your main points. Not every slide requires a visual. Slides 1 and 7 in Figure 10.2, for example, do not use visuals.
- Include only one graphic per slide; otherwise, your text will be more difficult to read.
- Avoid complicated tables, elaborate flow charts, or busy diagrams.
- Use clip art sparingly. Remember: Less is more. Clip art must be functional, not distracting. Each icon in Figure 10.2 is practical, easy to understand.
- Incorporate animation, sound effects, or movie/video clips only when they are persuasive, relevant, and undeniably professional.
- Don't bother with borders; they do not make a slide clearer.

Quality Check

- Be sure your spelling, grammar, names, dates, costs, and sources are correct.
- Double-check all math, tables, equations, and percentages.

Using Noncomputerized Visuals in Your Presentations

You can expect to incorporate visuals in a variety of other media besides PowerPoint presentations. There will be situations in which you may have to use a flip chart (where large pieces of white paper are anchored on an easel and flipped over like pages in a notebook or tablet) or an overhead projector to make a presentation instead of using a computer. You will almost surely be asked to prepare handouts that include text, visuals, or a combination of these to accompany your presentation, and to distribute them either before or during your talk.

Regardless of the medium you use, make sure your visuals are

- easy to see
- easy to understand
- self-explanatory
- relevant
- accurate

Getting the Most from Your Noncomputerized Visuals

The following practical suggestions will help you get the most from your visuals when time and space may prohibit using computer setups.

1. **Do not set up your visuals before you begin speaking.** The audience will be wondering how you are going to use them and so will not give you their full attention. When you are finished with a visual, put it away so that your audience will not be distracted by it or tempted to study it instead of listening to you.

2. **Firmly anchor any maps or illustrations.** Having a map roll up or a picture fall off an easel during a presentation is embarrassing.

3. **Never obstruct the audience's view by standing in front of your visuals**. Use a pointer or a laser pointer (a pen-sized tool that projects a bright red spot up to 150 feet) to direct the audience's attention to your visual.

4. **Avoid crowding too many images onto one visual**. Use more than one flip chart page, slide, or transparency instead.

5. **Do not put a lot of writing on a visual**. Elaborate labels or wordy descriptions defeat your reason for using the visual. Your audience will spend more time trying to decipher the writing than attempting to understand the visual itself. If any writing must appear on one of your visuals, enlarge it so your audience can read it quickly and easily.

6. **Be especially cautious with a slide projector**. Check beforehand to make sure all your slides are in order and are right side up. Most important, make sure the projector is in good working order. Practice changing from one transparency to another. And test your tape recorder if you are using one as part of your presentation.

Rehearsing Your Presentation

Don't skip rehearsing your presentation thinking it will save you time. Rehearsing will actually help you become more familiar with your overall message, building your confidence. Here are some strategies to use as you rehearse your speech.

- Know your topic and the various parts of your talk.
- Speak in front of a full-length mirror for at least one rehearsal to see how an audience might view you.
- Talk into an audio recorder to determine whether you sound friendly or frantic, poised or pressured. You can also catch and correct yourself if you are speaking too quickly or too slowly. A rate of about 120 to 140 words a minute is easy for an audience to follow.
- Time yourself so that you will not exceed your allotted time or fall far short of your audience's expectations.
- Practice with the presentation software, visuals, or equipment that you intend to use in your speech for valuable hands-on experience.
- Monitor the type of gestures (neither too many nor too few) you can use for clarity and emphasis in your talk. (See page 371.)
- Record a video of your final rehearsal and show it to a colleague or instructor for feedback.

Delivering Your Presentation

Poor delivery can ruin a good presentation. You will be evaluated by your style of presentation just as you are in your written work. Concentrate on the image you project: how you look, how you talk, and how you move (your body language). Do you mumble into your notes, never looking at the audience? Do you clutch the lectern? Do you shift nervously from one foot to the other? All those actions betray your nervousness and detract from your presentation.

First impressions are crucial. Research shows that people decide what they think of you in the first two or three minutes of your presentation. The way you dress is important but so is your body language. In fact, seventy-five percent of your audience's impressions are influenced by your body language. The nonverbal signals you send affect how your audience will regard your leadership abilities, your sales performance, even your sincerity. Pay attention to gestures, hand movements, how you stand, and so on. No matter how many hours you have worked to get your message ready, if your nonverbal presentation is misleading, awkward, or inappropriate, the impact of what you say will be lost.

The following suggestions on how to deliver a presentation will help you to be a well-prepared, poised speaker.

Settling Your Nerves Before You Speak

Being nervous before your presentation is normal—a faster heartbeat, sweaty palms, shaking. But don't let nerves stand in your way of delivering a highly successful talk.

Here are a few ways to calm yourself before you deliver your presentation and to get some healthy doses of confidence.

- Give yourself plenty of time to get there. The more you have to rush, the more anxious you will be.
- Don't bring anything with you that is likely to spill, such as coffee or a soft drink.
- Avoid caffeine for a few hours before your talk if it makes you jittery.
- Take some deep breaths and then hold your breath while you count to 10. Then exhale. This will slow your heart rate and lower your blood pressure.
- Remind yourself that you have spent hours preparing your presentation. Your hard work will pull you through.
- Try to chat with one or two members of the audience ahead of time and relax. See your audience as friends—people who can help your career.

Guidelines for Making Your Presentation

Everyone is nervous before a talk. Accept that fact and even allow a few seconds for "panic time." Then put your nervous energy to work for you. Chances are, your audience will have no idea how anxious you are; they cannot see the butterflies in your stomach. Again, see your audience as helpful colleagues, not enemies. Remember to do the following:

1. **Establish eye contact with your listeners.** Look at as many people in your audience as possible to establish a relationship with them. Never bury your head in notes or keep your eyes fixed on a computer screen or keyboard. You will only signal your lack of interest in the audience or your fear of public speaking. In a small PowerPoint presentation, try to establish rapport with each person in the room.

2. **Adjust to audience feedback.** Watch your listeners' reactions and respond appropriately—nodding to agree, pausing a moment, paraphrasing to clarify a confusing point. Know your material well so that if someone asks you a question or

wants you to return to a point, you are not fumbling through your notes or trying feverishly to locate the right screen or slide.

3. **Use a friendly, confident tone**. Speak in a natural, pleasant voice, but avoid verbal tics ("you know," "I mean") and fillers ("um," "ah," "er") repeated several times each minute. Such nervous habits will make your audience nervous and your speech less effective. Use pauses instead.

4. **Vary the rate of your delivery**. Use your natural speaking voice. Vary your rate and inflection to help you emphasize key points and make transitions. Talk slowly enough for your audience to understand you, yet quickly enough so that you don't sound as if you are belaboring or emphasizing each word.

5. **Adjust your volume appropriately**. Talking in a monotone, never raising or lowering your voice, will lull your audience to sleep, or at least into inattention. Talk loudly enough for everyone to hear, but be careful if you are using a microphone. Your voice will be amplified, so if you speak loudly, you will boom rather than project. Every word with a *b*, *p*, or *d* will sound like an explosive in your listener's ear. Watch out for the other extreme—speaking so softly that only people in the first two rows can hear you.

6. **Watch your posture**. Don't shift from one foot to another. But do not slouch or look wooden, either. If you stand motionless, looking as if *rigor mortis* has set in, your speech will be judged as cold and lifeless, no matter how lively your words are. Be natural yet dynamic; smile, nod your head. Refer to an object on the screen by touching it, or use a pointer to emphasize something on an overhead projector.

7. **Use appropriate body language**. Be natural and consistent. Do not startle an audience by suddenly pounding on the lectern or desk for emphasis. Avoid gestures that will distract or alienate your audience. For example, don't fold your arms as you talk, a gesture that signals you are unreceptive (closed) to your audience's reactions. Also, avoid the nervous habits that can divert the audience's attention: clicking a ballpoint pen, scratching your head, rubbing your nose, twirling your hair. Nor do you have to remain still or step with robotlike movements. The remote control for a PowerPoint presentation allows you to casually walk around the room as you click and change screens.

8. **Dress professionally**. Do not wear clothes or clanking jewelry (such as a necklace or bracelet charms) that call attention to themselves. Follow your company's dress code. Unless it specifies otherwise (e.g., casual Fridays), wear clothes that are the business norm. Women should wear a businesslike dress or suit; men should choose a dark suit or sports coat, white or blue shirt, and a tasteful tie.

When You Have Finished

Don't just sit down after your presentation or, worse yet, march out of the room. Thank your listeners and give them a chance to ask questions, as Marilyn Claire Ford does in Figure 10.2.

Evaluating Presentations

A large portion of this chapter has given you information on how to construct and deliver a formal presentation. As a way of reviewing that advice, study Figure 10.3—an evaluation form similar to those used by instructors in classes in oral communication. Note that the form gives equal emphasis to the speaker's performance or delivery and to the organization, content, and sequence of the presentation as well as the visuals that accompanied that presentation.

Figure 10.3   An evaluation form for a presentation.

Name of Speaker: ——————————   Date of Presentation: ——————

Title of Presentation: ——————————   Length of Presentation: ——————

PART I: THE SPEAKER'S DELIVERY
(Circle the appropriate number using the scale below.)

1. Appearance	1 unprofessional	2	3	4	5 well groomed
2. Eye contact	1 poor	2	3	4	5 effective
3. Voice	1 monotonous	2	3	4	5 varied
4. Posture	1 poor	2	3	4	5 natural
5. Gestures	1 distracting	2	3	4	5 appropriate
6. Self-confidence	1 nervous	2	3	4	5 poised

PART II: THE PRESENTATION ITSELF
(Circle the appropriate number: 1 = poor; 5 = superior.)

1. Made sure topic was relevant and timely	1	2	3	4	5
2. Began with a clear statement of purpose	1	2	3	4	5

3. Followed a logical organization	1	2	3	4	5
4. Gave audience cues for transitions between sections	1	2	3	4	5
5. Provided convincing supporting evidence	1	2	3	4	5
6. Concluded with a summary of main points	1	2	3	4	5
7. Allowed time for questions	1	2	3	4	5

PART III: USE OF VISUALS (SLIDES OR OTHER GRAPHICS)
(Circle the appropriate number: 1 = poor; 5 = superior.)

1. Ensured visuals were relevant	1	2	3	4	5
2. Carefully timed the sequence of visuals	1	2	3	4	5
3. Made sure audience could see visuals clearly	1	2	3	4	5
4. Referred to visuals and indicated why they were important	1	2	3	4	5

Revision Checklist

☐ Anticipated audience's background, interest, or even potential resistance, and their questions about message of both informal and formal presentations.

☐ Prepared outline and identified and corrected any weak or redundant areas.

☐ Prepared introduction to provide a "road map" of the presentation and to arouse audience interest.

☐ Started with interesting and relevant statistics, a question, an anecdote, or similar "hook" to capture audience attention.

☐ Limited body of presentation to main points.

☐ Sequenced main points logically and made connections among them.

☐ Used supporting examples and illustrations appropriate to audience and message.

☐ Made sure conclusion contains summary of main points of my presentation and/or specific call for action.

☐ Designed visuals that are clear, easy to read, and relevant for the audience.

☐ Experimented successfully with presentation software and applications before including them in a presentation.

☐ Used appropriate number of slides, made sure they were readable.

☐ Showed slides in right, carefully timed sequence.

☐ Prepared transparencies and handouts in case of computer trouble.

☐ Rehearsed presentation thoroughly to become familiar with its content, organization, and visuals.

☐ Monitored volume, tone, and rate to vary delivery and to emphasize major points.

☐ Rehearsed gestures to make them relevant and nonintrusive.

☐ Timed presentation, complete with visuals, to run close to allotted time.

Exercises

1. Prepare a three- to five-minute presentation explaining how a piece of equipment that you use on your job works. If the equipment is small enough, bring it with you to class. If it is too large, prepare an appropriate visual or two for use in your talk.

2. You have just been asked to talk about the students at your school or the employees where you work. Narrow the topic and submit an outline to your instructor, showing how you have limited the topic and gathered and organized evidence. Use two or three appropriate visuals (tables, photographs, maps, charts, icons, or even videos) to give a PowerPoint presentation. Follow the format of the presentation in Figure 10.2.

3. Prepare a 10-minute presentation on a controversial topic that you would present before a civic group—the PTA, the local chapter of an organization, a post of the Veterans of Foreign Wars, a synagogue, a mosque, or a church club.

4. Using the information contained in the long report on multinational workers in Chapter 9 (pages 338–354), prepare a short presentation (five to seven minutes) for your class.

5. Using the evaluation form in Figure 10.3, evaluate a speaker—a speech class student, a local politician, or a co-worker delivering a report at work. Specify the time, place, and occasion of the speech. Pay special attention to any visuals the speaker uses.

6. Deliver a formal presentation (15 to 20 minutes) on one of the following topics. Restrict your topic, and divide it into four key issues or parts, as in Figure 10.2. Use at least three visuals with your talk. Submit an outline to your instructor.

   a. a new piece of equipment at work
   b. the use of the Internet to provide health care in rural areas
   c. a major change in housing or traffic control in your city
   d. a paper or report you may have written in school or on your job
   e. "greening" your school's student center
   f. the budgetary allocations or pending cuts planned for your department or your town's school district for a given year
   g. applying for and receiving student financial aid

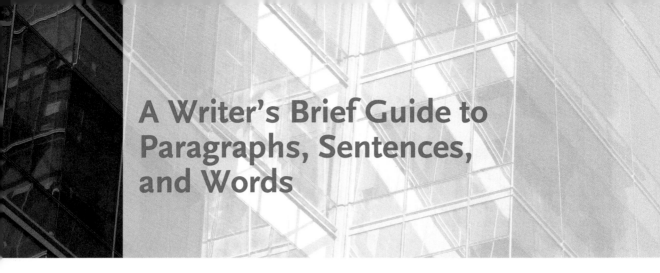

A Writer's Brief Guide to Paragraphs, Sentences, and Words

To write successfully, you must know how to create effective paragraphs, write and punctuate clear sentences, and use words correctly. This guide succinctly explains some of the basic elements of clear and accurate writing.

Paragraphs

Writing a Well-Developed Paragraph

A paragraph is the basic building block for any piece of writing. It is (1) a group of related sentences (2) arranged in a logical order (3) supplying readers with detailed, appropriate information (4) on a single important topic.

A paragraph expresses one central idea, with each sentence contributing to the overall meaning of that idea. The paragraph does that by means of a *topic sentence*, which states the central idea, and *supporting information*, which explains the topic sentence.

Supply a Topic Sentence

The topic sentence is the most important sentence in your paragraph. Carefully worded and restricted, it helps you to generate and control your information. An effective topic sentence also helps readers grasp your main idea quickly. As you draft your paragraphs, pay close attention to the following three guidelines.

1. Make sure you provide a topic sentence. In their rush to supply readers with facts, some writers forget or neglect to include a topic sentence. The following paragraph, with no topic sentence, shows how fragmented such writing can be.

No topic sentence: Sensors found on each machine detect wind speed and direction and other important details such as ice loading and potential metal fatigue. The information is fed into a microprocessor in the nacelle (or engine housing). The microprocessor then automatically keeps the blades turned into the wind, starts and stops the machine, and changes the pitch of the tips of the blades to increase power under varying wind conditions. Should any part of the wind turbine suffer damage or malfunction, the microprocessor will immediately shut the machine down.

Only when a suitable topic sentence is added—"The MOD-2 wind turbine is programmed to run according to the latest technology."—can readers understand what the technical details have in common.

2. Put your topic sentence first. Place your topic sentence at the beginning—not the middle or end—of your paragraph because the first sentence occupies a commanding position. Burying the key idea in the middle or near the end of the paragraph makes it harder for readers to comprehend your purpose or act on your information.

3. Be sure your topic sentence is focused and discusses only one central idea. A broad or unrestricted topic sentence leads to a shaky, incomplete paragraph for two reasons.

- The paragraph will not contain enough information to support the topic sentence.
- A broad topic sentence will not summarize or forecast specific information in the paragraph.

The following example of a carefully constructed paragraph contains a clear topic sentence in an appropriate position (highlighted in color) and adequate supporting details.

> Fat is an important part of everyone's diet. It is nutritionally present in the basic food groups we eat—meat and poultry, dairy products, and oils—to aid growth or development. The fats and fatty acids present in those foods ensure proper metabolism, thus helping to turn what we eat into the energy we need. Those same fats and fatty acids also act as carriers for important vitamins like A, D, E, and K. Another important role of fat is that it keeps us from feeling hungry by delaying digestion. Fat also enhances the flavor of the food we eat, making it more enjoyable.

Three Characteristics of an Effective Paragraph

Effective paragraphs have **unity**, **coherence**, and **completeness**.

(1) Unity
A unified paragraph sticks to one topic without wandering. Every sentence, every detail, **supports**, **explains**, or **proves** the central idea. A unified paragraph includes only relevant information and excludes unnecessary or irrelevant comments.

(2) Coherence
In a coherent paragraph all sentences flow smoothly and logically to and from each other like the links of a chain. Use the following techniques to achieve coherence.

a. Use transitional words and phrases. Some useful connective, transitional words, along with the relationships they express, are listed in Table A.1.

TABLE A.1   Transitional, or Connective, Words and Phrases

Addition	again	besides	moreover
	additionally	first, second, third	next
	along with	furthermore	together with
	also	in addition	too
	and	many	what's more
	as well as		
Cause/effect	accordingly	consequently	on account of
	and so	due to	since
	as a result	hence	therefore
	because of	if	thus
Comparison/ contrast	but	in contrast	on the other hand
	conversely	in the same way	similarly
	equally	likewise	still
	however	on the contrary	yet
Conclusion	all in all	in brief	on the whole
	altogether	in conclusion	to conclude
	as we saw	in short	to put into perspective
	at last	in summary	to summarize
	finally	lastly	to wrap up
Condition	although	granted that	provided that
	depending	if	to be sure
	even though	of course	unless
Emphasis	above all	for emphasis	of course
	after all	indeed	surely
	again	in fact	to repeat
	as a matter of fact	in other words	to stress
	as I said	obviously	unquestionably
Illustration	for example	in other words	that is,
	for instance	in particular	to demonstrate
	in effect	specifically	to illustrate
Place	across from	below	over
	adjacent to	beyond	there
	alongside of	here	under
	at this point	in front of	where
	behind	next to	wherever
Time	afterward	formerly	previously
	at length	hereafter	soon
	at the same time	later	simultaneously
	at times	meanwhile	subsequently
	beforehand	next	then
	currently	now	until
	during	once	when
	earlier	presently	while

Paragraph with
connective words:
Advertising a product on the radio has many advantages over using television. *For one thing*, radio rates are much cheaper. *For example*, a one-time 60-second spot on local television can cost $5,000. *For that money*, advertisers can purchase nine 30-second spots on the radio. *Equally attractive* are the low production costs for radio advertising. *In contrast*, television advertising often includes extra costs for models and voice-overs. *Another* advantage radio offers advertisers is immediate scheduling. *Often* the ad appears during the same week a contract is signed. *On the other hand*, television stations are *frequently* booked up months in advance, so it may be a long time *before* an ad appears. *Furthermore*, radio gives advertisers a greater opportunity to reach potential buyers. *After all*, radio follows listeners everywhere—in their homes, at work, and in their cars. *Although* television is very popular, it cannot do that.

b. Use pronouns and demonstrative adjectives. Words like *he, she, him, her, they, their*, and so on contribute to paragraph coherence and increase the flow of sentences.

Paragraph with
pronouns:
Traffic studies are an important tool for store owners looking for a new location. These studies are relatively inexpensive and highly accurate. They can tell owners how much traffic passes by a particular location at a particular time and why. Moreover, they can help owners to determine what particular characteristics the individuals have in common. Because of their helpfulness, these studies can save owners time and money and possibly prevent financial ruin.

c. Use parallel (coordinated) grammatical structures. Parallelism means using the same *kind of* word, phrase, clause, or sentence to express related concepts.

Orientation sessions accomplish four useful goals for trainees. First, they introduce trainees to key personnel in accounting, IT, maintenance, and security. Second, they give trainees experience logging into the database system, selecting appropriate menus, editing core documents, and getting off the system. Third, they explain to trainees the company policies affecting the way supplies are ordered, used, and stored. Fourth, they help trainees understand their ethical responsibilities in such sensitive areas as computer security and use.

Parallelism is at work on a number of levels in the paragraph above, among them

- The four sentences about the four goals start in the same way grammatically ("...they introduce/give/explain/help...") to help readers categorize the information.
- Within individual sentences, the repetition of *present participles* (logg*ing*, select*ing*, edit*ing*, gett*ing*) and of *past participles* (order*ed*, us*ed*, stor*ed*) helps the writer to coordinate information.
- Transitional words—*first, second, third, fourth*—provide a clear-cut sequence.

(3) Completeness

A complete paragraph provides readers with sufficient information to **clarify, analyze, support, defend,** or **prove** the central idea expressed in the topic sentence. The reader feels satisfied that the writer has given necessary details.

Skimpy paragraph:	Farmers are turning their crops and farm wastes into cost-effective fuels. Much grown on the farm is being converted to energy. This energy can have many uses and save farmers a lot of money in operating expenses.
Fully developed paragraph:	Farm crops and wastes are being turned into fuels to save farmers' operating costs. Alcohol can be distilled from grain, sugar beets, and corn. Converted to ethanol (90 percent gasoline, 10 percent ethanol), this fuel runs such farm equipment as irrigation pumps, feed grinders, and tractors. Similarly, through a biomass digestion system, farmers can produce methane from animal or crop wastes as a natural gas for heating and cooking. Finally, cellulose pellets, derived from plant materials, become solid fuel that can save farmers money in heating barns.

Sentences

Constructing and Punctuating Sentences

The way you construct and punctuate your sentences can determine whether you succeed or fail in the world of work. Your sentences reveal a lot about you. They tell readers how clearly or how poorly you can convey a message. And any message is only as effective and as thoughtful as the sentences of which it is made.

What Makes a Sentence

A sentence is a complete thought, expressed by a subject and a verb that can make sense standing alone.

subject	verb	
Websites	sell	products.

The Difference Between Phrases and Clauses

The first step toward success in writing sentences is learning to recognize the difference between phrases and clauses. A **phrase** is a group of words that does not contain a subject and a verb; phrases cannot make sense standing alone. Phrases cannot be sentences.

in the park	No subject:	Who is in the park?
	No verb:	What was done in the park?
for every patient in intensive care	No subject:	Who did something for every patient?
	No verb:	What was done for every patient?

A **clause** does contain a subject and a verb, but *not every clause is a sentence.* Only **independent** (or **main**) **clauses** can stand alone as sentences. Here is an example of an independent clause that is a complete sentence.

subject   verb   object

The president closed the college.

A **dependent** (or **subordinate**) **clause** also contains a subject and a verb, but it does not make complete sense and cannot stand alone. Why? A dependent clause contains a subordinating conjunction—*after, although, as, because, before, even though, if, since, unless, when, where, whereas, while*—at the beginning of the clause. Such conjunctions subordinate the clause in which they appear and make the clause dependent for meaning and completion on an independent clause.

After
Before
Because   } the president closed the college
Even though
Unless

"After the president closed the college" is not a complete thought but a dependent clause that leaves us in suspense. It needs to be completed with an independent clause telling us what happened "after."

dependent clause	independent clause		
(Not a sentence)	subject	verb	phrase
After the president closed the college,	we	played	in the snow.

Avoiding Sentence Fragments

An incomplete sentence is called a **fragment**. Fragments can be phrases or dependent clauses. They either lack a verb or a subject or have broken away from an independent clause. A fragment is isolated: It needs an overhaul to supply missing parts to turn it into an independent clause or to glue it back to an independent clause to have it make sense.

To avoid writing fragments, follow these rules. **Note that incorrect examples are preceded by a minus sign, correct revisions by a plus sign.**

1. Do not use a subordinate clause as a sentence. Even though it contains a subject and a verb, a subordinate clause standing alone is still a fragment. To avoid this kind of sentence fragment, simply join the two clauses (the independent clause and the dependent clause containing a subordinating conjunction) with a comma—*not* a period or semicolon.

– Unless we agreed to the plan. (What would happen?)
– Unless we agreed to the plan; the project manager would discontinue the operation. (A semicolon cannot set off the subordinate clause.)
+ Unless we agreed to the plan, the project manager would discontinue the operation.

– Because safety precautions were taken. (What happened?)

+ Because safety precautions were taken, 10 construction workers escaped injury.

Sometimes subordinate clauses appear at the end of a sentence. They may be introduced by a subordinate conjunction, an adverb, or a relative pronoun (*that, which, who*). Do not separate these clauses from the preceding independent clause with a period, thus turning them into fragments.

– An all-volunteer fire department posed some problems. Especially for residents in the western part of town.

+ An all-volunteer fire department posed some problems, especially for residents in the western part of town. (The word *especially* qualifies *posed*, referred to in the independent clause.)

2. Every sentence must have a subject telling the reader who does the action.

– Being extra careful not to spill the solution. (Who?)

+ The technician was being extra careful not to spill the solution.

3. Every sentence must have a complete verb. Watch especially for verbs ending in *-ing*. They need another verb (some form of *to be*) to make them complete.

– The machine running in the computer department. (Did what?)

You can change that fragment into a sentence by supplying the correct form of the verb.

+ The machine *is running* in the computer department.

+ The machine *runs* in the computer department.

Or you can revise the entire sentence, adding a new thought.

+ The machine running in the computer department processes all new accounts.

4. Do not detach prepositional phrases (beginning with *at, by, for, from, in, to, with*, and so forth) **from independent clauses.** Such phrases are not complete thoughts and cannot stand alone. Correct the error by leaving the phrases attached to the sentence to which they belong.

– By three o'clock the next day. (What was to happen?)

+ The supervisor wanted our reports by three o'clock the next day.

Avoiding Comma Splices

Fragments occur when you use only bits and pieces of complete sentences. Another common error that some writers commit involves just the reverse kind of action. They weakly and wrongly join two complete sentences (independent clauses) with a comma as if those two sentences were really only one sentence. Such an error is called a **comma splice.** Here is an example.

– Gasoline prices have risen by 15 percent in the last month, we will drive the car less often.

Two independent clauses (complete sentences) exist:

+ Gasoline prices have risen by 15 percent in the last month.
+ We will drive the car less often.

A comma alone lacks the power to separate independent clauses.

As the preceding example shows, many pronouns—*I, he, she, it, we, they*—are used as the subjects of independent clauses. A comma splice will result if you place a comma instead of a semicolon between two independent clauses where the second clause opens with a pronoun.

– Maria approved the plan, she liked its cost-effective approach.
+ Maria approved the plan; she liked its cost-effective approach.

However, relative pronouns (*who, whom, which, that*) are preceded by a comma, not a period or a semicolon, when they introduce subordinate clauses.

– She approved the plan. Which had a cost-effective approach.
+ She approved the plan, which had a cost-effective approach.

Four Ways to Correct Comma Splices

1. Remove the comma separating two independent clauses and replace it with a period. Then, capitalize the first letter of the first word of the new sentence.

+ Gasoline prices have risen by 15 percent in the last month. We will drive the car less often.

2. Insert a coordinating conjunction (*and, but, or, nor, so, for, yet*) after the comma. Together, the conjunction and the comma properly separate the two independent clauses.

+ Gasoline prices have risen by 15 percent in the last month, so we will drive the car less often.

3. Rewrite the sentence (if it makes sense to do so). Turn the first independent clause into a dependent clause by adding a subordinate conjunction; then, insert a comma and add the second independent clause.

+ Because gasoline prices have risen by 10 percent in the last month, we will drive the car less often.

4. Delete the comma and insert a semicolon.

+ Gasoline prices have risen by 15 percent in the last month; we will drive the car less often.

Of the four ways to correct the comma splice, sentences 3 and 4 are equally suitable, but sentence 3 reads more smoothly and so is the better choice.

The semicolon is an effective and forceful punctuation mark when two independent clauses are closely related, that is, when they announce contrasting or parallel views, as the two following examples reveal.

+ The union favored the new legislation; the company opposed it. (contrasting views)
+ Night classes help the college and the community; students can take more credit hours to advance their careers. (parallel views)

How *Not* to Correct Comma Splices

Some writers mistakenly try to correct comma splices by inserting a conjunctive adverb (*also, consequently, furthermore, however, moreover, nevertheless, then, therefore*) after the comma.

− Gasoline prices have risen by 15 percent in the last month, consequently we will drive the car less often.

Because the conjunctive adverb (*consequently*) is not as powerful as the coordinating conjunction (*and, but, for*), the error is not eliminated. If you use a conjunctive adverb—*consequently, however, nevertheless*—you still must insert a semicolon or a period before it, as the following examples show.

+ Gasoline prices have risen by 15 percent in the last month; consequently, we will drive the car less often.
+ Gasoline prices have risen by 15 percent in the last month. Consequently, we will drive the car less often.

Avoiding Run-on Sentences

A **run-on sentence** is the opposite of a sentence fragment. The fragment gives the reader too little information, the run-on too much. A run-on sentence forces readers to digest two or more grammatically complete sentences without the proper punctuation to separate them.

Run-on: The Internet has become a primary source of information and students and other researchers are right to call it a virtual library this library is not like the collections of hard-copy books and magazines that are carefully shelved always waiting for students to check and recheck them too often a website disappears or changes considerably and without a backup file or a hard copy the researcher has no document to quote from and no exact citation to prove that he or she consulted an authentic source.

Revised: The Internet has become a primary source of information. Students and other researchers are right to call it a virtual library, although this library is not like the collections of hard-copy books and magazines that are carefully shelved, waiting for students to check and recheck them out. But too often a website disappears, is under construction, or changes considerably. Without a backup file or a hard copy of the site, the researcher has no document to quote from and no exact citation to prove that he or she consulted an authentic source.

As the revision shows, you can repair a run-on sentence by (1) dividing it into separate, correctly punctuated sentences and (2) by adding coordinating conjunctions (*and, but, yet, for, so, or, nor*) between clauses.

Making Subjects and Verbs Agree in Your Sentences

A subject and a verb must agree in number. A singular subject takes a singular verb, whereas a plural subject requires a plural verb.

Singular Subject	Plural Subjects
the engineer calculates	engineers calculate
a report analyzes	reports analyze
a policy changes	policies change

You can avoid subject-verb agreement errors by following several simple rules.

1. Disregard any words that come between the subject and its verb.

Faulty: The customer who ordered three parts want them shipped this afternoon.
Correct: The <u>customer</u> who ordered three parts <u>wants</u> them shipped this afternoon.

2. A compound subject (two parts connected by *and*) **takes a plural verb.**

Faulty: The engineering department and the safety committee prefers to develop new guidelines.
Correct: The <u>engineering department</u> and the <u>safety committee</u> <u>prefer</u> to develop new guidelines.

3. When a compound subject contains *neither…nor* or *either…or*, the verb agrees with the subject closest to it.

Faulty: Either the residents or the manager are going to file the complaint.
Correct: Either the residents or the manager is going to file the complaint.
Correct: Either the manager or the residents are going to file the complaint.

4. Use a singular verb after collective nouns (like *committee, crew, department, group, organization, staff, team*) **when the group functions as a single unit.**

Correct: The crew was available to repair the machine.
Correct: The committee asks that all recommendations be submitted by Friday.

but

Correct: The staff were unable to agree on the best model. (The staff acted as individuals, not a unit, so a plural verb is required.)

5. Use a singular verb with indefinite pronouns (such as *anyone, anybody, each, everyone, everything, no one, somebody, something*).

Each of the programmer<u>s</u> has completed the seminar.
Somebody usually volunteer<u>s</u> for that duty.

Similarly, when *all, most, more,* or *part* is the subject, it requires a singular verb.

Most of the money is allocated.
Part of the equipment was salvageable.

6. Words like *scissors* and *pants* are plural when they are the true subject.

Faulty: A pair of trousers were available in his size. (*Pair* is the singular subject.)
Correct: The trousers were on sale.

7. Some foreign plurals (*curricula, data, media, phenomena, strata, syllabi*) **always take a plural verb.**

> The data conclusively <u>prove</u> my point.
> The media <u>are</u> usually the first to point out a politician's weak points.

8. Use a singular verb with fractions.

> Three-fourths of her research proposal was finished.

Writing Sentences That Say What You Mean

Your sentences should say exactly what you mean, without double talk, misplaced humor, or nonsense. Sentences are composed of words and word groups that influence each other.

Writing Logical Sentences

Sentences should not contradict themselves or make outlandish claims. The following examples contain errors in logic; note how easily the suggested revisions solve the problem.

> Illogical: Steel roll-away shutters make it possible for the sun to be shaded in the summer and to have it shine in the winter. (The sun is far too large to shade; the writer meant that a room or a house, much smaller than the sun, could be shaded with the shutters.)
> Revision: Steel roll-away shutters make it possible for owners to shade their living rooms in the summer and to admit sunshine during the winter.

Using Contextually Appropriate Words

Sentences should use the combination of words most appropriate for the subject.

> Inappropriate: The members of the Nuclear Regulatory Commission saw fear radiated on the faces of the residents. (The word *radiated* is obviously ill advised in this context; use a neutral term.)
> Revision: The members of the Nuclear Regulatory Commission saw fear reflected on the faces of the residents.

Writing Sentences with Well-Placed Modifiers

A **modifier** is a word, phrase, or clause that describes, limits, or qualifies the meaning of another word or word group. A modifier can consist of one word (a *blue* car), a prepositional phrase (the man *in the toll booth*), a relative clause (the woman *who won the marathon*), or an *-ing* or *-ed* phrase (*walking three miles a day*, the student was in good shape; *seated in the first row*, we saw everything on stage).

A **dangling modifier** is one that cannot logically modify any word in the sentence.

– When answering the question, his calculator fell off the table.

One way to correct the error is to insert the right subject after the *-ing* phrase.

+ When answering the question, he knocked his calculator off the table.

You can also turn the phrase into a subordinate clause.

+ When he answered the question, his calculator fell off the table.
+ His calculator fell off the table as he answered the question.

A **misplaced modifier** illogically modifies the wrong word or words in the sentence. The result is often comical.

− Hiding in the corner, growling and snarling, our guide spotted the frightened cub. (Is the guide growling and snarling in the corner?)
− All travel requests must be submitted by employees in red ink. (Are the employees covered in red ink?)

The problem with both of those examples is word order. The modifiers are misplaced because they are attached to the wrong words in the sentence. Correct the error by moving the modifier to where it belongs.

+ Hiding in the corner, growling and snarling, the frightened cub was spotted by our guide.
+ All travel requests by employees must be submitted in red ink.

Misplacing a relative clause (introduced by relative pronouns like *who, whom, that, which*) can also lead to problems with modification.

− The salesperson recorded the merchandise for the customer that the store had discounted. (The merchandise was discounted, not the customer.)
− The salesperson recorded the merchandise that the store had discounted for the customer. (The salesperson recorded for the customer; the store did not discount for the customer.)
+ The salesperson recorded for the customer the merchandise that the store had discounted.

Always place the relative clause immediately after the word it modifies.

Correct Use of Pronoun References in Sentences

Sentences will be vague if they contain a faulty use of pronouns. When you use a pronoun whose **antecedent** (the person, place, or object the pronoun refers to) is unclear, you risk confusing your reader.

Unclear: After the plants are clean, we separate the stems from the roots and place them in the sun to dry. (Is it the stems or the roots that lie in the sun?)

Revision: After the plants are clean, we separate the stems from the roots and place the stems in the sun to dry.

Unclear: The park ranger was pleased to see the workers planting new trees and installing new benches. This will attract more tourists. (The trees or the benches or both?)

Revision: The park ranger was pleased to see the workers planting new trees and installing new benches, because additional trees and benches will attract more tourists.

Words

Spelling Words Correctly

Your written work will be judged in part on how well you spell. A misspelled word may seem like a small matter, but on an employment application, an e-mail, an incident report, a letter, a short or long report, or a PowerPoint slide, it can make you look careless or, even worse, uneducated to a client or a supervisor. Readers will inevitably question your other skills if your spelling is incorrect.

The Benefits and Pitfalls of Spell Checkers

Spell checkers can be handy for flagging potential problem words. But beware! Spell checkers recognize only those words that have been listed in them. A proper name or new, infrequently used word may be flagged as an error even though the word is spelled correctly. Moreover, a spell checker will not differentiate between such homonyms as *too* and *two* or *there* and *their*. A spell checker identifies only misspelled words, not misused words. In short, do not rely exclusively on spell checkers to solve all your spelling and word-choice problems.

Consulting a Dictionary

Always have a dictionary handy. Two useful online dictionaries to consult are *Merriam-Webster OnLine* at *http://www.Merriam-webster.com* and *http://www.dictionary.com*.

Using Apostrophes Correctly

Apostrophes cause some writers special problems. Basically, apostrophes are used for four reasons: (1) contractions, (2) possessives, (3) plurals, and (4) abbreviations. The following guidelines will help you to sort out those uses.

1. In a **contraction**, the apostrophe takes the place of the missing letter or letters: *I've = I have*; *doesn't = does not*; *he's = he is*; *it's = it is*. (*Its* is a possessive pronoun—the dog and *its* bone—not a contraction. There is no such form as *its'*.)

2. To form a **possessive**, follow these rules.

a. If a singular or plural noun does not end in an -*s*, add *'s* to show possession.

Mary's locker	the woman's jacket
children's books	the women's jackets
the staff's dedication	the company's policy

b. If a singular noun ends in -*s*, add *'s* to show possession.

the class's project	the boss's schedule

c. If a plural noun ends in -*s*, add just the ' to indicate possession.

employees' benefits	computers' speed
lawyers' fees	stores' rates

d. If a proper name ends in *-s*, add *'s* to form the possessive.

Jones's account   Keats's poetry
the Williams's house   James's contract

e. If it is a compound noun, add an ' or *'s* to the end of the word.

brother-in-law's business   Ms. Melek-Patel's order

f. To indicate shared possession, add just *'s* to the last name.

Rao and Kline's website   Juan and Anne's major

g. To indicate separate possession, add *'s* to each name.

Juan's and Tia's transcripts   Shakespeare's and Byron's poetry

3. To form the plural of numbers and capital letters used as nouns, including abbreviations without periods, just add *s*. To avoid misreading some capital letters, however, you may need to add the apostrophe.

during the 1980s   all perfect 10s
their SATs   several local YMCAs
the 3 Rs   straight A's

4. For abbreviations with periods and for lowercase letters used as nouns, form the plural by adding *'s*.

his *p*'s and *q*'s   Q and A's   Ph.D.'s   I.R.S.'s

Using Hyphens Properly

Use a hyphen (-, as opposed to a dash, —) for

- **compound words**
  four-part lecture   heavy-duty machine   long-term prospects
- **most words beginning with** *self*
  self-starting   self-defense   self-regulating   self-governing
- **fractions used as adjectives**
  at the three-quarter level   two-thirds majority   three-dimensional drawing

Using Ellipses

Sometimes a sentence or passage is particularly useful, but you may not want to quote it fully. You may want to delete some words that are not really necessary for your purpose. These omissions are indicated by using an *ellipsis* (three spaced dots within the sentence to indicate where words have been omitted). Here is an example.

Full Quotation: "Diet and nutrition, which researchers have studied extensively, significantly affect oral health."

Quotation with Ellipsis: "Diet and nutrition … significantly affect oral health."

Using Numerals Versus Words

Write out numbers as words rather than numerals

■ **to begin a sentence**

Nineteen ninety-nine was the first year of our recruitment drive.

■ **to list the first number when two numbers are used together**

The company needed eleven 9-foot slabs.

But use numerals, not words,

■ **with abbreviations, percentages, symbols, units of measurement, dates**

17%	11:30 a.m.	70 ml
Dec. 3, 2003	$250.00	50 K

■ **for page references**

pp. 56–59

■ **for large numbers**

3,000,000	23,750	1,714

Use both numerals and words when you want to be as precise as possible in a contract or a proposal.

> We agreed to pay the vendor an extra twenty-five dollars ($25.00) per hour to finish the job by the eighteenth (18th) of May.

For information on conventions of writing and using numbers for an international audience, see Chapter 4.

Matching the Right Word with the Right Meaning

The words in the following list frequently are mistaken for one another. Some are true homonyms; others are just similar in spelling, pronunciation, or usage. The part of speech is given after each word. Make sure you use the right word in the right context.

accept (v) to receive, to acknowledge: *We accept your proposal.*
except (prep) excluding, but: *Everyone attended the meeting except Neelou.*

advice (n) a recommendation: *I should have taken Xi's advice.*
advise (v) to counsel: *Our lawyers advised us not to sign the contract.*

affect (v) to change, to influence: *Does the detour on Route 22 affect your travel plans?*
effect (n) a result: *What was the effect of the new procedure?*
effect (v) to bring about: *We will try to effect a change in company policy.*

allot (v) to distribute, to assign: *The manager allotted the writing team two weeks to complete the report.*
a lot (n) a quantity: *They bought a lot of supplies for the trip.*

all ready (adj) two-word phrase *all + ready*; to be finished; to be prepared: *We are all ready for the inspector's visit.*
already (adv) previously, before a given time: *Our webmaster had already updated the sites.*

altar (n) central place of worship: *The bride met the groom at the altar.*
alter (v) to change, to amend: *The tailor altered the trousers.*

ascent (n) move upward: *We watched the space shuttle's ascent.*
assent (n) agreement: *She won the teacher's assent.*
assent (v) to agree: *The committee asked the company to assent to the new terms.*

attain (v) to achieve, to reach: *We attained our sales goal this month.*
obtain (v) to get, to receive: *You can obtain a job application on their website.*

bimonthly (adv) every other month: *The boss wanted marketing studies done bimonthly.*
semimonthly (adv) twice a month: *We were paid semimonthly—on the 15th and 30th.*

cite (v) to document: *Please cite several examples to support your claim.*
site (n) place, location: *They want to build a parking lot on the site of the old theater.*
sight (n) vision: *His sight improved with bifocals.*

coarse (a) rough: *The sandpaper felt coarse.*
course (n) subject of study: *Sharonda took a course in calculus this fall.*

complement (v) to add to, enhance: *Her graphs and charts complemented my proposal.*
compliment (v) to praise: *The customer complimented us on our courteous staff.*

continually (adv) frequently and regularly: *This answering machine continually disconnects the caller in the middle of the message.*
continuously (adv) constantly: *The air-conditioning is on continuously during the summer.*

council (n) government body: *The council voted to increase salaries for all city employees.*
counsel (n) advice: *She gave the trainee pertinent counsel.*

defer (v) To put off until later: *His student loan was deferred while he finished his degree.*
differ (v) to disagree, to be different: *The committee differed among its members about the bond issue.*

discreet (adj) showing respect, being tactful: *The manager was discreet in answering the complaint letter.*
discrete (adj) separate, distinct: *Put those figures into discrete categories for processing.*

dual (adj) double: *A clock-radio serves a dual purpose.*
duel (n) a fight, a battle: *The argument almost turned into a duel.*

eminent (adj) prominent, highly esteemed: *Dr. Felicia Rollins is the most eminent neurologist in our community.*
imminent (adj) about to happen: *A hostile takeover of that company is imminent.*

envelop (v) to surround: *The major feared that fog would envelop the city.*
envelope (n) covering for a letter: *Always send letters in an envelope with the company logo on it.*

fair (n) convention, exhibition: *The technological fair featured a DVD-CD home theater with five satellite speakers.*
fair (adj) honest: *Their price was fair.*
fare (n) cost for a trip: *She was able to get a discount on a round-trip fare.*
fare (n) food: *They ate East Asian fare.*

foreword (n) preface, introduction to a book: *The foreword outlined the author's goals and objectives in her study of new global markets.*
forward (adv) toward a time or place; in advance: *We moved the time of the visit forward on the calendar so we could meet the overseas manager.*
forward (v) to send ahead: *We forwarded her e-mail to her new server.*

imply (v) to suggest: *The supervisor implied that the mechanics had taken too long for their lunch break.*
infer (v) to draw a conclusion: *We can infer from these sales figures that the new advertising campaign is working.*

it's (pronoun + verb) contraction of *it* and *is*: *Do you think it's too early to tell?*
its (adj) possessive form of *it*: *That old printer is on its last legs.*

knew (v) (past tense of *know*): *She knew the new regulations.*
new (a) never used before: *The subwoofer was new.*

lay/laid/laid (v) to put down: *Lay aside that project for now. He laid aside the project. He had already laid aside the project twice before.*
lie/lay/lain (v) to recline: *I think I'll lie down for a while. He lay there for only a few minutes before the firefighter rescued him. She has lain out in the sun too often.*

lean (adj) thin, skinny: *She asked for a lean slice of roast.*
lean (v) to rest against: *The shovel leaned against the fence.*
lien (n) a claim against: *There was a lien against his property for back taxes.*

lose (v) to misplace, to fail to win: *Be careful not to lose my calculator. I hope I don't lose my seat on the planning board.*
loose (adj) not tight: *The printer ribbon was too loose.*

miner (n) individual who works in a mine: *His uncle was a miner in West Virginia.*
minor (n) someone under legal age: *The law forbids the sale of tobacco to minors.*

overdo (v) to exceed, to do in excess: *The coach did not want her players to overdo their practice time.*
overdue (adj) past due: *The quarterly bill was overdue by three weeks.*

pare (v) to cut back: *Sandoval pared the skin from the apple.*
pair (n) a couple: *They offered a pair of resolutions.*
pear (n) a fruit: *Alphonso ate a pear with lunch.*

passed (v) went by (past tense of *pass*): *He passed me in the hall without recognizing me.*
past (n) time gone by: *We've never used their services in the past.*

peace (n) absence of war or conflict: *Joaquin enjoyed the peace he found in the new branch office.*
piece (n) a fragment, portion: *Each day care child received a piece of Wanda's birthday cake.*

personal (adj) private: *The manager closes the door when she discusses personal matters with one of her staff.*
personnel (n) staff of employees: *All personnel must participate in the 401(k) retirement program.*

perspective (n) viewpoint: *From the customer's perspective, we are an honest and courteous company.*
prospective (adj) expected, likely to happen or become: *E-mail the prospective budget to district managers.*

plain (adj) simple, not fancy: *He ate plain food.*
plane (n) airplane: *The plane for Dallas leaves in an hour.*
plane (v) to make smooth: *The carpenter planed the wood.*

precede (v) to go before: *A slide show will precede the open discussion.*
proceed (v) to carry on, to go ahead: *Proceed as if we had never received that letter.*

principal (adj) main, chief: *Sales of new software constitute their principal source of revenue.*
principal (n) the head of a school: *She was a high school principal before she entered the business world.*
principal (n) money owed: *The principal on that loan totaled $32,800.*
principle (n) a policy, a belief: *Sales reps should operate on the principle that the customer is always right.*

quiet (adj) silent, not loud: *He liked to spend a quiet afternoon surfing the Net.*
quite (adv) to a degree: *The officer was quite encouraged by the recruit's performance.*

stationary (adj) not moving: *Miguel rides a stationary bicycle for an hour every morning.*
stationery (n) writing supplies, such as paper and envelopes: *Please stop off at the stationery store and buy some more address labels.*

than (conj) as opposed to (used in comparisons): *He is a faster keyboarder than his predecessor.*
then (adv) at that time: *First she called the vendor; then she summarized their conversation in an e-mail to her boss.*

their (adj) possessive form of *they*: *All the lab technicians took their vacations during June and July.*
there (adv) in that place: *Please put the printer in there.*
they're (pronoun + verb) contraction of *they* and *are*: *They're our two best customer service representatives.*

to (prep): *They invited us to their new facility.*
too (adv) also, excessive: *The painters put too much enamel on the railings.*
two (n) the number: *Two new notebooks arrived today.*

waiver (n) international relinquishment of a right, claim, or privilege: *The company issued a waiver so that additional liability insurance would not have to be secured.*
waver (v) to shake, to move: *Our company would not waver in its commitment to safety.*

who's (pronoun + verb) contraction of *who* and *is*: *Who's up next for a promotion?*
whose (adj) possessive form of *who*: *Whose idea was that in the first place?*

you're (pronoun + verb) contraction of *you* and *are*: *You're going to like their decision.*
your (adj) possessive form of *you*: *They agree with your ideas.*

Index